Carolus Clusius

History of Science and Scholarship in the Netherlands, volume 8

The series *History of Science and Scholarship in the Netherlands* presents studies on a variety of subjects in the history of science, scholarship and academic institutions in the Netherlands.

Titles in this series

1. Rienk Vermij, *The Calvinist Copernicans. The reception of the new astronomy in the Dutch Republic, 1575-1750*. 2002, ISBN 90-6984-340-4
2. Gerhard Wiesenfeldt, *Leerer Raum in Minervas Haus. Experimentelle Naturlehre an der Universität Leiden, 1675-1715*. 2002, ISBN 90-6984-339-0
3. Rina Knoeff, *Herman Boerhaave (1668-1738). Calvinist chemist and physician*. 2002, ISBN 90-6984-342-0
4. Johanna Levelt Sengers, *How fluids unmix. Discoveries by the School of Van der Waals and Kamerlingh Onnes*. 2002, ISBN 90-6984-357-9
5. Jacques L.R. Touret and Robert P.W. Visser, editors, *Dutch pioneers of the earth sciences*, 2004, ISBN 90-6984-389-7
6. Renée E. Kistemaker, Natalya P. Kopaneva, Debora J. Meijers and Georgy Vilinbakhov, editors, *The Paper Museum of the Academy of Sciences in St Peterburg (c. 1725-1760), Introduction and Interpretation*, 2005, ISBN 90-6984-424-9, ISBN DVD 90-6984-425-7, ISBN Book and DVD 90-6984-426-5
7. Charles van den Heuvel, *'De Huysbou'. A reconstruction of an unfinished treatise on architecture, town planning and civil engineering by Simon Stevin*, 2005, ISBN 90-6984-432-X
8. Florike Egmond, Paul Hoftijzer and Robert Visser, editors, *Carolus Clusius. Towards a cultural history of a Renaissance naturalist*, 2007, ISBN 978-90-6984-506-7

Editorial Board

K. van Berkel, University of Groningen
W.Th.M. Frijhoff, Free University of Amsterdam
A. van Helden, Utrecht University
W.E. Krul, University of Groningen
A. de Swaan, Amsterdam School of Sociological Research
R.P.W. Visser, Utrecht University

Carolus Clusius

Towards a cultural history of a Renaissance naturalist

Edited by
Florike Egmond
Paul Hoftijzer
Robert Visser

Koninklijke Nederlandse Akademie van Wetenschappen, Amsterdam 2007

Copyright © 2007 Royal Netherlands Academy of Arts and Sciences.
No part of this publication may be reproduced, stored in a retrieval system
or transmitted in any form or by any means, electronic, mechanical, photocopying, recording or otherwise, without the prior written permission of
the publisher.

Information and orders:
Edita KNAW
P.O. Box 19121, 1000 GC Amsterdam, the Netherlands
T + 31 20 551 07 00
F + 31 20 620 49 41
E edita@bureau.knaw.nl, www.knaw.nl/edita

ISBN 978-90-6984-506-7

The paper in this publication meets the requirements of ∞ ISO-norm 9706
(1994) for permanence.

Illustration cover: Drawing of a daffodil, accompanying a letter from Carolus
Clusius in Leiden to Mattea Caccini in Florence, dated 10 October 1608.
Universiteitsbibliotheek Leiden, BLP 2414/14b. See also colour plate 2.

Contents

Preface   1

PART I   CLUSIUS' NETWORK AND EXCHANGES   7

Florike Egmond
   Clusius and friends: Cultures of exchange in the circles of European naturalists   9

Marie-Elisabeth Boutroue
   French manuscript sources on Carolus Clusius   49

Gillian Lewis
   Clusius in Montpellier, 1551-1554: A humanist education completed?   65

Josep L. Barona
   Clusius' exchange of botanical information with Spanish scholars   99

PART II   CLUSIUS AND INDIVIDUAL CORRESPONDENTS: TWO CASE STUDIES   117

Dóra Bobory
   'Qui me unice amabat.' Carolus Clusius and Boldizsár Batthyány   119

Kjell Lundquist
   Lilies to Norway and cloudberry jam to the Netherlands – On the relationship, correspondence and exchange of naturalia between carolus Clusius and Henrik Høyer, 1597-1604   145

PART III   CLUSIUS' TRANSLATIONS AND ILLUSTRATIONS: PROCESSING INFORMATION   171

José Pardo Tomás
   Two glimpses of America from a distance: Carolus Clusius and Nicolás Monardes   173

Peter Mason
   Americana in the *Exoticorum libri decem* of Charles de l'Écluse   195

Sachiko Kusukawa
   Uses of pictures in printed books: The case of Clusius' *Exoticorum libri decem*   221

PART IV   IDEAS AND INFLUENCE OF CLUSIUS   247

Irene Baldriga
   The influence of Clusius in Italy. Federico Cesi and the Accademia dei Lincei   249

Andrea Ubrizsy Savoia
   Some aspects of Clusius' Hungarian and Italian relations   267

Sabine Anagnostou
   The international transfer of medicinal drugs by the Society of Jesus (sixteenth to eighteenth centuries) and connections with the work of Carolus Clusius   293

Bibliography   313

List of illustrations   333

About the contributors   337

Index   341

Preface

The botanical renaissance of the sixteenth century has for a long time been regarded first and foremost as an event of an epistemological nature. Generations of historians of science praised pioneers like Bock, Brunfels, Clusius, Dodoens and Fuchs for their introduction of empiricism in botanical research. As long as the history of science focused primarily on the history of ideas this was generally regarded as the essential element of the revolution in botany during the sixteenth century.

During the last twenty years perspectives have changed. History of science has opened up to include social and cultural aspects of science, and a much richer picture of natural history has started to emerge. The new history of natural history has directed our attention to the importance of processes of professionalization and changes in the concept of professionalism, to the circulation, exchange and reception of knowledge, the interaction between experts within and outside the academic world, the role of universities and learned societies of the informal kind, the intertwining of natural history with collections (institutional and private), gardens and herbaria, the role of the market, and the confrontation with non-European systems of knowledge. Thus, human relations, politics, religion, commerce, culture, locations, objects, practices, reception have all found a place in this new history of science. In close connection with these developments, the interest in the visual aspects of science – from book illustration to styles of visual representation – has grown extensively. Images are no longer either left to art historians or regarded as less important, but have become an object of study in their own right. It is becoming increasingly clear that the developments in these and other areas are of no less relevance than the empiricist turn to understand the botanical renaissance of the sixteenth century.

During the second part of the century Carolus Clusius (1526-1609) (Ill. 1) was a key figure in this multi-faceted innovation of botanical science. Clusius' publications show a remarkable degree of originality. While most of his predecessors focused their studies on regional flora's, his field of research knew, at least in principle, no geographical boundaries. Plants from all over the world

Ill. 1. Portrait of Carolus Clusius, at the age of 59. Canvas, painted in Vienna in 1585 by an unknown artist, possibly Jacob de Monte. (See also colour plate 1).

were Clusius' subject. No other sixteenth-century botanist has described so many new exotic species as he did. His extensive correspondence was an important instrument in realizing his encyclopaedic ambitions. It gave him access to plants, their representations or written information about them, on an unprecedented scale: Clusius' network of correspondents brought the world within his reach. Herbaria, botanical gardens and illustrations formed other essential research instruments. Unlike most of his colleagues Clusius did not confine his research to the isolated plant. His publications show a striking interest in the plant's natural relations with other species in its habitat – heralding a science that much later came to be called ecology. Other innovative elements in his research were, for instance, frequent fieldtrips, the complete absence of any interest in the emblematic contextualising of natural objects, the special attention for popular plant names and local expertise with regard to the medicinal use of plants. Last but not least, Clusius was one of the first scientists of the early modern period to study botany primarily for its own sake and not as the handmaid of medicine. In all of these ways he contributed to the emergence of botany as an independent scientific discipline.

The nature of his scientific achievements forms only one of the many reasons for new historical research concerning Clusius and natural history. This is made both possible and especially attractive by the unique sources available for such an enterprise, which also allow us to address many of the topics mentioned above. In the course of his life Clusius exchanged letters with more than 300 correspondents from all over Europe. His correspondence comprises approximately 1300 letters addressed to him and some 300 letters written by Clusius to various correspondents. The vast majority of these letters can be found in Leiden University Library. They form a source of major importance for the history of sixteenth-century natural history in general and for botany in particular. Nonetheless, no major historical study has been devoted to Clusius during the past 60 years. He shares this fate with most of the innovative naturalists of his time.

Under the auspices of the Scaliger Institute of Leiden University an attempt is being made to change this situation. In 2004 the Scaliger Institute and Leiden University Library started the Clusius Project with the digitization of the circa 1300 Clusius letters in Leiden. This part of the project was concluded by making the high quality scans of all letters worldwide available in a database on the internet. The database can be found via the site of the Scaliger Institute at Leiden University Library (see Clusius Project). Access is free, the images can be downloaded, and it is hoped that easy accessibility will stimulate new research by scholars anywhere in the world.

A second part of the Clusius Project (2005-2009) concerns research along the lines described in the introductory paragraph of this preface. The Netherlands Organization for Scientific Research (Nederlandse Organisatie voor Wetenschappelijk Onderzoek, NWO) has generously provided grants for two Ph.D. students, respectively working on 'Clusius and botany in the context of Habsburg court culture' and 'Exchange and language in Clusius' European network of botanists', and a post doctoral researcher who will write a synthetic monograph provisionally entitled 'Natural history in the making. Carolus Clusius and the European community of naturalists'. A third facet of the Clusius Project is to organize international workshops bringing together European scholars from many disciplines with relevant expertise.

The first of these workshops, with the title 'Clusius in a New Context', was organized by the Scaliger Institute in Leiden in September 2004. Its main aims were to present the digitized Clusius letters, create a European network for the then still future research part of the project, and bring together researchers from a wide range of disciplines – from art history, and the history of collecting, to the history of science, botany, garden architecture, and book history – who were dealing with Clusius, in order to further dialogue between them and establish the state of the art of Clusius research. The present volume is largely based on the papers presented at this conference. A substantial part of it is devoted to the Clusius correspondence network, his correspondents, and the exchange of knowledge, but further topics include his influence on the history of botany, the reception of his work after his death, Clusius' ways of organizing information, and the relevance of visual material in his research and work. We regard all of these contributions as building stones towards a cultural history of a Renaissance naturalist.

The opening chapter by Florike Egmond presents us with the general characteristics of this network and its functioning. The most striking feature is its democratic nature. Clusius corresponded with a substantial number of non-academics. His correspondents came from a wide range of social backgrounds and included a significant number of women. The analyses of Clusius' French connections deal with his years in Montpellier, that played a formative role in his development as a naturalist (Lewis), and with Clusius' letters and manuscripts in French libraries, notably his annotations in De Thou's *Historia sui temporis* (Boutroue). His contacts with Spanish physicians and naturalists enabled him to become a key figure in the dissemination of knowledge of the Spanish flora in the rest of Europe and to collect information about South American plants and animals for his magnum opus: the *Exoticorum libri decem* (Barona).

Two of Clusius' correspondents are examined in more detail in the second part of the volume. Clusius knew both of them personally, face to face, which was by no means always the case with his correspondents. Lundquist discusses the correspondence with the Norwegian Hendrik Høyer, which was in essence about the simple exchange of specimens and information. Clusius' contacts with the Hungarian nobleman Boldizsár Batthyány greatly differed from those with Høyer. They were much more versatile and intensive, and Bobory's contribution indicates that Batthyány was not only Clusius' patron but also his friend. The connection with Batthyány was one way in which Clusius influenced the development of botanical studies in Hungary.

The third part of the book focuses on Clusius' use of illustrations in obtaining and disseminating knowledge, especially of exotic plants and animals. Pardo Tomás argues that Clusius considerably modified the organisation of Monardes' *Historia medicinal* by adding new illustrations and that these efforts were guided by his intention to incorporate it in his *Exoticorum*. Mason documents the considerable attention paid by Clusius to the quality of his illustrations. Clusius demonstrated a continuous ambition to enhance the credibility of his descriptions, especially in cases where he lacked first-hand information about the objects. Kusukawa too discusses the *Exoticorum* and concentrates on the pictorial representations of plants. She stresses the point that Clusius used pictures as an aid to strengthen the veracity of his claims about the identity of the plants.

The last part of the book opens with a contribution by Baldriga about Clusius (non)-relations with the Italian Accademia dei Lincei. It raises the question why he ignored the invitation to become a corresponding member of this famous Accademia, and discusses his possible influence on its botanical activities and interest in the naturalia of South America. Ubrizsy adds a new publication to her impressive list of books and articles on the impact of Clusius on Italian and especially Hungarian botany. Anagnostou closes the book with a chapter in which she investigates Clusius' influence on the worldwide efforts of the Jesuits to distribute knowledge about medicinal drugs.

The editors wish to thank various persons and institutions for their support in making the first phase of the Clusius Project possible. The staff of Leiden University Library was very helpful in providing the technical and personal means to make the Clusius correspondence digitally accessible. In particular we would like to thank Paul Gerretsen, former librarian of the University Library, André Bouwman, keeper of the Department of Western manuscripts, and his assistant-keeper Jan Vellekoop, as well as Maarten Steenhuis and Marlon Domingus of the Information technology section. Kasper van Ommen, coordinator of the Scaliger Institute, was of great assistance, both during the conference and the editorial work for this book. He also was responsible for making a number of photographs for the illustrations. Peter Mason, one of the authors of this volume, checked the English of some contributions.

Institutional sponsorship was received from various organisations. We are grateful to the Netherlands Research Foundation (NWO), the Dr. C. Louise Thijssen-Schoute Stichting and the J.J. Jurriaanse Stichting for their generous financial support in organizing the conference and publishing the transactions. Finally, we are pleased that this volume could appear in the series 'History of Science and Scholarship in the Netherlands' of the Koninklijke Nederlandse Academie van Wetenschappen in Amsterdam. We thank Yola de Lusenet for her professional advice and efficiency.

PART I

Clusius' network and exchanges

Clusius and friends: Cultures of exchange in the circles of European naturalists[1]

Florike Egmond

Clusius and Darwin

Some four hundred years ago friendship and friendly gift exchanges were at the core of the relations between the naturalist Carolus Clusius (Arras 1526 – Leiden 1609) and his network of European correspondents, who shared his fascination with nature (Ill. 2). Clusius acted like a spider in a web of communications and it is this aspect of his life and work on which I will focus in this essay. Partly thanks to improved transport and postal connections outside Europe, his famous successors of the eighteenth and nineteenth centuries Linnaeus and Darwin could operate on a grander geographical scale. All of Clusius' correspondents lived in Europe, even if some of them had travelled widely and most of them also exchanged information about non-European exotica. Nonetheless, much of what Janet Browne, one of Darwin's most important biographers, states below applies to Clusius as well:

> He relied on these letters for every aspect of his evolutionary endeavour, using them not only to pursue his investigations across the globe, but also to give his arguments the international spread and universal application that he and his colleagues regarded as essential footings for any new scientific concept. They were his primary research tool. [...] If there was any single factor that characterised the heart of Darwin's scientific undertaking it was this systematic use of correspondence.[2]

---

[1] Several earlier versions of (parts of) this essay have been tried out in papers at conferences in Florence (European University Institute), Paris (Gulbenkian Foundation, European Science Foundation) and Leiden (Scaliger Institute), and I have enjoyed and tried to profit from the discussions and comments. I would like to thank in particular Peter Mason, Sabine Anagnostou, Irene Baldriga, Francisco Bethencourt and José Pardo Tomás for their thoughtful comments and for our ongoing and stimulating exchanges about these topics. Special thanks go to Giuseppe Olmi and Lucia Tongiorgi Tomasi for the continuing inspiration of their work.

[2] J. Browne, *Charles Darwin. The power of place. Volume II of a biography* (London, 2002), 11-12. Cf. J. Browne, *Charles Darwin. Voyaging. Volume I of a biography* (London, 1995).

Ill. 2. Portrait of Carolus Clusius. Engraving by Martinus Rota, sixteenth-century.

Clusius, like Darwin (to limit this brief comparison to the man furthest apart from him in time), used correspondence both as a means to obtain and exchange information and as an instrument to disseminate his views in the wider community of naturalists – naming but two of the most obvious professional functions. Yet, there seem to be some essential differences between the networks of exchange in which Darwin and Clusius were involved. Clusius' network of correspondents appears to have formed more of a virtual community than Darwin's in the sense that many of Clusius' correspondents were themselves spiders in their own circle of correspondents, while their networks often overlapped. Thus, many of Clusius' contacts were also in exchange with each other, and the density of exchanges seems to have differed from those in Darwin's world of correspondence. Many of the latter's contacts were mobilised only for specific reasons and did not become friends. The closely related theme of reciprocity is at the core of the second difference between the networks of these naturalists. According to Janet Browne, Darwin used his vast correspondence network in order to obtain information from a wide geographic and social range of informants, but usually did not reciprocate, either with ideas or gifts.[3] Clusius operated in a less one-sided manner, exchanging gifts for gifts and information for information. This was the case not just during the early stages of his career, but right up to his final days. Nor did he follow this course of action only with respect to those persons who might be most important to him in either an intellectual or a social sense. A first impression of exchanges among other members of his circle indicates, moreover, that he was by no means the only one to do so. That fact alone should prevent us from immediately attributing this difference between Clusius and Darwin to personal characteristics, such as liberality or stinginess. As will be argued below, both these differences and some interesting parallels between the two men may well be connected with structural aspects of the discipline of European natural history.

Those parallels concern the professional background of the two men and the complicated issue of professionalism. Neither Clusius nor Darwin had studied natural history at university. In Clusius' time it did not yet exist as a specialisation, as we shall see in more detail below, while by Darwin's time natural history had become a slightly schizophrenic subject: a highly popular topic that was eminently suitable for amateur-enthusiasts, on the one hand, and a technical discipline already fragmented into several specialisations yet at the same time fraught with philosophical and theological questions, on the other hand. Both men started out by studying medicine and only later made natural history their profession; both thus started out as 'amateur' natural historians. Interestingly, in spite of the three centuries that separate them, each did so in a social context in which nature and natural history were in high fashion, among the elites as well as the middle social strata. Both reached the international top of their profession, while at the same time defining or redefining it and its standards.

It would be fascinating to undertake a thorough comparison of the two men along these lines and analyze the implications concerning the social history of natural science during the centuries concerned, but this is not a comparative essay. The initial comparison between Clusius and Darwin has been made simply to bring the questions that are central to this article about 'Clusius and Friends' into sharper focus. The contrast between their respective networks of correspondents and the notion of a (virtual) community leads us to pay special attention to the key notions of *friends* and *friendship*. It is not by

---

[3] Browne, *Charles Darwin. The power of place*, here esp. 9-14.

chance that expressions referring to friendship can be found innumerable times throughout Clusius' correspondence. What does it precisely mean? And who were these friends of Clusius? Other key terms are *pleasure, service,* and *gift.* They should make us pay special attention to questions of professionalism and amateur enthusiasm, to rivalry and friendships, the modalities of exchange in this network, possible tensions between 'free' gift exchange and commercial exchanges, and therefore to the roles of intermediaries or brokers. Finally we will come back to the issue of the connections between some of the main social and cultural characteristics of the community of naturalists (and their exchanges) in Europe and the formation of a new field of expertise.

The new cultural history of science and the Clusius correspondence

Most of the above questions have been inspired by the theoretical framework of the *new cultural history of science*, which has become increasingly important since the mid 1980s.[4] Three elements of this approach are especially relevant to this essay. The first requires the relinquishing of an evolutionist perspective – and therefore of the notion of permanent scientific progress – as well as of an internalist approach, focused almost exclusively on the history of ideas. The second requires a continuous rethinking of the concept of science itself, along less anachronistic (modernist) lines, creating room to subsume, for instance, alchemy and physics or botany, art and anatomy under the same heading and study them as mutually relevant. Finally, and closely intertwined with the former, the notion of the *social construction* of knowledge presupposes that social and cultural contexts are *constitutive* elements in the formation of ideas and knowledge. All three points may be especially relevant to the natural or 'life' sciences, since these did (and perhaps still do) combine scientific, theoretically informed notions with expertise obtained in practice by men and women who were not professional scientists. For this reason the natural sciences – or those sections that have remained open to practical knowledge – have often found themselves on the borderline between 'real', respectable science and a 'mere' field of expertise.[5]

---

[4] To name only some especially inspiring examples: M. Biagioli, 'Scientific revolution, social bricolage, and etiquette', in R. Porter and M. Teich (eds.), *The scientific revolution in national context* (Cambridge, 1992), 11-54; N. Jardine, J.A. Secord and E.C. Spary (eds.), *Cultures of natural history* (Cambridge, 1996); P. Findlen, *Possessing Nature: Museums, collecting and scientific culture in early modern Italy* (Berkeley, 1994); and G. Olmi, *L'Inventario del mondo. Catalogazione della natura e luoghi del sapere nella prima età moderna* (Bologna, 1992).
[5] Obviously, this distinction between 'mere' expertise and 'real' science is a cultural construction as well, while the borderline shifts, moreover, over time. For an illuminating essay concerning hierarchies of scientific disciplines see J.A. Secord, 'The crisis of nature', in Jardine, Secord and Spary (eds.), *Cultures of natural history*, 447-459.

Until recently the focus in the *new cultural history of science* has been principally on the seventeenth to nineteenth centuries; learned societies and academies; medicine and the domain of the 'hard' sciences, such as mathematics, physics, and astronomy; and on their most famous representatives (such as Galilei, Newton, Hooke, or Boyle).[6] The history of collections has, moreover, emerged as an exciting, new and related specialisation. In the field of early-modern *natural* history, attention has mainly focused on Italy, while some inspiring collections of essays have discussed European natural history from the Renaissance to the twentieth century.[7] During the past thirty to forty years, however, hardly any major studies have been devoted to the innovative sixteenth-century naturalists from Central- and (North-)Western Europe – Carolus Clusius, Rembert Dodoens, Pierre Belon, Guillaume Rondelet.[8] The few synthetic studies concerning early natural history are of good quality but old; they date back to the 1930s or 1970s, and thus to before the period when the new cultural history of science started raising a whole range of new questions.[9] Surprisingly, especially given the fact that the sixteenth century is known as a period of major innovation in botany or even as the (first phase of) the 'botanical renaissance', there seem to be no monographs at all which discuss sixteenth-century natural history in connection with the scientific revolution.

Next to his near contemporaries, the Swiss polymath, physician, bibliographer and natural historian Conrad Gessner (1516-65), and the Italian physician, collector and naturalist Ulisse Aldrovandi (1522-1605), Clusius may well be the most interesting European botanist to study, given our interest in the questions mentioned above. The reasons are simple. He has left one of the

---

[6] This focus can hardly have been determined alone by the availability of sources, but probably also reflects an implicit hierarchy of prestige of the subject, which itself deserves more attention.

[7] For major synthetic studies on Italy see Findlen, *Possessing nature*; and Olmi, *L'Inventario del mondo*. Examples of inspiring volumes of essays discussing a wide range of centuries and areas are Jardine, Secord and Spary (eds.), *Cultures of natural history*, and A. Ellenius (ed.), *The natural sciences and the arts. Aspects of interaction from the Renaissance to the twentieth century* (Uppsala, 1985).

[8] A recent German dissertation focuses on Clusius' *Exoticorum libri decem*, and provides an updated biography, but in terms of perspective has no connection with new approaches in the history of science. See: A. Fetzner, *Carolus Clusius und seine* Libri Exoticorum (Marburg, 2004).

[9] Among the best of the older synthetic studies are P. Delaunay, *La zoologie au seizième siècle* (Paris, 1962); K. Reeds, *Botany in medieval and Renaissance universities* (New York, 1991), which was in fact written decades earlier; A. Arber, *Herbals, their origin and evolution; a chapter in the history of botany, 1470-1670* (Cambridge, 1938²; 1st edn. 1912). A wide-ranging and more recent exhibition catalogue with interesting essays is F. de Nave and D. Imhof (eds.), *Botany in the Low Countries (end of the 15th century – ca. 1650)* (Antwerp, 1993). Good surveys can also be found in Z. Mirek and A. Zemanek (eds.), *Studies in Renaissance botany* (Kraków, 1998) [Polish botanical studies, Guidebook series, 20]. It looks as if the forthcoming B. Ogilvie, *The science of describing. Natural history in Renaissance Europe* (Chicago, 2006) may be the first modern synthetic study to fill this gap.

most fascinating, massive and wide-ranging correspondences of all sixteenth-century naturalists, and must be regarded as one of the key figures in European natural history of this age because of his innovative work and his crucial role in collecting information and disseminating ideas, information and plants. In fact, our thesis – to be further researched in coming years – is that the exchanges in the network of which Clusius was the central figure contributed in a major way to the creation of a new European community of experts on nature in the course of sixteenth century, and that this community (which developed its own modes of exchange and cooperation) played a crucial part in the development of a new discipline or field of expertise. Clusius' European stature rested on three pillars: his travel and investigations in several European countries; his publications and innovative approach to botany; and his wide-flung network of correspondents. Clusius was born in Arras in the Southern Netherlands. He studied both there and in Germany and France, travelled and did botanical field research in Spain, Portugal, the Southern Netherlands, Austria, Hungary, Germany and England, lived, studied and worked in the Southern Netherlands, at the universities of Paris, Montpellier and some German towns, at the Habsburg court in Vienna, on aristocratic estates in Hungary, in Frankfurt and at the university of Leiden. He maintained friendly exchanges by letter with a large network of friends, collectors, fellow experts and others for half a century, at least from the early 1560s until his death in Leiden in 1609.

Natural history in the sixteenth century and the role of Clusius will be discussed in some more detail in the following section. First, the importance of his correspondence needs some further explanation, since it is its very nature which enables us to investigate the themes mentioned above. Some 1500 letters of the Clusius correspondence have been preserved, of which the bulk of more than 1300 are kept in Leiden University Library, while about 200 are held by the library at Erlangen in Germany.[10] Of the total of about 1500 letters the majority (some 1200) were sent to Clusius; the remaining 300 or so are letters from Clusius to various correspondents. An as yet unknown number of letters, which probably will run to the dozens rather than hundreds, are scattered throughout libraries and archives all over Europe. Next to those of Gessner and Aldrovandi, the Clusius correspondence may well be the most valuable collection of correspondence in the field of early European natural history. The reason why far more letters *to* Clusius than letters written *by* him to others have been preserved is a simple one. While the letters sent by Clusius were

---

[10] The numbers mentioned cannot be exact; there are still double counts and uncertain identifications and datings.

dispersed all over Europe, the letters sent to him centred on one person. Clusius must have taken very good care of them – perhaps because they formed not only a permanent and rich source of information for his own research and the manifestation of important friendships, but also the material evidence of the growth of a virtual European community of naturalists and of his own central position in that community. Clusius' correspondence seems to be exceptional in this respect: most early scientific or scholarly correspondences of which larger numbers of letters have been preserved consist of letters from a famous person to others, implying that the sender considered himself important enough to keep copies. Clusius was clearly a different kind of person, and that fits with the evidence concerning his generosity and lack of pretensions.

Between about 1560 and 1609 Clusius corresponded with at least 300 different persons. Clusius' correspondents lived all over Europe, from England to Hungary and Austria, from Greece and Italy to Poland, and from Spain and Portugal to the Northern Netherlands, France, Germany and Norway. In a geographical sense his network was truly European, and his correspondence collection is unique in the sense that it shows what was happening in these parts of the world. Even the limited preliminary research which has been done so far shows that most of Clusius' correspondents – of whom a considerable number were expert botanists, collectors, and garden owners – had access in their turn to large and partly overlapping networks of friends, acquaintances and fellow experts with whom they maintained relations of exchange. If we try to imagine the range of information and informants to which he could have access if he wished, it is no exaggeration to claim that he directly or indirectly could have access to all the then relevant naturalists in Europe. Clusius' curiosity, his desire for knowledge, and the way in which he maintained friendships and meticulously kept the letters sent to him, allow us both to reconstruct a virtual community and to obtain glimpses of life and the development of natural history in these diverse parts of Europe in the second half of the sixteenth century.

The range of Clusius' correspondence network was wide in a social sense as well. He exchanged information with social equals, but also with people of both a higher and a lower social position. Among Clusius' correspondents are aristocratic collectors, princes, courtiers and rich patrons (some of whom have been mentioned above), such as the Count of Arenberg, Lamoraal van Egmont, and Charles de Saint Omer in the Southern Netherlands, Sir Philip Sydney and Lord Zouche in England, Ludwig I Duke of Wurttemberg and Ludwig VI Elector of the Palatinate in Germany, the Hungarian Count Batthiány, Princess Marie de Brimeu, famous humanists such as Benito Arias Montano and Justus Lipsius, fellow physicians or botanical experts such as Felix Platter, Joachim Camerarius, Matteo Caccini, Ulisse Aldrovandi, or

Simon de Tovar, printer-publishers and artists such as Christoph Plantin, Franciscus Raphelengius and Anselmus de Boodt, diplomats such as Ogier de Busbeq, and apothecaries such as the Garet family, Jean Mouton, Christian Porret and Hugh Morgan. But this list also includes relatives and many others. The better known Clusius became as a leading botanical expert, the more his epistolary contacts proliferated. His network snowballed.

Given this wide social and geographic range of Clusius correspondence and the fact that is has been known to exist and to have been preserved intact since shortly after his death,[11] it is amazing that it has not been studied as a corpus, and that almost no systematic use has been made of the whole, except by his biographer F.W.T. Hunger.[12] Apart from its bulk and the practical difficulties involved in deciphering the diverse handwritings and languages, I can only think of one reason for this relative neglect: the relatively low status of natural history and the concomitant lack of interest in it among general historians and historians of science. This is borne out by the fact that access to correspondences of many other important early modern European naturalists is often equally difficult. While lists and (partial) editions of the correspondence of scholars or scientists such as Lipsius, Scaliger, Grotius and Huygens are available, for the contents of the naturalists' letters we have to rely in most cases on old (generally nineteenth-century and early twentieth-century, sometimes much older) editions of small selections of their correspondence or on the original letters themselves which can be found in libraries and archives all over Europe. Even surveys summing up or just listing the letters and correspondents of individual naturalists are scarce.[13] Clusius' famous counterpart Conrad Gessner, whose influence on both learned and popular European culture can hardly be overestimated, forms an exception – at least up to a point. Many of his letters were already published in the late sixteenth century, shortly after his death, and additional material was published during the 1950s and 1970s.[14] The situation

---

[11] Thanks to his Leiden colleague Bonaventura Vulcanius.

[12] F.W.T. Hunger, *Charles de l'Escluse (Carolus Clusius) Nederlandsch kruidkundige, 1526-1609*, 2 vols. (The Hague, 1927-43). The first attempt at a survey of the whole collection was by Ans Berendts, 'Carolus Clusius (1526-1609) and Bernardus Paludanus (1550-1633). Their contacts and correspondence', *Lias* 5 (1978), 49-64.

[13] In this respect internet search instruments specializing in manuscripts and letters, and the lists or transcriptions of letters by naturalists put on internet by European libraries (for example, Bologna, Aldrovandi) promise important changes.

[14] Yet publications of more than 400 years old are by now almost as inaccessible as the original letters, and Gessner's correspondence is certainly worth a new look. See R.J. Durling, 'Konrad Gessner's Briefwechsel', in R. Schmitz and F. Krafft (eds.), *Humanismus und Naturwissenschaften* (Boppard, 1980) [Beiträge zur Humanismusforschung, 6], 101-112; G. Rath, 'Die Briefe Conrad Gessners aus der Trewschen Sammlung', *Gesnerus*, 7 (1950), 140-70, and 8 (1951), 195-215; and C. Longeon, *Conrad Gesner. Vingt lettres à Jean Bauhin fils (1563-1565)* (Saint Etienne, 1976).

with respect to the Clusius correspondence has long been far worse. Given its size, it is understandable that the bulk has never been published, but even a decent survey listing the individual letters was lacking until the start of the present Clusius project in 2005.[15] In so far as letters to him have been published, this concerns only a relatively limited number, and with a few exceptions, such as the correspondence between Clusius and his Spanish contacts, these publications date back to the seventeenth, late nineteenth or early twentieth century.[16]

The general situation sketched above was, however, also the result of the dominant research questions of the time, and in its turn did not facilitate the emergence of new ones. Older publications about (or editions of) the correspondence of sixteenth-century naturalists were generally inspired by a history of science which focused on the biographies of great scientists in the context of an evolutionist approach. Simplified, the agenda of most publications could be summed up as: the progress of science and what my hero contributed to it. Thus only those letters (or parts of them) were deemed of importance which provided information of a basically biographical kind about the naturalist himself (or one or some of his correspondents) or about their scholarly ideas and (in this case botanical) knowledge. Much of Clusius' correspondence, for instance, was consulted by his biographer F.W.T. Hunger before World War II, and the letters sent to Clusius that have appeared in print are almost all letters from correspondents who were famous in their own right.[17] Thus, the focus has tended to be on a very limited part of the contents of such letters and not on exchanges or on the larger network. Interest in networks, information exchange, or the ways in which botany, and for instance collecting or anatomy overlapped and may have formed part of a single field of expertise could only emerge once the perspective had shifted with the emergence of the new cultural history of science – which was at least partly triggered from outside the circles of historians of science. The consequence is a very different approach

---

[15] Scans of all Clusius' letters in Leiden University Library (further Leiden UL) are now available on internet to all (for further information see the introduction of the present volume).

[16] Exceptions are: J.L. Barona and X. Gómez i Font, *La correspondencia de Carolus Clusius con los científicos españoles* (Valencia, 1998); and G. Olmi, 'Lettere di fra Gregorio da Reggio, cappuccino e botanico del tardo rinascimento', in M. Beretta, P. Galluzzi, C. Triarico (eds.), *Musa Musaei, studies on scientific instruments and collections in honour of Mara Miniati* (Florence, 2003), 117-139. Among the main older editions are: G. Istvánffi, *Études et commentaires sur le code de l'Escluse* (Budapest, 1900); the letters published by Clusius' biographer Hunger in the second volume of his *Charles de l'Escluse* (1942); G.B. de Toni, *Il carteggio degli Italiani col botanico Carlo Clusius nella bibliotheca Leidense* (Modena, 1911); and P. Ginori Conti, *Lettere inedite di Charles de l'Escluse (Carolus Clusius) a Matteo Caccini, Floricultore Fiorentino. Contributo alla storia della botanica* (Florence, 1939).

[17] Hunger's biography (*Charles de l'Escluse*), of which the first volume is written in Dutch and the second (mainly presenting sources) in German, is chronologically organized and in many ways old-fashioned, but it is based on thorough research and is still very readable.

to both the letters themselves and their contents. The former have to be studied as part of a collection and can no longer be seen as disembodied carriers of information: they are objects in their own right, and their very survival testifies to the fact that they were regarded as precious collector's items.[18]

Nor can the contents of the letters any longer be sifted for 'relevant' material according to a pre-set group of criteria which derive directly from modern notions of what is relevant information concerning botany or natural history. In many ways, the latter has become exactly the question: what was botany or natural history at a time when it was just beginning to emerge as a field? If a sixteenth-century correspondent deemed it relevant to write about stolen bulbs, the best ways of planting daffodils, religious persecution and the layout of a garden as if it were a tapestry, that information is highly pertinent to botany as a *historical* field of expertise. And it is as important to look at what is written in these letters about strictly botanical matters as it is to look at the ways in which these people exchanged knowledge, their backgrounds, forms of presentation, conflicts, and the other types of information they regarded as interesting.

Fashionable nature: A botanical renaissance

Clusius' interest in nature originated in a period during which a wave of fascination with nature was reaching its crest all over Europe. In fact, in his old age Clusius told the Florentine flower expert Matteo Caccini in one of his letters that he had already been very interested in plants as a boy.[19] Nature had become increasingly fashionable from the early decades of the sixteenth century, and this interest had been greatly stimulated by the discovery of the New World and the transportation of exotic naturalia to Europe from all over the world. While the Middle East (especially Persia) had been famous for its sophisticated garden culture since medieval times, it is not generally known that the New World was more than just a source of exciting new botanical discoveries and formed an example to the Europeans as well. Spanish conquerors

---

[18] Interesting observations on the place of twelve volumes with letters to the seventeenth-century Roman Jesuit and collector Athanasius Kircher, which seem to have functioned partly as a visitor's book to his collection and may have been on display as part of that collection, can be found in Paula Findlen, 'Un incontro con Kircher a Roma', in E. Lo Sardo (ed.), *Athanasius Kircher. Il Museo del Mondo* (Rome, 2001), 39-47, here 42-43. With thanks to Irene Baldriga for this information. For a volume discussing many aspects of early modern correspondence see F. Bethencourt and F. Egmond (eds.), *Correspondence and cultural exchange in early modern Europe* (Cambridge, 2007).
[19] '[…] et come da fanciullo ho sempre havuto grande piacere nelle piante […].' See the letter from Clusius to Caccini, dated 29 September, 1606, when Clusius was eighty years old. Ginori Conti, *Lettere inedite di Charles de l'Escluse*, 37.

were amazed and impressed by the Aztec gardens and menageries in and around Tenochtitlán which they encountered upon their first arrival in 1521. Plants were grown there for pleasure as well as for medicinal and ritual use.[20] The great voyages of discovery to the New World and Far East and explorations in the Middle East led to the introduction of exotic plants or drugs based on them (such as potatoes, cacao, tomatoes, bulbs, quinine) in Europe, which in the long run would have enormous effects on food, medicine and Europe's physical appearance. Clusius was personally involved in the introduction and scientific description of both the potato and various types of tulip in the Low Countries. As a more short-term effect, the very existence of the strange plants and animals discovered in newly explored regions demonstrated the incompleteness of the knowledge of natural historians from Antiquity. It triggered new investigations both in and outside Europe.[21]

The European interest in nature focused mainly on plants and gardens, and to a somewhat lesser extent on animals, which were much more difficult and expensive to transport and keep. Three major manifestations were the creation of numerous private gardens all over Europe; the proliferation of botanical and zoological motifs in decorative art (such as tapestries, frescos); and the collection of naturalia (such as plants, shells, and parts of animals) and of their visual representations in drawings, watercolours, woodcuts, engravings or illustrations from printed works. In the rich aristocratic and princely collections of the period, naturalia were often part of much larger *Kunst- und Wunderkammern*.[22] Almost every self-respecting nobleman owned a garden and showed off exotic flowers to his visitors.[23] Some even owned a maze or a private

---

[20] See D. Heyden, 'Jardines botánicos prehispánicos', *Arqueología Mexicana*, 57 (2000), 18-23; and A.M.L. Velsaco Lozano, 'El jardín de Itztapalapa', *Arqueología Mexicana*, 57 (2000), 26-35.

[21] On the connections between the voyages of discovery, the New World and natural history see G. Olmi, *L'Inventario del mondo*, esp. 211-255; and H. Lowood, 'The New World and the European catalog of nature', in K. Ordahl Kupperman (ed.), *America in European consciousness, 1493-1750* (Chapel Hill/London, 1995), 295-323.

[22] There is an extensive literature on this subject. Classic studies are: J. von Schlosser, *Die Kunst- und Wunderkammern der Spätrenaissance. Ein Beitrag zur Geschichte des Sammelwesens* (Brauschweig, 1923¹; rev. edn. 1978); O. Impey and A. Macgregor (eds.), *The origins of museums. The cabinet of curiosities in sixteenth- and seventeenth-century Europe* (Oxford, 1985); A. Schnapper, *Le géant, la licorne et la tulipe. Collections et collectionneurs dans la France du XVIIe siècle* (Paris, 1988); and A. Lugli, *Naturalia et mirabilia. Les cabinets de curiosités en Europa* (Paris, 1998; a French edition of the Italian 1st edition, Milan, 1983 with an updated bibliography).

[23] On gardens of this period see, for instance, the excellent R. Strong, *The Renaissance garden in England* (London, 1979); and U. Härting (ed.), *Gärten und Höfe der Rubenszeit. Im Spiegel der Malerfamilie Brueghel und der Künstler um Peter Paul Rubens* (Munich, 2000). For science at court see for example (for Mantua) D.A. Franchini et al., *La scienza a corte. Collezionismo eclettico natura e immagine a Mantova fra Rinascimento e Manierismo* (Rome, 1979); (for Hessen-Kassel) I. Dübber, *Zur Geschichte des Medizinal-*

menagerie. Quite a few of them took gardening and botany very seriously indeed and read as widely as possible on the subject. The fact that they were rich and sometimes acted as patrons to scholars and scientists as well as artists should not obscure the fact that they could also become real experts in their own right who conversed and corresponded about plants on a more or less equal footing with scholarly experts such as Clusius. Princess Marie de Brimeu and the Count of Arenberg in the Low Countries, Wilhelm IV of Hessen-Kassel, and the English Lord Zouche are good examples among Clusius' correspondents. An aristocrat with a particularly varied collection was Charles de Saint Omer, Lord of Moerkercke and Dranoutre, who lived near Bruges in the Southern Netherlands. His collection comprised books, tapestries, a menagerie, gardens, a maze, prints, paintings and coins, naturalia and other curiosa. He commissioned the famous botanical and zoological watercolours of the *Libri Picturati* collection now held in Kraków, and was one of Clusius' first patrons (Ill. 3).[24]

Cultural interests and the fashion of collecting and display often coincided or overlapped with scientific and commercial ones. The Spanish king Philip II, for instance, who sent out the physician Francisco Hernandez to investigate, describe and depict the flora of Mexico, and at whose court Clusius may briefly have been a guest during his trip to Spain and Portugal, was very much alive to the possibility of creating plantations in the New World to grow medicinal plants for commercial purposes.[25] Economic motives were even more prominent in the early explorations of the Dutch in the East Indies. The instruction to the commanders of the first Dutch ships sent out to the East Indies to explore and bring back herbs, fruits and spices was certainly of help to botanists (and Clusius was probably personally involved in it), but had also been inspired by the expectation of enormous profits in the spice trade.[26] Nor were

---

und Apothekenwesens in Hessen-Kassel und Hessen-Marburg von den Anfängen bis zum Dreissigjährigen Krieg (Marburg, 1969); and in particular Mario Biagioli, 'Le Prince et les savants. La civilité scientifique au 17me-siècle', *Annales E.S.C.*, 6 (1995), 1417-1453.

[24] See F. Egmond, 'Clusius, Cluyt, Saint Omer. The origins of the sixteenth-century botanical and zoological watercolours in Libri Picturati A. 16-30', *Nuncius*, 20 (2005), 11-67, with references to the further literature.

[25] About the possible stay of Clusius and his pupil Jacobus Fugger at Philip's court in 1564 see Fetzner, *Carolus Clusius*, 30. For the plantations see H. Kamen, *Philip of Spain* (New Haven/London, 1997), 91; and for Hernández: J. Pardo Tomás (with J.M. Lopez Pinero), *La influencia de Francisco Hernández (1515-1587) en la constitución de la botánica y de la materia médica modernas* (Valencia, 1996).

[26] See P. Baas, 'De VOC in Flora's lusthoven', in L. Blussé and I. Ooms (eds.), *Kennis en Compagnie: De Verenigde Oost-Indische Compagnie en de moderne wetenschap* (Amsterdam, 2002), 124-137. Cf. K. van Berkel, 'Een onwillige mecenas? De rol van de VOC bij het natuurwetenschappelijk onderzoek in de 17e eeuw', in J. Bethlehem and A.C. Meijer (eds.), *VOC en Cultuur: Wetenschappelijk onderzoek en culturele relaties tussen Europa en Azië ten tijde van de Verenigde Oostindische Compagnie* (Amsterdam, 1993), 39-58.

Ill. 3. A beautiful watercolour of a clematis in *Libri Picturati*, vol. A 23, f. 18v. (See also colour plate 3).

academic institutions interested in nature for exclusively scientific reasons. Many European universities claimed prestige and status by creating botanical gardens during this period. All of the oldest such gardens in Europe (e.g. Pisa, Padua, Bologna, Leipzig, Leiden) originated between 1520 and 1610.[27] They formed the visual proof that these universities took their teaching of medicine and medicinal herbs seriously, while at the same time demonstrating how these universities vied with each other and with aristocratic garden owners by growing rare exotic plants whose medicinal value was often irrelevant.[28]

From its earliest phase the fascination with nature and curiosities by no means affected only the rich or the learned in Europe. The origins of many smaller and more specialized collections comprising naturalia and their pictures can be traced to the decades between 1540 and 1590 and to men (and a few women) who belonged to the professional middle classes. They too could own a garden and grow new plants from seeds or cuttings obtained by barter.[29] The most important non-elite collectors and aficionados of natural history all over Europe were apothecaries, perfumers and druggists.[30] As such they have not been taken seriously enough by historians of science. Some well documented examples show that their interest in nature could go far beyond mere professional (medicinal) requirements and manifest itself in ways that equalled or came close to scholarly or elite forms of expertise. Some owned experimental gardens, grew exotics, compiled encyclopaedic surveys of information about nature, and had portraits of plants or animals painted. Apothecaries such as Hugh Morgan, James Garet Jr and Thomas Penny in England, or their Continental counterparts Peeter van Coudenberghe, Thomas de la Fosse, Jean Mouton, and Christiaan Porret (in the Low Countries) – all of them correspondents of Clusius – were famous for being the first to grow certain exotic plants in Europe and to experiment with highly prized new varieties of

---

[27] See the useful survey A. Zemanek, 'Renaissance botany and modern science', in Mirek and Zemanek (eds.), *Studies in Renaissance botany*, 9-47.

[28] That lines distinguishing 'scientific' academic gardens and the private ones of rich collectors should not be drawn too sharply is also clear from H. Brunon, 'Il bell' ordine della natura: spazio e collezioni nel giardino di villa Medici', in M. Hochmann (ed.), *Villa Medici. Il sogno di un cardinale* (Rome, 1999), 67-73, here especially 71.

[29] Whether this holds true as much for Mediterranean as for Northwest Europe has to be investigated further.

[30] There is a quantity of fragmented information on apothecaries and their interest in nature, but (to my knowledge) no monograph or synthetic essay. See P. Dilg, 'Apotheker als Sammler', in A. Grote (ed.), *Macrocosmos in Microcosmo. Die Welt in der Stube. Zur Geschichte des Sammelns 1450 bis 1800* (Opladen, 1994) [Berliner Schriften zur Museumskunde, 10], 453-474. An excellent regional study is Dübber, *Zur Geschichte des Medizinal- und Apothekenwesens in Hessen-Kassel*. For Jesuits as apothecaries see S. Anagnostou, *Jesuiten in Spanisch-Amerika als Übermittler von heilkundlichem Wissen* (Stuttgart, 2000) [Quellen und Studien zur Geschichte der Pharmazie, 78].

non-indigenous flowers. Some of them were also experimenting with the medicinal uses of exotic plants and involved in their commercial importation.[31] The gardens owned by these men and their upper-class counterparts were visited by foreign travellers, aristocratic collectors (such as Sir Philip Sidney or Lodovico Guicciardini), scholars (like Lipsius or Lobelius), and by many anonymous friends and relatives who all shared an interest in plants, animals, exotic naturalia, or 'curious simples' as they were often called at the time.[32] The comparatively democratic openness of these gardens to visitors of various classes and backgrounds (though they were certainly not public domains) should be compared with the situation of access to the *Kunst- und Wunderkammern*.

During the sixteenth century natural history thus formed an increasingly varied field of practice which was inextricably connected with collecting, gardening, the universities, and the great voyages of discovery as well as with the more mundane worlds of apothecaries and folk healers. A status problem was the consequence. Precisely because there was neither a demarcated discipline nor a formal training in natural history, all forms of expertise (whether based on medicinal practice, book learning, collecting, growing plants, voyages, or attempts at describing and classifying) were relevant to those scholars who were turning themselves into specialist naturalists. They could only develop their own expertise by drawing upon a vast range of different forms of knowledge. This comprised both learned (written and to a large extent classical) knowledge, the everyday practical expertise of those who were in daily touch with nature and *naturalia*, their own trial and error research and observations, information deriving from rich collectors, folklore, and many other sources. All experts, moreover – including those with a university training (which inevitably was *not* in natural history) – were to a large extent amateurs. The value and status of their respective types of expertise still had to be compared, weighed and established.

This process was not made easier by the fact that – typically, for a Renaissance way of thinking – 'low' activities involving both manual work and contact with lower-class persons meant low-quality knowledge unless it was transformed and elevated by scholarship. While hunting and the outlay of private gardens had by then long been considered pastimes suitable for princes and

---

[31] The Garets played an important role in this respect. See for instance R.S. Roberts, 'The early history of the import of drugs into England', in F.N.L. Poynter (ed.), *The evolution of pharmacy in Britain* (London, 1965), 165-186.

[32] See, for example, for short descriptions of Guicciardini's visits to the gardens of Charles de Saint Omer and Peeter van Coudenberghe (respectively near Bruges and Antwerp): L. Guicciardini, *Descrittione di tutti i paesi bassi*, ed. Dina Aristodemo (Amsterdam, 1994).

aristocrats, the arduous activities that could go with gardening and botany (getting muddy, digging up roots, pruning trees, gathering mushrooms, being out in the fields in all seasons and types of weather) were generally looked down upon and usually left to servants. It is no coincidence, therefore, that derogatory remarks abound in various botanical publications of the early modern period about information and plants obtained from rustics, old women, and other less elevated persons. In the preface to his *Cruijdt-boeck* (1554) the Southern Netherlandish botanist (and friend of Clusius) Rembert Dodoens states that the knowledge of plants and herbs had long been despised in the circles of physicians, who regarded it as a demeaning handwork and a domain only suited to lesser figures, such as 'apothecaries and other untrained persons, who search for herbs on a daily basis in the woods and fields, and that it would have been dishonourable for themselves and a useless burden to acquire and investigate the knowledge of herbs'.[33] The research methods of the new botanical experts required at least some physical contact with nature, however, and they must have been careful about defining their own position between lowly manual labourers and aristocratic garden owners. The well known botanizing trips by sixteenth-century French and Italian naturalists and their (medical) students consisted primarily of teaching and social expeditions during which most of the hard work seems to have been done by servants.[34] It would be interesting to know whether a man like Clusius was doing much manual work himself during his early field trips.[35] The evaluation of the status of botanical (or more generally natural historical) expertise and the attempts at elevating it would take a large part of the sixteenth century and were even then by no means concluded, although by then the hierarchy of the various types of naturalist expertise was fairly clear. In this sense natural history was unlike law or medicine, where the distinctions between university-trained physicians or lawyers of high status and other practitioners (such as surgeons or notaries) dated back to much earlier periods.

Many of the men who have become famous as sixteenth-century naturalists did so on the basis of their publications which presented descriptive surveys of all available information about nature and (in some cases) attempts to classify and systematize it. New illustrated botanical surveys were published from the mid-1530s (by Fuchs and Brunfels) and especially during the 'second wave' of interest during the 1560s-1600s (for example by Dodoens and Clusius). All of these men experimented with new ways of ordering and

---

[33] My translation from the Dutch text as quoted in A. Schierbeek, *Van Aristoteles tot Pasteur. Leven en werken der groote biologen* (Amsterdam, 1923), 41.
[34] Findlen, *Possessing nature*, 164-170.
[35] For some evidence see the contribution by Gillian Lewis in the present volume.

presenting nature.³⁶ Thousands of watercolours and woodcuts of plants and animals were produced during the 1530s-1600s, some of them made 'after nature', while printed works experimented with new forms of scientific illustration. Even though the use of such illustrations could be haphazard and their quality was by no means always good, this relatively new visual means directly contributed to the spread of information about nature, new forms of standardization, and the identification of plants and animals throughout Europe. It would be a mistake, however, to imagine that only the men whom we now regard as the great botanists of the sixteenth century were involved in this more 'abstract' or 'theoretical' attempt to order, classify, describe, and depict nature. Aristocrats such as Saint Omer and apothecaries such as James Garet Jr shared these interests and were involved in the same process of discovery and classification.³⁷ Their assistance, information and ideas were, moreover, indispensable to the scholars who further selected, channelled, organized and translated some of this knowledge and elevated it to the level of specialist expertise. The main difference between these three categories of experts – aristocrats with collections, gardens and varying degrees of erudition; apothecaries with great practical expertise, better access to folk knowledge and considerable erudition; and the scholars themselves – seems not to have been their type of interest or even their range of knowledge, but its results. The rich collectors created gardens and collections, only a few of which still remain. Most apothecaries generally did not write books and typically, those who did are usually classified as scholars rather than as apothecaries. But scholars published.

Naturally, the success of naturalists to proclaim themselves as experts depended not only on their ability to develop their own traditions, expand and define their expertise and community, but also upon their recognition as experts by others. Conditions were right at the time. Various institutions (such as universities and courts) and groups (rich collectors, aristocrats) needed experts in natural history, or thought they did given the fashion of nature. The growth of this new field of expertise may in fact well have been as much an effect of this demand as of the activities on the part of the experts themselves to widen their expertise and promote it. In good Renaissance fashion the fact that people of quality (i.e. high status) were interested in being seen as patrons

---

³⁶ On new classifications of nature see for instance S. Atran, *Cognitive foundations of natural history: Towards an anthropology of science* (Paris/Cambridge, 1989).

³⁷ To name but one example: a letter written by Guy Laurin to Clusius in Spain (dated 25 November 1564, Leiden UL, VUL 101) describes how Laurin and Saint Omer (both members of a small circle of rich Bruges humanists and patrons) were busy in the middle of winter at Moerkercke identifying 'simples' and arranging botanical watercolours 'according to the Dioscoridean method'.

of naturalists, owners of exotic gardens and collectors of naturalia in itself raised the status of this field of knowledge. Within little more than half a century (between about 1530 and 1590) natural history established itself as a respected scholarly and elite activity and a recognized field of expertise.[38]

Clusius was by no means the only naturalist to have trained as a physician. In fact, a large majority of the men who have become known as the principal experts in natural history of the sixteenth century in Europe north of the Alps – the most famous are Leonhart Fuchs, Otto Brunfels, and Hieronymus Bock during the 1530s-1540s; Pierre Belon, Guillaume Rondelet, Conrad Gessner, Carolus Clusius, Rembertus Dodoens, and Kaspar Bauhin during the 1550s-1610s – had studied medicine. Some of them were actually employed as general practitioners, or acted as private physicians to aristocrats or princes. In the latter case they were probably appointed to such posts partly because they could also offer advice on matters of botany, gardening and collecting naturalia. Most of these famous specialists spent the larger part of their working life outside universities, but it is interesting to see that botany (rather than natural history as a whole) had by the late sixteenth century not only managed to become a respectable elite activity but also gained a place as a specialisation within the academic domain of medicine. By that time several European universities had created (generally combined) chairs in anatomy and botany within the medical faculty.[39]

Given the medical background of many naturalists this might seem a natural setting, but the knowledge of medicinal plants (*materia medica*) had until this period been regarded as belonging much more to the domain of the practical experts – the apothecaries who generally did not go to university but learned their trade in practice – than to that of the learned physicians. The process by which university-based medicine appropriated botany and established it as a respectable form of scientific knowledge resembles in some respects the similar appropriation of anatomy and dissection, which in earlier times had belonged largely to the domain of the (non-university trained) surgeons.[40] These

---

[38] For the creation of a field of expertise see especially P. Findlen, 'The formation of a scientific community: natural history in sixteenth-century Italy', in A. Grafton and N. Siraisi (eds.), *Natural particulars* (Cambridge [Mass.], 1999), 369-400. See also her excellent *Possessing nature*; and cf. Reeds, *Botany in medieval and Renaissance universities*. This is not the place to go into the closely related phenomenon of the 'academies': generally private gatherings (especially in Italy) of members of the elite who studied subjects ranging from antiquity to botany, alchemy and astronomy. For the famous Roman Accademia dei Lincei see for example I. Baldriga, *L'occhio della lince. I primi Lincei tra arte, scienza, e collezionismo (1603-1630)* (Rome, 2002).

[39] See Reeds, *Botany in medieval and Renaissance universities*.

[40] See K. Park, 'The criminal and the saintly body: Autopsy and dissection in Renaissance Italy', *Renaissance quarterly*, 47 (1994), 1-33, with an extensive bibliography; R.K. French, 'Natural philosophy and anatomy', in J. Céard, M.M. Fontaine and J.C. Margolin (eds.), *Le corps à la Renaissance. Actes*

are, in fact, by no means unique examples in which the formation of a field of specialized knowledge went hand in hand with the increase in status of that particular type of knowledge, the development of jargon, and the exclusion of others from (continued) access to specialized knowledge. Precisely in the latter respect there was a crucial difference between dissection/anatomy and botany, however. Medical faculties all over Western Europe could and did impose a monopoly on dissection by means of closing off access to dead bodies. There was no way of doing the same with nature, and the contribution of practical (non-university based) expertise to botany, zoology and plant-based medicine continues to be important up to this day.

Even by the end of the sixteenth century natural history as a whole had not yet become a clearly demarcated domain of knowledge. In fact, precisely because of the continuous influx of practical expertise and the continued involvement of people with diverse backgrounds and types of knowledge, it continued to overlap with alchemy, collecting, art, and medicine during the whole of the early modern period. In consequence it suffered to some extent from boundary problems in defining its identity as a serious field of study. Interestingly, natural history may prove to be an example of an initially successful attempt to become a serious field of knowledge – a discipline – which in the very long run has turned out to be less than completely successful. As the situation during Darwin's days showed, natural history (unlike medicine) has remained relatively open to amateurs who could (and still today can) reach a very high level of expertise without advanced technology or formal training.[41]

Natural history also changed drastically in terms of contents and approach in the course of the sixteenth century. The attempts of various scholars to invent new systems of organization and classification were cause for continuous debate, which triggered new research concerning the best criteria for classification and systems of naming.[42] Both the practical uses (as food and medicine) and the symbolic or religious relevance of plants and animals continued to be important throughout the sixteenth and seventeenth centuries. Yet, nature was

---

du XXXe Colloque de Tours 1987 (Paris, 1990), 447-460; F. Egmond, 'Execution, dissection, pain and infamy – a morphological investigation', in F. Egmond and R. Zwijnenberg (eds.), *Bodily extremities. Preoccupations with the human body in early modern European culture* (Ashgate, 2003), 92-128; M. Pelling and C. Webster, 'Medical practitioners', in C. Wester (ed.), *Health, medicine and mortality in the sixteenth century* (Cambridge, 1979), 165-235, esp. 174-78; and a major study which addresses the issue of the relations between surgeons and physicians specifically: E. Huizenga, *Tussen autoriteit en empirie. De Middelnederlandse chirurgieën in de veertiende en vijftiende eeuw en hun maatschappelijke context* (Hilversum, 2003).

[41] Concerning hierarchies of scientific disciplines see Secord, 'The crisis of nature'.

[42] Two centuries before Linnaeus, Clusius and some of his contemporaries were, for instance, experimenting with binomial (Latin) nomenclature. Some of the plant names used by Clusius are still in use in more or less the same form.

increasingly studied for its own sake, in non-emblematic and matter of fact ways.[43] Observation, practical experience, experiment, standardized description, classification and a non-emblematic approach to the object of study – all characteristic of a scientific attitude often labelled as 'modern' and 'rational' – dramatically gained in importance during the sixteenth century. Clusius both embodied these developments and was a key figure in them. His many publications and letters demonstrate his life-long emphasis on precise observation and description, interest in field observation and ecology, and attempts to expand the knowledge of 'exotic' nature.

In spite of the fact that the term 'scientific revolution of the seventeenth century' originated as part of a whiggish and evolutionist style of historiography and has given rise to much unfortunate reification, it is tempting to describe these changes in the approach to natural history during the sixteenth century as a first phase of the scientific revolution or even as an earlier revolution in the life sciences. It is intriguing to note that the term 'botanical *renaissance* of the sixteenth century' (which seems to have been coined in the twentieth century)[44] couches the important changes in this field in cultural rather than scientific terms. We may well ask whether this could not itself be a result of the eventually unsuccessful attempt of the non-mathematical and non-mechanistic life sciences to raise their status. If the botanical renaissance should indeed be regarded as an early scientific revolution in the life sciences – and the topic certainly deserves much more research – that would open up a new range of questions with respect to the history and status of the life sciences themselves and to changing early modern attitudes to nature.

Expert women

A point of special interest with respect to the growth of a community of experts in natural history and the development of a new discipline concerns the

---

[43] About an emblematic worldview see W.B. Ashworth Jr., 'Natural history and the emblematic worldview', in D.C. Lindberg and R.S. Westman (eds.), *Reappraisals of the Scientific Revolution* (Cambridge, 1990), 303-332. Cf. K. Thomas, *Man and the natural world* (Harmondsworth, 1983), and pace M. Foucault, *Les mots et les choses* (Paris, 1966).

[44] The term crops up regularly, but has to my knowledge not been clearly defined; I have also come across references to the 'botanical renaissance of the 19th century'. According to A. Zemanek, historians of science have assigned various dates to the different stages of the botanical renaissance; such as A. Morton, who located its beginning in 1483 with the publication of Latin translations of works by Theophrastus, and 1623 (appearance of Caspar Bauhin's *Pinax theatri botanici*) as its end. See Zemanek, 'Renaissance botany and modern science', 11-13. For an interesting discussion of the innovative character of Renaissance botany in comparison with other scientific disciplines see F.D. Hoeniger, 'How plants and animals were studied in the mid-sixteenth century', in F.D. Hoeniger and J.W. Shirley (eds.), *Science and the arts in the Renaissance* (Cranbury/London, 1985), 130-148.

role of women. Clusius never married, and we know of no affairs, lovers or children. Among his close friends and correspondents were several women, with some of whom he exchanged letters for decades. For instance, ten letters from Anna von Aicholtz Starzerin to Clusius survive for the years 1588-92: she was the wife of one of the professors at the University of Vienna and Clusius' landlady during most of his long stay in Vienna. The 25 letters from the rich Viennese garden owner Anna Maria von Heusenstain Starhemberg to Clusius likewise testify to a longstanding friendship spanning the years 1588-1606. His best-known and almost lifelong female friend, however, was the notorious Netherlandish princess Marie de Brimeu (Ill. 4).[45] Twenty-seven of her letters to Clusius have been preserved for the period 1571-1607. She was the owner of several superb gardens and an enthusiastic amateur grower of tulips and other exotic flowers and shrubs.

In all, letters from 24 different women to Clusius survive. Among them were Yzabeau van Arkel, Ottavia Peverella de Bruti, princess Louise de Coligny, Louise de Boisot, Eva Ungnadin baroness von Sonnegk, Anna de Lalaing domina de Marquette, Conillemette madame de Haeften, and Aleidis Wyetfleet, and two female relatives of Clusius: Catherine Pacquet de l'Escluse and Genevieve le Comte de l'Escluse.[46] Nearly all of them seem to have possessed a garden and most of them – as far as we know at present – were ladies who belonged to the upper ranges of society in terms of both wealth and family. The numbers of their remaining letters to Clusius are, however, much smaller than in the above mentioned three cases. The fact that Clusius kept the letters of these women is important in itself, as is the fact that the number of surviving letters from Marie de Brimeu (27), Anna Marie von Heusenstain Starhemberg (10) and Anna von Aicholtz Starzerin (25) is greater than that of the large majority of his (roughly 275) male correspondents. Longstanding correspondences with men of which more than 10 letters survive are actually only few: Justus Lipsius (22), Jacques Plateau (19), Giovanni Vincenzo Pinelli (about 80!), Joachim Venerius (18), Jean Boisot (12), Joannes de Castaneda (14), Joannes Crato (36), Johannes van Hoghelande (11), Jean de Maes (18), Filippo di Monte (12), and Adolphus Pantius (11). The most outstanding case is the correspondence with Joachim Camerarius II (1534-1598). Only a small number of the latter's letters to Clusius can be found in the Leiden collection, but some

---

[45] See J.L. van der Gouw, 'Marie de Brimeu. Een Nederlandse prinses uit de eerste helft van de tachtigjarige oorlog', *De Nederlandsche leeuw*, 64 (1947), 5-49, which also uses her correspondence with Clusius. She deserves a full-length biography.

[46] Some of these women, such as Marie de Brimeu and Louise de Coligny, were famous persons in their own right; about others we know as yet virtually nothing. This is one of the points of special attention in the Clusius Project.

Ill. 4 Portrait of princess Marie de Brimeu. Anonymous engraving from Jacques de Bie, *Livre contenant la généalogie et descente de ceux de la mayson de Croy* (Antwerp, c. 1615).

195 letters of Clusius to Camerarius covering the period 1573-1598 survive in the original in Erlangen (Germany) and in copy in Leiden. We do not know yet what all of this means. Were friendships with women especially important to Clusius? Was Camerarius more than just a friend? May we infer that Clusius wrote such large numbers of letters to other friends as well, which, however, have not survived? Will the remaining letters in Leiden tell us more about such aspects of his life, or about the reasons why he may have kept certain letters and not others? Since the Clusius collection of letters has not suffered losses since it came into the possession of Leiden University soon after Clusius' death, may we infer that this collection of correspondence contains only those letters that Clusius wanted (or allowed) to be seen after his death?

None of Clusius' correspondence with women friends has been published or even described or analyzed, but even a quick perusal of the letters written to him by Marie de Brimeu and Anna Maria von Heusenstain makes clear that Clusius did not only discuss household or family matters with them. In fact, a large part of their letters to him concern plants, the best ways of growing them in their gardens, and exchanges of seeds, cuttings, bulbs, and fruits. By such means both garden owners and botanical experts increased their knowledge and experience in this field. Nor did they discuss only common garden flowers or plants for kitchen gardens. Like the male garden owners in Clusius' network, these women provided Clusius with information about plants and exotic flowers, grew precious exotica such as white oleanders, tulips and Spanish lilies, and took pleasure in showing these (plus their close connection with Clusius as expert) to their neighbours and friends. In 1592 Anna Maria van Heusenstain, for instance, sent an exotic flower which she had received by courier from Constantinople to Clusius in Frankfurt: 'With this [letter] I send Sir two little flowers which someone sent me from Constantinople by courier'.[47]

Her letters demonstrate moreover, that there was a certain rivalry among these garden owners as to who received most botanical rarities from Clusius:[48]

I have real sorrow because I think that Sir is unfriendly towards me because he does not want anything at all from me and writes me in the last but one letter that I have anyhow so much more than Sir and do not give anything to anyone. I just believe that

---

[47] 'Hiemiet schicke ich dem Herrn 2 Blümlein, die hat mir einer von Konstantinopel geschickt mit einem Kurier', dated 20 September 1592 from Starhemberg to Frankfurt, Leiden UL, VUL 101.

[48] 'Ich habe rechten Kummer (weil ich denke) der Herre sei mir feindlich (gesinnt), weil er so gar nichts von mir begehrt und mir im nächstsvorhergegangenen Brief schreibt, ich hätte sowieso viel mehr Sachen als der Herr und gebe niemandem etwas. Ich glaube nur es seien Leute, die mich gerne mit dem Herrn verfeinden willten, damit der Herr nichts schickt. Aber ich hoffe der Herre wird es nicht tun und nicht glauben dass ich jedermann gebe; dat tue ich nicht, es leistet mein Garten nichts', dated 15 August 1591(?) from Vienna to Frankfurt, Leiden UL, VUL 101.

there are some people who want to create enmity between me and Sir, so that he will not send me anything any more. But I hope Sir will not do this and will not believe that I give things to just anyone; I do not do that and it would not serve my garden.

As Marie de Brimeu indicates in one of her letters to Clusius, this rivalry concerned not only gifts but also the general reputation of a garden; clearly there was considerable competition between experts in this field, which she interestingly denotes as a 'profession':[49]

[…] thanking you very affectionately for your instruction about how to bring out the best in them which will cause my garden to start to acquire a name, even to the point of provoking ordinary envy (but without danger) among those of this profession.

Clusius discussed plants and gardening with many of his female correspondents on the same footing as with some of his male ones. With respect to the formation of disciplines and expert knowledge it is intriguing that botany, plants and gardens were one of the few (semi-)public domains in which women could gain expertise and be addressed as experts in these relatively private exchanges.[50] From the letters of Marie de Brimeu it is clear that she was not only well aware of the exceptional status of Clusius as expert in his domain, but also discussed him with (and was regarded as a discussion partner by) men of equally high position such as Lipsius. After receiving a letter from Clusius accompanied by a large basket full of bulbs, seeds and roots, Marie de Brimeu wrote the following to him:[51]

---

[49] 'Vous en remerciant fort affectueusement et de linstruction pour les faires valoir qui sera cause que mon iardin comencera aussij a avoir quelque nom iusques mesmes a en estre subiect a envije ordinaire (mais sans danger) entre ceux de ceste profession', dated 24 May 1593 from Leiden to Clusius in Frankfurt, Leiden UL, VUL 101.

[50] H. Bots and F. Waquet, *La République des Lettres* (Paris, 1997), here 96-98, discuss women as absent from the Republic of Letters with a few exceptions and do not mention their role in botany. Cf. *Les Correspondances. Leur importance pour l'historien des sciences et de la philosophie. Problèmes de leur édition* (Paris, 1976) [Revue de synthèse, Série générale, vol. CVII, 3rd. ser., 81-82]. Although the longstanding connection between women and botany seemed such an obvious one, the literature I have been able to find as yet concentrates almost completely on the eighteenth and nineteenth centuries, with the exception of a small number of seventeenth-century Dutch and German female artists (such as Maria Sybilla Merian, Alida and Maria Withoos, and Rachel Ruysch, the daughter of a well-known Dutch collector) who depicted naturalia.

[51] 'Je le recevraij cependant dentiere affection tant pour la rarete dicelluij que pour loccasion que dites avoir eu a me lenvoier a scavoir le plaisir que jaij de jardiner le quel certes je vous confesse estre grand mais de joli et non encores de science pour navoir jusques ores eu moien de mij emploier a bon escient comme je voudrois ny eu maistre pour my introduire et assister comme vous que jaij bien entendu par monsr Lipsius estre le pere de tous les beaux jardins de ce paijs et ce tant pour sa cognoissance quaves des simples que pour la liberalite dont aves use envers plusieurs, et maintenant envers moij […]', dated 18 September [year unknown, probably 1591] from Leiden to Clusius in Frankfurt, Leiden UL, VUL 101.

I will receive them certainly with complete affection, as much because of their rarity as for the opportunity that you say you have had to send them to me knowing how much I enjoy gardening. That pleasure, I confess to you, is certainly great but is as yet more enjoyment than science because I have neither many hours to spend on it and become more expert as I would want nor a good instructor who can introduce and assist me in this subject, like you who – as I have understood from monsieur Lipsius – are the father of all the beautiful gardens in this country, on account of both your knowledge of simples and the liberality which you have shown towards many and now towards me [...].

Modes of exchange: languages and informality

It is fascinating that none of the undoubtedly well educated women of high social position who corresponded with Clusius did so in Latin. They used either French, German or Dutch/Flemish, while Clusius probably responded in the same language. It is not enough in this respect to point out that women at the time did not have access to universities or other forms of higher education, since many of Clusius' male correspondents likewise preferred the vernacular. Their choice of language seems to have been only partly related to their social or educational status. For instance, the fact that some Polish or Hungarian correspondents wrote in Latin is simple enough to explain in practical terms. Most of the humanists (such as Lipsius, or the Laurin brothers, who were core members of the Bruges humanist circle) as well as several German, French, Dutch and English physicians wrote in Latin, although they could have used their respective vernacular languages if they had wanted to. Some of the learned Italians (such as Ferrante Imperato, Hieronimo Calzolari, and Giacomo Cortuso, but not Ulisse Aldrovandi who used Latin) did indeed prefer Italian, while quite a few other men who must have known Latin preferred French. Dutch, German and possibly Spanish seem to have been used mainly by correspondents who did not know Latin. A few correspondents used more than one language, either in one and the same letter or successively, changing back and forth. This was the case with the rich bourgeois garden owner Johannes van Hoghelande from Leiden, who mixed French and Latin, Christian von Ecgk, who wrote in French for some years, then changed to German and later turned back to French, and Joannes Lewenklau, who alternated between Latin and Italian.

Judging from the information available at the present time about the bulk of the letters sent to Clusius, about half of them were written in Latin, a third in French, while the others were in Italian, German, Spanish or Dutch (in that descending order). Clusius himself was a polyglot. He read and wrote an impressive number of languages: besides Latin, Dutch and French, he was fluent in German, Italian and Spanish. He could probably read and understand Portuguese, read (classical) Greek, and may have understood some Hungarian. One of the few languages he did not manage to master well was the present-day lingua franca, English. Although he visited Great Britain several times, learnt some English, and obviously did manage to communicate with its inhabitants (probably some of the time in French), no letters in English to him (or by him) survive. We may infer that Clusius' correspondents could generally write to him in the language of their choice. Clusius himself did not invariably adapt to his correspondents' choice of language, however. In the correspondence (1550-1578) with Hubertus Languetus, a friend from his student days in Paris, Languetus always used Latin, while Clusius consistently wrote back in French – perhaps for nostalgic reasons.

The multiple language use in this correspondence should make us wonder. Latin was perhaps slightly less of a lingua franca in the circles of *studiosi* in early modern Europe than we might have thought. Should we interpret the possibility offered by Clusius to his correspondents to use the language of their choice as part of a code of civility? It may certainly have had this effect on some of his less educated correspondents, who obviously felt at home writing to Clusius in their vernacular (or dialect) and in handwriting that had very little to do with the beautiful and legible hand in which humanists were supposed to write. But could the fact that Clusius learned so many different languages (whatever practical reasons may have existed) also imply that Latin was not enough for his purposes, and that communication in the vernacular with informants from other countries was essential to his type of research? Did this apply to other natural historians too? In a more general sense, could this point to a difference in modes of communication between naturalists and other scholars of the early modern age?[52]

These are issues that cannot be answered without further investigation of the languages used in correspondence by other European naturalists and of their correspondents' backgrounds. It certainly looks, however, as if the fact that natural history was (as yet) less closed off by scientific disciplinary boundary markers than some other branches of the sciences had its effects on the use of language. In comparison with the mathematical sciences or even the medical sector, a specialized jargon developed relatively late and only slowly in natural history. This may have happened precisely because botanical expertise

---

[52] Mario Biagioli's work has inspired many of the topics raised in this essay. See especially his: 'The anthropology of incommensurability', *Studies in the history and philosophy of science*, 21 (1990), 183-209; 'Etiquette, interdependence, and sociability in seventeenth-century science', *Critical inquiry*, 22 (1996), 193-238; and the abovementioned 'Le Prince et les savants', and 'Scientific revolution, social bricolage, and etiquette'.

was based to a considerable extent on local knowledge and first-hand observation. Communication with non-elite practical experts and other non-university trained persons could not be conducted in Latin. Yet they possessed important information, as Clusius and others well realized. From both his printed works and his correspondence, it is evident that Clusius was seriously interested in 'folk' nomenclature of plants, its possible relevance to their classification, and in local information about the natural 'habitat' of plants. His own multilingualism and the fact that he corresponded in so many different languages may therefore well be clues to his innovative research methods and, vice versa, to the way in which the vernacular languages and 'practical expertise' were making their way into (and contributed to) the new discipline of botany.

The considerable number of Clusius' correspondents who did not belong to the range of university-trained scholars or even the social elite do not quite fit the usual image of members of the republic of letters. Yet, some of his correspondents explicitly spoke of a *republic* to describe the community of naturalists. Nor do the styles of exchange characteristic of the letters written to Clusius conform to the stereotypic image of humanist letters. In fact, quite a few of the letters written to Clusius look as if no particular care had been taken about composition, lay-out, or handwriting. The contents usually match their appearance. Clusius was clearly held in great respect and affection by most of his correspondents, but they rarely go out of their way to address him in a formally deferential way. Clusius must either have been a very easy person to approach, even in his later years when his reputation had reached its peak, or the rules of deference in the circles of naturalists appear to have been less strict than among humanists. Moreover, anarchy appears to reign in the correspondence concerning the topics that could be discussed and the order in which to do so. Apart from the fact that information about previous letters and sent packages can nearly always be found in the first few lines, while greetings to and from family members, friends and other naturalists are usually reserved for the last ones, no order seems to have been prescribed. Requests for seeds, information about new plants and bad growing seasons, new publications, complaints about servants, information about political events, gossip about affairs, illness, marital problems, or financial misfortunes of acquaintances, a request for some new shoes to be sent, news about the great voyages to the East and West Indies, hints at religious controversy and so on are mentioned indiscriminately. Ciceronian models of written eloquence are a long way off.

Such elementary characteristics are revealing. Most of these exchanges were largely informal and matter of fact. They belonged to an ongoing exchange between people who shared the same interests, so that many things did not need to be explained. Clusius was neither a prince nor an apothecary, but he corresponded with members of both social groups about nature, recognizing a shared interest in nature, special expertise and a real pleasure in it which could suspend (if not obliterate) social differences that would probably have been very hard to ignore under any other circumstances in this age. Among the most interesting general characteristics of these letters are their liveliness, practical orientation, and above all, the fact that they were obviously *not* meant for publication or even circulation on a larger scale.[53] They neither look nor read like the semi-public type of correspondence which was written to be passed on or read aloud, and which belongs rather to the circles of Grotius, Lipsius, Mersenne or Erasmus.[54]

The rules of friendship

The theme of friendship is explicitly mentioned in almost every single letter in the Clusius correspondence. Could there be a link between the informal character of the exchanges between these naturalists, the frequent use of the vernacular, the apparent ease by which significant status differences were set temporarily aside, on the one hand, and the formation of a European community of scholars and the establishment of a new field of expertise, on the other hand? Friendship did have a much wider meaning during the early modern period than it does now, however, and could include the notions of mutual assistance and kinship. It played an important part in humanist correspondence of the seventeenth century as well as in the more mundane exchanges of, for instance, early modern merchant families, while it implicitly or explicitly carried overtones of the classical rhetorical concept of friendship.

In the Clusius correspondence the notion of friendship is closely linked with the idea of disinterested services to a friend and of gift exchange.[55] The

---

[53] The fact that these letters were definitely *not* meant for publication undermines the often implicit assumption in discussions about early modern correspondence that earlier examples of correspondence were generally meant for 'public' use while later ones had a more 'private' character. That assumption seems to be based on teleological models of historical development and the idea that 'the private' and 'privacy' (much like 'the individual') were niches carved out of a larger public domain only during the Renaissance.

[54] See Bots and Waquet, *La République des Lettres*; and cf. *Les correspondances*. C.M.G. Berkvens-Stevelinck, H. Bots and J. Häseler (eds.), *Les grands intermédiaires culturels de la République des Lettres. Etudes de réseaux de correspondance du XVIe au XVIIIe siècles* (Paris, in press), had not yet come out by the time this essay was finished.

[55] Present day discussions about gift exchange by historians generally go back to issues raised in M. Mauss, *The gift. The form and reason for exchange in archaic societies* (London, 1990; originally 'Essai sur le don. Forme et raison de l'échange dans les sociétés archaïques', *L'Année sociologique*, 2nd ser., 1 (1923-24); cf. N.Z. Davis, *The gift in sixteenth-century France* (Oxford, 2000).

letter which the Flemish merchant Hendrik Bloeme sent to Clusius in 1601 is typical in this respect:[56]

I send you in this box three leaves of my Indian fig[57] and two of my aloë americana plants. If I had any larger ones I would send them to you, but since I have shared all of them with friends who also desired them I don't have any more at the moment [...] I thank you very warmly for the gift that it has pleased you to make to me of both your Historia plantarum and the book of the late monsieur de St. Aldegonde[58] and consider it a great favour. I have always regarded myself as very much obliged to you for the many courtesies that I have received from you over such a long period and have often wished for an occasion that would enable me to do something in return, but continuing to burden myself with new favours (without using any ceremony) I ask you affectionately whether there is anything with which I can serve you, or whether there is anything from here that would be pleasant to you; please ask me directly [...].

Clusius' old friend Jean de Brancion also wrote to him about their friendship:[59]

[...] but I beg you not to use the formalities that you put in your letter, because you know well that I am not taken in by them and that I am and always will be a true and devoted friend and that on my side I need nothing but a reciprocation of that friendship [...].

Sharing was seen as a virtue, as a means of giving pleasure, and as a sign of friendship. Liberality and generosity – to share ánd be seen to share – were very important in the virtual community of sixteenth-century naturalists, both for men and women. The apothecary Thomas de la Fosse wrote to Clusius in

---

[56] '[...] vous envoije en ceste cassetta 3 feuilles de mon figuijer d'inde avec deux plantes de mon Aloë Americana, si je le eusse en plus grandes je les vous envoijeroije, mais les ayant tous participe avec amis qui le desiroijent n'en aij pour le present aultres [...] Je vous remercie bien affectueusement de la presentation qu'il vous a pleu faire tant de votre Historia plantarum que du livre de feu monsieur de St. Aldegonde, l'estimant un tres grand faveur. Je m'aij toussiours estime estre fort oblige envers vous pour tant de courtoisies des long temps receu et en aij souvente fois desire quelque occasion pour aulcunement vous en pouvoir gratifier, mais me chergeant continuellement de nouvelles faveurs (sans user de ceremonies) vous prie tres affectueusement s'il ia perdeca en quoij je vous pourroij servir, ou que desiriez quelque chose dici que vous seroit agrable me le vouloir ainsi promptement demander [...]', dated 13/23 April, 1601 from Frankfurt to Clusius in Leiden, Leiden UL, VUL 101.
[57] An opuntia cactus.
[58] This is the Southern Netherlandish nobleman Philip de Marnix de St. Aldegonde (1540-1598).
[59] 'Mais je vous prie n'user plus des ceremonies que mettez en v-re l-re, car vous scavez bien que je n'en suijs refaict et que je vous suijs et seraij toujours vraij et entier amij et qu'il ne fault en mon endroit aultre recognoissance sinon la reciprocque', dated 3 August 1571, from Malines to Clusius in Paris, Leiden UL, VUL 101.

1596 that he hoped to share the special plant 'lenticus' with as many friends as possible who were worthy and emphasized that he would be pleased to be of service to Clusius because he valued his friendship.[60] Clusius' own career is illuminating in this respect. He started out as a young, Protestant physician from the Southern Netherlandish nobility. In the course of his professional life he built up an extensive knowledge of European botany by personal observation and fieldtrips (in France, Spain, Portugal, Hungary, Austria, Germany and England). For various reasons he did not travel to Italy himself, however, nor did he ever leave Europe. For information about plants and animals from Asia and the New World as well as for expertise concerning plants grown in Italy, the Balkans, Greece, Poland, and the Middle East he was thus completely dependent on his correspondents. Moreover, his many friends who owned gardens all over Europe continued to provide him during the whole of his life with information about the best ways to grow certain types of plants – especially those imported from different climates – how to multiply them, sow or prune them, and how to develop new varieties.[61]

As indicated in the introduction, this was by no means a one way system. Clusius reciprocated. He shared not only plants, roots, bulbs, seeds and (probably to a lesser extent) ideas, but was also generally regarded as a man who did so. In some cases he initiated new friendships by such gifts. There are very few letters from his correspondents in which he is not thanked by them for the latest box of seeds and cuttings and offered new ones in return. Until shortly before his death Clusius maintained his complex pattern of give and take, keeping up friendships and gift exchange relations via correspondence. The amount of time and effort it must have taken him to do so is a direct reflection of the importance of such exchanges to Clusius, in terms of both friendship and the maintenance of his position as leading European expert. If Clusius could not become a leading figure in this new field of expertise without relying strongly on a large network of informants (as well as on important patrons), we may conclude that no naturalist could gain prestige or even become a real expert without the help of others.

---

[60] 'J'aij espoir de le sy bien conserver que je le poldraij tenir et en participer avecq le tans les bons amys qui en sont dignes', dated 12 July 1596, from Middelburg to Clusius in Leiden, Leiden UL, VUL 101.

[61] Vice versa, this wide network of correspondence itself may have contributed to new notions of European geography and the concept of Europe itself – a theme I have developed a little bit in F. Egmond, 'A European community of scholars: Exchange and friendship among early modern natural historians', in D. Curto and A. Molho (eds.), *Rethinking the history of Europe. Images, symbols, discourses* (Florence, in press). The implications of the fact that Clusius could also check the results of the cultivation of certain plants in different climates and conditions by correspondence deserve further attention (with thanks to Irene Baldriga).

The emphasis on disinterested friendship and a 'free' exchange of gifts may have been especially strong in the European ('virtual') community of naturalists which had, after all, only started to develop from the 1530s on. This was very much a new community of experts that tried to carve out its own niche in a social and cultural domain in which many different parties operated – such as university academics, physicians, surgeons, aristocratic garden owners, apothecaries, explorers and peasants, local healers – and where different economies overlapped and clashed. 'Free' gift exchange was in everyone's interest because it helped raise the general quality of the available expertise and thus the status and honour of the discipline (and by extension those of its members). Thus the idiom of friendship both expressed and underpinned the value of free exchange, which itself helped to create a 'virtual' community of natural history experts throughout Europe.

We should not be naïve, however, about how 'free' such exchanges actually were. The Clusius correspondence contains many clues to the informal but very important rules of behaviour governing these apparently liberal exchanges. Once a person was recognized as 'a friend' – that is, a fellow member of the network of naturalists – it was not done to withhold information or lie about it, refuse to give gifts in return, be stingy, steal bulbs (or have them stolen by servants), publish some else's results or discoveries without any form of recognition, bribe agents or brokers in order to obtain rare naturalia that were destined for someone else, plagiarize (although the definition and the notion of authorship were not identical with modern ones), et cetera.

For instance, the apothecary James Garet Sr wrote to Clusius about the rich bourgeois garden owner Johannes van Hoghelande in whose garden he had seen some beautiful flowers that were a present from Clusius himself: 'he is so tight-fisted that I could not get any out of his hands, patience'.[62] A remark by James' son Pieter in a letter to Clusius also reveals that sharing was certainly not done indiscriminately:[63]

Paludanus[64] implored me to give it [i.e. a special type of gum used by Indians to waterproof their canoes] to him or at least part of it, but I refused, hearing that he

---

[62] '[…] hij es zoe vasthoudende ic en conster niet van crijghen pacientia', dated 4 April 1592 from London to Clusius in Frankfurt, Leiden UL, VUL 101.

[63] 'Paludanus heefter mij seer om gebeden ofte ten minsten een stuck daervan dwelck ick geweijgert hebbe hoorende dat hij het wilde bescrijven, dan hebbe hem geseijt soo het UL geliefde dat ghij hem een stuck soude willen vereeren', dated 30 January, 1602, from Amsterdam to Clusius in Leiden, Leiden UL, VUL 101.

[64] Bernardus Paludanus (1550-1633), a physician and collector, lived in the town of Enkhuizen in North Holland. He had a private botanical garden and his collection of curiosities was famous throughout Europe north of the Alps. Many foreign guests of high rank visited him. He travelled widely, but from his international correspondence very few letters survive, while he published very little

wanted to describe it, and I told him that if you wanted him to have part of it you would make a present of some to him.

In other words, when Pieter Garet discovered that Paludanus wanted to 'steal' the honour from Clusius of being the first to describe or publish a special gum, he protected Clusius – and thereby his own good relationship with him – also leaving the option open for Clusius to show his generosity and give some of the gum to Paludanus. Transgressions of these rules of civility and courtesy resulted in conflict, broken friendships, denial of access to further information, or loss of reputation and status in the virtual community of scholars.[65] Ultimately, an offending person could be informally ostracized from this virtual community.

How much irritation, jealousy and even exclusion from the 'exchange circle' the breaking of these rules could cause, emerges from a letter from Jacques Plateau, a rich garden owner living at Tournay in the Southern Netherlands, to Clusius. He describes the visit of Jehan Robin, Parisian apothecary, creator of the botanical garden of Paris (in 1597) and 'hortulanus' of the French king, to the Low Countries and to Tournay, where Robin visited all of the gardens (Ill. 5):[66]

(botanical entries for the *Itinerario* by his friend and fellow townsman Jan Huygen van Linschoten, many of which were based on Garcia de Orta's *Colóquios dos simples*). There is no monograph on him. See H.D. Schepelern, 'Natural philosophers and princely collectors: Worm, Paludanus and the Gottorp and Copenhagen collections', in O. Impey and A. MacGregor (eds.), *The origins of museums: the cabinets of curiosities in sixteenth and seventeenth-century Europe* (Oxford 1985), 121-127; and J. Drees, 'Die 'Gottorfische Kunst-Kammer'. Anmerkungen zu ihrer Geschichte nach historischen Textzeugnissen', in H. Spielmann and J. Drees (eds.), *Gottorf im Glanz des Barock. Kunst und Kultur am Schleswiger Hof 1544-1713*, 4 vols. (Schleswig, 1979), vol. II, 11-28.

[65] See Bots and Waquet, *La République des Lettres*, 124-126; and for a later period cf. Anne Goldgar, *Impolite learning. Conduct and community in the Republic of Letters, 1680-1750* (New Haven/London, 1995). For a comparison and differences between northern Europe and Italy in this respect see Findlen, 'The formation of a scientific community', esp. 383-389.

[66] 'Il a visite tous les jardins de la ville tant de gens d'église que d'apothicaire, desques il a amasse grande nombre de plantes communes. Je luy dit pourquoy il s'obligoit a tant de personne voyant que une seule personne luy pouvoit furnir. Il me disoit quil alloit de Tournay a Gand puis a Bruges, Anvers, Malines, Bruxelles, Mons, Valenchienne, Cambray et la a Paris. Jestime quil aura faict le mesure ausdictes villes comme en Tournay s obligeant a une infinite de personnes. Depuis son retour a Paris j'ay receu deux lettres le messager quoy me delivre sa lettre me dit quil avoit bien 8 ou 10 stant seulement pour Tournay [...] pour plantes nouvelles je ne pense point quil en aura eu beaucoup quant est de moy il n'en a eu une seul. Il y a plus de 15 ou 16 ans et davantaige que je n'ay receu plante de luy aussi je ne me demande estre oblige a luy pour cause quil demande si grande nombre de plante a une fois que lon en est degoute, sil en encore une il en demandere 10 ou 12 voyant quil n'a sceu avoir de moy ce quil desiroit, il m'a prie luy en vouloir vendre aucunes, Je luy ay respondu quelles nestoient a vendre depuis je n'ay eu des ses novelles. [...] Jay peu de fiance aux Franchois ce me seroit pas le premier, ne seconde fois quil m'auroient trompe', dated 8 February, 1602 from Tournay to Clusius in Leiden, Leiden UL, VUL 101.

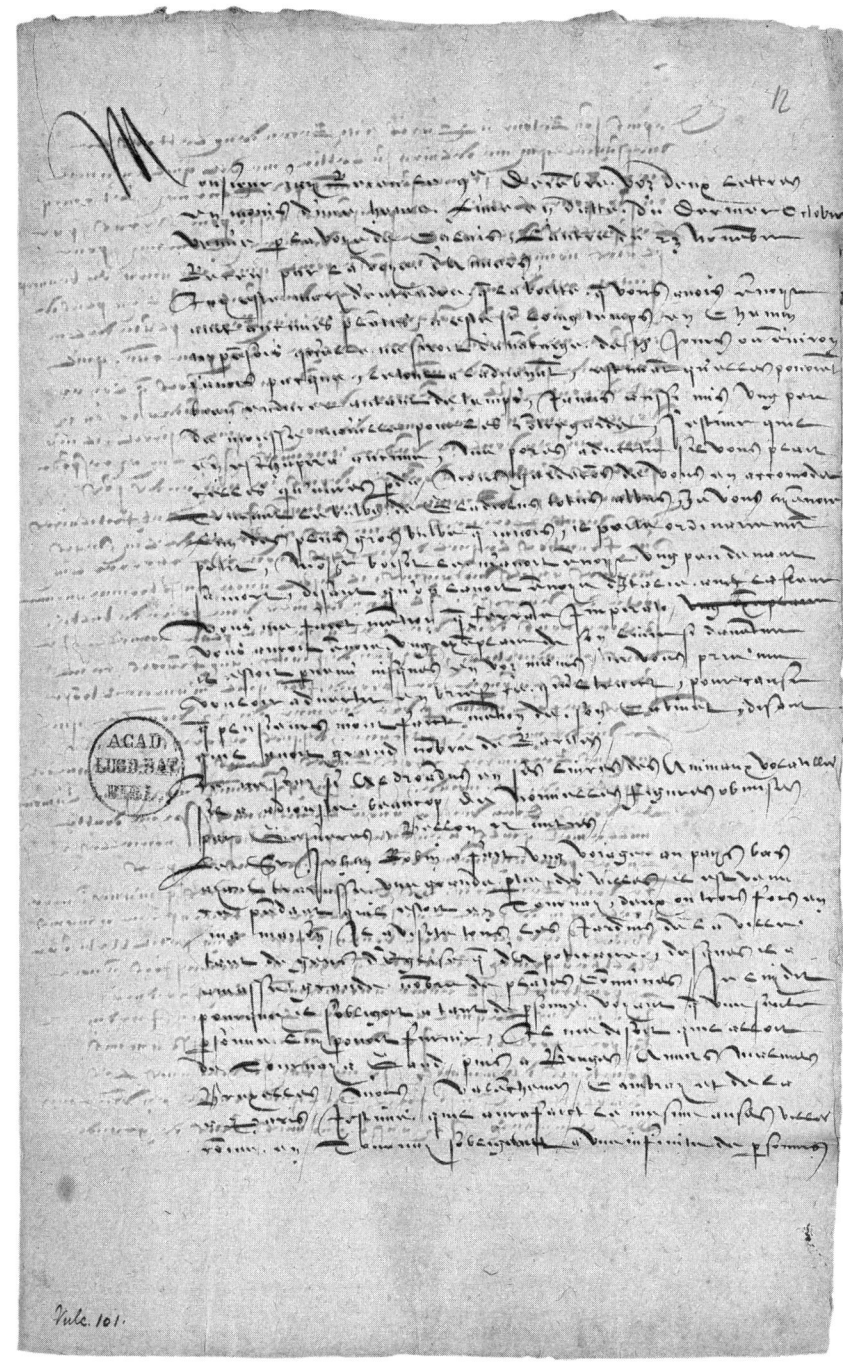

Ill. 5. First page of the letter from Jacques Plateau in Tournay to Clusius in Leiden, quoted in the text (8 February 1602).

He visited all the gardens of the town, both those of the church dignitaries and those of the apothecaries, from where he amassed a large quantity of ordinary plants. I asked him why he was putting himself under an obligation to so many people when he could have had the same plants from a single person. He told me that he would go from Tournay to Ghent, then to Bruges, Antwerp, Malines, Brussels, Mons, Valenciennes, Cambray and from there to Paris. I expect that he did the same thing there as he did in Tournay, putting himself under an obligation to an infinite number of persons. Since his return to Paris I have received two letters; the messenger told me that there were as many as 8 or 10 letters for Tournay alone [...] in which he asked for new plants. I don't think he will have received many. From me he had not a single one. For more than 15 or 16 years or more I have not received a single plant from him. I also do not feel obliged to him, because he asks for such a large number of plants all at once that it makes one feel disgusted; if he still has one he asks for 10 or 12. Seeing that he did not get from me what he desired, he has asked whether I wanted to sell him some. I have replied to him that they were not for sale. Since then I have had no news from him. [...] I have little trust in the French, it would be neither the first nor the second time that I have been deceived by them [...].

It was bad enough that Robin asked for too much and gave nothing in return. The penultimate sentence of this letter shows how he made matters even worse by offering to buy plants from Plateau, who disgustedly said that they were not for sale. This whole episode nicely illustrates the point that the type of 'free' exchange of naturalia as described in these examples existed exclusively *within* the community of naturalists who regarded each other as friends and were bound by the rules of civility and honour.

While these types of 'free' exchanges persisted, by the later sixteenth century European society was becoming geared to merchant enterprise, the market, and commodities. Quite a large number of naturalia were not only of academic interest, but also formed part of the market economy as (potential) drugs, spices, or valuable rarities. The fact that gardens were regularly plundered and bulbs were frequently stolen from both gardens and packages of Clusius and his correspondents indicates that many people had an eye for their commercial value.[67] The market value of plants would become much more evident some decades later, during the early seventeenth century, with the famous tulip craze of the 1630s when prices of tulips reached astronomical proportions, the growing market for medicinal herbs, and the fortunes made in the spice trade (nutmeg, pepper, ginger and cinnamon). During the 1580s and 1590s the persons who probably best realised the economic potential of

---

[67] References to the plundering of gardens (including Clusius') can be found in letters of Marie de Brimeu and many other correspondents.

special plants were those naturalists who (also) acted as brokers and were involved in commerce themselves.

The members of the Garet family are a case in point.[68] James Sr and his sons James Jr and Pieter worked as apothecaries and spice traders. They originated in Antwerp, fled the Southern Netherlands in 1569-70, no doubt for religious as well as economic reasons, and moved to London, where they joined the Dutch Reformed Church at Austin Friars. Pieter later moved to Amsterdam. The Garets combined scholarly and economic interests. The London Garets grew European and exotic plants in their gardens, experimented with them and exchanged information with fellow botanists, druggists, perfumers, and spice traders. They received and corresponded with many learned guests from the Continent, besides forming part of a highly specialized circle of botanists, physicians and apothecaries in London, including Hugh Morgan, Thomas Penny, John Gerard, Thomas Moffet, Jacob Cole (Ortelianus), Richard Garth, Mathias de L'Obel, and John Rich, many of whom likewise corresponded with Clusius. Both the younger Garets developed a regular passion for exotic naturalia. James (also known as Jacques) Jr was especially interested in new varieties of tulips and lilies. In 1589-90 he also grew potatoes in his garden, a most exotic plant that had only just been brought to England from Peru. He was closely in touch with the famous overseas adventurers of this period, such as Sir Francis Drake and Thomas Cavendish, and it was through him that Clusius gained access to some of the newly discovered drugs and plants brought from Roanoke and Virginia.[69] Together with his fellow London apothecary Hugh Morgan, 'royal druggist' and likewise correspondent of Clusius, James Garet Jr pioneered the importation of drugs and the cultivation of plants from the New World in Britain.[70]

In Amsterdam his brother Pieter Garet also owned a garden where he grew exotica and in his turn maintained close contacts with the first Dutchmen who explored the East Indies, such as Wybrant van Warwijck and Jacob van Neck. After 1602 he may even have had a contract with the Dutch East Indies Company (VOC) to provide the outward bound ships with drugs for the medicine chest. Although he was a merchant himself, Pieter complained bitterly in a letter to Clusius when he discovered that upon the return of several ships from

---

[68] I am preparing an essay specifically dealing with the Garets, their letters and biographies, based on extensive archival research. They also figure in D.E. Harkness, '"Strange" ideas and "English" knowledge. Natural science exchange in Elizabethan London', in P.H. Smith and P. Findlen (eds.), *Merchants and marvels. Commerce, science and art in early modern Europe* (New York/London, 2002), 137-160.

[69] See L.G. Matthews, *The royal apothecaries* (London, 1967).

[70] See Pelling and Webster, 'Medical practitioners', 178; and Roberts, 'The early history of the import of drugs'.

the East Indies, buyers from the emperor's court in Vienna or Prague were ready to board the ships in the harbour of Amsterdam, in the middle of the night if necessary, and pay enormous sums for exotica – thus preventing him from obtaining anything special for Clusius or for himself.[71] Thus, at this personal level, an economy of free exchange and barter among experts overlapped and sometimes clashed with a market economy. The two were certainly not mutually exclusive, but belonged to different 'registers': non-market exchange going with hospitality, gift, generosity, friendship and reputation; market exchange with the money economy, capital, profit and commodification.[72] No wonder that Plateau was so offended by Robin's behaviour!

Friendship and pleasure: a shared fascination

Friendship was at the same time a *topos* and practical reality. Up to a point friendship and sharing the same passion for nature were able to efface (or perhaps rather temporarily overrule) some of the social and gender differences that undeniably existed between Clusius and quite a few of his correspondents. This could work in more than one direction. The Garets stood below Clusius in terms of status and education, although probably not in terms of wealth. None of them had gone to university, while as businessmen they belonged to the world of trade rather than study. This is also apparent from the respectful tone of the Garets' letters and their frequent demonstrations of willingness to be of assistance. Nonetheless, the Garets and Clusius were friends and that term was not a merely rhetorical one. Their (gift) exchanges were mutual. Respect and liking obviously existed on both sides. Clusius visited the Garets in London and recognized them as experts in their own right. They are regularly mentioned as friends and experts in letters from other correspondents of Clusius too, such as Joachim Camerarius II and Richard Garth, as well as in Clusius' own *Exoticorum libri decem* (Antwerp, 1605). The fact that three members of one family living in different countries and belonging to two generations maintained a correspondence from 1583 (or even before) to 1605 also points in the direction of mutual and longstanding friendship.

In terms of social status and wealth Clusius himself, in his turn, stood definitely below correspondents such as princess Marie de Brimeu, Charles de Saint Omer, the Count of Arenberg and other members of the high aristocracy.

---

[71] Pieter Garet (in Dutch), 9 February, 1605 from Amsterdam to Clusius in Leiden, Leiden UL, VUL 101.
[72] P.H. Smith and P. Findlen (eds.), *Merchants and marvels. Commerce, science and art in early modern Europe* (New York/London, 2002) largely ignores the fact that these two types of economy could co-exist and were not necessarily mutually exclusive (and that most naturalists participated in both).

Education and his growing prestige as a scholar probably acted to some extent as a balance. Clearly the differences in status between patrons and their protégés could be overcome up to a point by friendship, shared affinities and the sheer pleasure they took in discoveries and the beauty of nature. Moreover, botany was not only a passion shared by those who already were friends, but it could and did create new friendships between people from different countries and different classes, as is clear from the Clusius correspondence. In the continuous search throughout Europe for new discoveries, friends of friends continually recommended each other as new contacts for exchange. Frequently new friendships grew out of such contacts, thus promoting the growth of the virtual community of naturalists.

The Clusius correspondence also provides ample evidence that pleasure, the enjoyment of nature and emotional responses to it were felt long before the Romantic era. It cannot be a coincidence that Clusius' near-contemporary, the Swiss naturalist Conrad Gessner, was probably the first European to write (1541) about the beauty and majesty of the Swiss mountains, which he explored during herbalizing trips.[73] Such herbalizing excursions and journeys which became part of training for many students of natural history during the later part of the sixteenth century may have had an important effect on the appreciation of nature in an aesthetic, non-utilitarian and also non-religious and non-symbolic way.[74] Expressions of enjoyment can be found not only in the letters from wealthy and aristocratic owners of private botanical gardens, such as princess Marie de Brimeu, but also in those from more modest middle-class garden owners. A touching example of the owner's emotional attachment to his garden comes from two letters written to Clusius by the apothecary Thomas de la Fosse about his deceased uncle Jean Mouton, likewise an apothecary. The uncle had transmitted his pleasure in plants to the nephew, as De la

---

[73] In a treatise about milk and milk products: *Libellus de lacte et operibus lactariis, philologus pariter ac medicus. Cum epistolae ad Jacobum Avienum de montiium admiratione [...]* (Zurich, 1543). See H. Wellisch, 'Conrad Gessner: A bio-bibliography', *Journal of the Society for the Bibliography of Natural History*, 7 (1975), 151-247, especially 159 and 180-181. Cf. J. Spicer, 'Roelandt Savery and the 'Discovery of the Alpine waterfall', in E. Fučíková et al. (eds.), *Rudolf II and Prague. The imperial court and residential city as the cultural heart of Central Europe* (Prague/London/Milan, 1997), 146-156, where a letter from Gessner on this topic is quoted.

[74] Naturally I am not arguing here that (some of) these naturalists were a-religious. The point is that people started writing (and, so it seems, feeling) about nature in terms of beauty and not necessarily of religion or symbolism. For other examples see e.g. Findlen, *Possessing nature*, 180-185, and Baldriga, *L'occhio della lince*, 1. Baldriga (personal communication) indicates that there was a clear difference for the Linceans between the savage experience of nature (suitable to the dignity of the scholar) and the mere practice of gardening (more suitable to apothecaries and simplicisti). This is something to be further investigated for Clusius and friends.

Fosse said retrospectively: 'who is the cause of my total delight and pleasure in plants'. Mouton had for many years been one of Clusius' regular correspondents and shared the latter's love of plants down to the last day of his life. Even in his final months, when he was extremely weak and paralysed from the neck down, one of Mouton's main delights and preoccupations in life was his garden: 'still he was taking care of his garden and had plants moved because of the cold and other things done as if he were not ill at all [...].'[75]

This emotional involvement with nature and gardens could have a complex background and for at least some men and women was deeply connected with the experience of danger. Clusius' own passionate interest in plants and animals may well have been fed by a wish to escape from the contemporary world of religious persecution and warfare. Not only Clusius himself but many of his friends and relatives had suffered losses on those accounts.[76] For many expert naturalists in the sixteenth century 'nature' formed a safe enclave and a hide-out. Giuseppe Olmi argues, for instance, that Ulisse Aldrovandi's shift from medicine and philosophy to natural history after his imprisonment by the Inquisition and during the Counter-Reformation was decisively influenced by the uncontroversial nature of natural history.[77] Quite a few naturalists were Protestants or even Calvinists, and nearly all had personally experienced the effects of warfare and persecution in a Europe that was deeply divided by wars, conspiracies and religious controversy.[78]

As they could see in their immediate surroundings, many fields of knowledge were dangerous. Theology and alchemy literally cost some experts their lives and apothecaries in particular could be suspected of knowing too much for their own good about drugs and poisons.[79] The case of William Turner, the English naturalist and author of *Libellus de Re Herbaria* (1538), was particularly dramatic. He joined the Reformation, was banished from England, lived

---

[75] 'Qui est cause que j'ay entierement delectassion et plaisir aulx simples.' 'Encore avoit il soing de son jardin faisant transporter les plantes pour la froidure et faire autres chozes comme s'il n'eut este aucunement malade' (quotations from letters dated respectively 12 July 1596 from Middelburg to Clusius in Leiden, and 31 August 1589 from Tournay to Clusius in Frankfurt).

[76] See Hunger, *Charles de l'Escluse*.

[77] Giuseppe Olmi, *Ulisse Aldrovandi. Scienza e natura nel secondo cinquecento* (Trento, 1976), 45 and 57.

[78] See Reeds, *Botany in medieval and Renaissance universities*, 13-14. In this essay I have not gone into the interesting topic of connections between religion and botany (or its scientific culture), because it is a much more complex one than might seem, and has not yet been researched for Clusius and his network.

[79] See B.M.I. White, *A cast of ravens. The strange case of Sir Thomas Overbury* (London, 1965) for information about the murder by poison of Overbury (1615) in the Tower of London in which, it seems, the son (Paul de Lobel, an apothecary) of the botanist Mathias de Lobel was indirectly involved. Lobel was examined but not put on trial.

in Italy and Switzerland around 1542-43 (where he became a close friend of Conrad Gessner), and travelled in the Low Countries and Germany. His works were banned by Henry VIII, but he was able to return to England upon the accession of Edward VI. He lost his job and almost his life during the first year of Queen Mary's reign and had to flee once more to the Continent. His works were banned a second time, and he was only able to return when Elizabeth I became queen in 1558.[80] A rare example in which this connection between dangerous times and the image of natural history and botany as a safe domain was explicitly made at the time occurs in a letter from the Southern Netherlandish nobleman Philip de Marnix de Saint Aldegonde to Clusius: 'The French king himself, who is a great lover of horticulture, desires nothing but leisure after the turmoil of these civil wars, so that he can establish botany in France with the help of such men as I have described to you.'[81]

Conclusion

Correspondence formed only one of the means of exchange among early modern experts in natural history: there were face to face meetings, travel, social gatherings, the participation in a court or aristocratic setting as either patron or protégé, printed books and treatises, herbalizing expeditions, diplomatic missions, voyages of discovery, meetings of academies, university gatherings and so on. Yet, correspondence has the unique advantage of being a means of conveying information *about* all of these other aspects of exchange as well as a means of information exchange itself *and* an instrument in the creation of both a community of scholars and a new field of expertise. We can only start to interpret correspondence in this way, however, once we have shifted from the individual, biography, and the history of ideas – which has underlain many older publications about correspondence among scholars and scientists – to communities, group culture, and modes of exchange.

Between 1530 and about 1600 experts in natural history lacked the prestige of the old and established disciplines. They were scattered throughout Europe, moreover, and only rarely had access to organisations or locations where they could meet or deposit their information.[82] This situation changed – up to a

---

[80] See W. Jones, *William Turner: Tudor naturalist, physician, divine* (London, 1988), esp. 17-25.
[81] 'Rex certe ipse qui hortulariae rei est amantissimus nihil nisi ab his bellorum intestinorum furoribus otium exoptat ut eiusmodi virorum qualem te praedicavi opera, rem herbariam in Galliae possit instaurare', 8 November 1590 from Souburg in the Dutch province of Zeeland to Clusius in Frankfurt. It is clear from the rest of the letter that Aldegonde had Clusius himself in mind for such a job, since he enquired whether Clusius might be interested in a connection with the French court.
[82] See Biagioli, 'Etiquette, interdependence, and sociability'.

point – in the course of the seventeenth century, when in many European countries academies and royal societies were created that helped to channel research in natural history, but also established and demanded new loyalties that may have become less international and European than in the preceding century. During most of the sixteenth century, however, experts in natural history could only gain by working together and sharing information. No naturalist could acquire prestige or even become a real expert without the help of others. 'Free' gift-exchange was in everyone's interest: it helped increase expertise and raise the general status of the discipline – and by extension that of its members. The idiom of friendship gains a new meaning in this way. In terms of function it both expressed and underpinned the value of free exchange which itself helped to create a virtual community of natural history experts throughout Europe. As we have seen, the correspondence about natural history in Europe – of which Clusius' huge collection forms only a small part – belonged to an ongoing exchange which comprised both material objects (plants, seeds, pictures of plants and books) and immaterial gifts, such as pleasure, assistance, knowledge, respect, time, friendship. Perhaps the most valuable gift of all expressed and conveyed via these letters was the feeling of belonging to a virtual but very real and wide-flung community of friends and naturalists. Even now, these letters help us to reconstruct that community of exchanges. At the time they both embodied and helped to constitute that community.

French manuscript sources on Carolus Clusius

Marie-Elisabeth Boutroue

On the whole, French libraries hold few documents concerning Carolus Clusius.[1] A consultation of Michel Popoff's general index of manuscripts in French public libraries[2], for example, yields only a small number of documents: some letters, autographs or copies, and some commentaries, in particular on the *Historia sui temporis* by Jacques Auguste de Thou. Taken together they constitute a limited, though interesting ensemble. This article is intended to establish the coherence of this ensemble from the perspective of Clusius' incoming letters, most of which are preserved in Leiden University Library. In Paris, the manuscripts related to Clusius are mostly preserved in the Dupuy collection of the Bibliothèque Nationale. To this series can be added some documents in the *Nouveau fonds français* originating from the correspondance of Peiresc.[3]

---

[1] In the quotations the following conventions have been adopted: on the whole the French orthography is respected; missing letters are indicated by square brackets; consonantal u and i have been systematically replaced by v and j. With regard to Latin texts, the u and i have been retained in all instances. The original punctuation is preserved, even when it departs considerably from modern practice. Abbreviations are given in full.

[2] Michel Popoff, *Index général des manuscrits décrits dans le Catalogue général des manuscrits des bibliothèques publiques de France* (Paris, 1993). A search through the digital version of the *Iter Italicum* (P. Kristeller [comp.], *Iter Italicum: A finding list of uncatalogued or incompletely catalogued humanistic manuscripts* [Leiden, 1996]) does not yield other French material. On the other hand, and independently of the collections which today are kept in Leiden University Library, the search results include the documents written by Clusius and preserved in the Bibliotheca Ambrosiana in Milan: letters by Pinelli to Clusius preserved in Ms. 1934, as well as letters of a predominantly botanical interest. The correspondence of Pinelli has been the subject of quite a few studies; the part which concerns the Dupuy brothers has been published by Anna-Maria Raugei (introd. and ed.), *Pinelli, Gian Vincenzo – Dupuy, Claude: Une correspondance entre deux humanistes* (Florence, 2001).

[3] The manuscripts are the following: Paris, BnF, Ms. Dupuy 632: collection of letters, memoirs and observations concerning the *Historiae sui temporis* of president Jacques Auguste de Thou; followed by epitaphs in his honour, 17th century, 228 ff. This collection is composed of documents of various sizes, as well as the minutes of letters addressed to or written by de Thou; Ms. Dupuy 699: collection of letters by important persons and scholars of the 16th and 17th centuries, modern binding in red morocco, 273 ff. The manuscript contains a letter by Clusius to Paul Reneaulme;

Clusius' annotations on the Historia sui temporis

The manuscript Dupuy 632 of the Bibliothèque Nationale de France contains the annotations by Clusius to one of the most important works for historians of seventeenth-century France: Jacques Auguste de Thou's *Historia sui temporis*. Simply entitled 'Notae Karoli Clusii', this text is a commentary on various passages of the *Historia sui temporis* in the form of a letter addressed to the author himself. The commentary deals with a number of contemporary events of which Clusius had either been a direct witness or had been informed by personal acquaintances. It is on the basis of this privileged position that he is able to correct some of the statements made by de Thou. The notes refer to an edition in octavo; the text is not dated, but the following letter mentions the year 1607. Moreover, given that the commentary by Clusius is in Latin, that he cites the Latin version of de Thou's memoirs, that the edition was published in octavo, and that Clusius died in 1609, the only edition of the *Historia* that Clusius can have used is that of 1604[4], i.e. one of the two first editions of the work.

The manuscrit Dupuy 632 belongs to the category of comments that Clusius has left here and there in manuscripts or in the margins of his books. Although strictly speaking outside French territory, similar annotations can be found in the margins of a copy of the *Observations* by Pierre Belon, kept in the Plantin-Moretus Museum in Antwerp.[5] In both cases it concerns marginal notes to a work of which the subject is unfamiliar to him, but on which he claims to have spent much time at the request of the authors themselves. In the case of the *Historia* of de Thou, Clusius' comments concern persons as well as events. With regard to persons, the remarks by Clusius give us access to some detail

---

Ms. Dupuy 836: collection of letters written in Latin to president de Thou, 16th and 17th centuries, 303 ff. The manuscript contains a letter by Clusius, dated 20 May, 1607; Ms. Dupuy 951: collection of Latin poems from the sixteenth century, almost entirely copied by Daniel Rogers. The collection contains an elegy by P. Lotichius Secundus, who is mentioned in Clusius's correspondence, addressed to Guillaume Rondelet and dated February, 1554. This piece of poetry also contains a dedication written by Théodore de Bèze to Clusius; Ms. n.a.f. 5172, 17th century, 229 ff.: part of a collection regarding the correspondance of Peiresc. The manuscript contains minutes or copies of letters written by Peiresc to various correspondents. The order of the letters is alphabetical.

[4] Jacques Auguste de Thou, *Historiarum sui temporis libri XVIII* (Paris, 1604), 2°. Other editions in Latin and French would soon follow. On the historical publications of de Thou, see Samuel Kinser, *The works of Jacques-Auguste De Thou* (The Hague, 1966); A.-Joseph Rance-Bourrey, *Jacques-Auguste de Thou, son Histoire universelle et ses démêlés avec Rome* (Paris, 1881); John Collinson, *The life of Thuanus, with some account of his writings, and a translation of the preface to his history* (London, 1807).

[5] An exemplar of Pierre Belon's *Observationes*, published in one volume with Clusius' own *Exoticorum libri decem* by Franciscus II Raphelengius of the Leiden Officina Plantiniana in 1605, is preserved in the Plantin-Moretus Museum in Antwerp, shelf-mark A 1330.

in the lives of some of his contemporaries which is indicative of the person's general conduct or just noteworthy. For example, in the case of Francisco de Erasso,[6] Clusius points out that he was secretary to king Philip II of Spain, but his allusion to the man's status at court borders on disloyalty:

Supremus Regis Philippi secretarius is fuit; et in magna apud eum gratia etiam quum in Hispaniam viuerem: atque nonnulli mihi significabant adeo gratiosum apud regem esse, quod valde formosam uxorem haberet.
[Francisco de Erasso was the first secretary to King Philip; he enjoyed great credit with the king at the time when I was in Spain; yet some have informed me that his position was as strong as his wife was beautiful.][7]

Still, such precisions relating to the persons named in the *Historia* and known to Clusius are only rarely marked by these expressions of irony. Most often it is the desire for precision that is his motive, and his indications and suggestions for correction do not suggest any malevolence on his part. Thus he writes on the death of Sebastian Münster, which, according to de Thou, had taken place at Heidelberg:

Ubi agit de munsterii obitu, qui acciderit Heydelbergae, Basileae reponendum puto. Nam ibi commorabatur. Eum sane anno 1550, dum Francofurto Mompelium proficiscerer isthic salutaui, comite Petro Lotichio II, viro mihi amicissimo summo Germaniae poeta.
[On the page where there is mention of the death of Sebastian Münster, which supposedly had taken place at Heidelberg, this should be changed to Basle. Because this is where he was staying. While I was travelling from Frankfurt to Montpellier, I met him there in 1550, in the company of Petrus Lotichius II, my dear friend and a great poet of Germany.][8]

Another well-known figure, Guillaume Pellicier, hardly needs any introduction. Bishop of Montpellier and a close adviser to Francis I, he also was a keen collector of ancient manuscripts and a knowledgeable expert on natural history. His correspondence is ample evidence of his taste for ancient manuscripts, and it is well known that Rondelet acknowledged his help in the composition

---

[6] On this secretary, who was one of the key figures in the diplomacy of Philip II, see Alonso de Castillo Solórzano, *Jornadas alegres a D. Francisco de Erasso, Conde de Humanes, señor de las Villas de Mohernando y el Cañal* (Madrid, 1907).
[7] Ms. Dupuy 632, f. 78. The note refers to p. 657 in the 1604 edition of the *Historia sui temporis*.
[8] The elegies by Petrus Lotichius (1528-1560), which are now available in an electronic edition for instance at http://www.uni-mannheim.de/mateo/camautor/lotic.html, have been the subject of several studies, for exampe: Ulrike Auhagen and Eckart Schäfer (eds.), *Lotichius und die römische Elegie* (Tübingen, 2001) [NeoLatina, 2]; Stephen Zon, *Petrus Lotichius Secundus: Neo-Latin poet* (Bern, 1983).

of *De piscibus*. He also wrote a commentary on the *Historia naturalis* of Pliny.[9] Clusius, whose knowledge of Pellicier dates from his own Languedoc period, was clearly appreciative of Pellicier's scientific interest, but he also discusses another, less pleasant episode of his life: that of his arrest on the orders of the parliament of Toulouse, and his imprisonment in the castle of Beaucaire. The evocation of the friendship between the cleric and the naturalist is also the occasion to refer to the allegations that the real author of the treatise on fishes was Pellicier rather than Rondelet. Clusius categorically denies this allegation: '[…] mais le fait que Rondelet aurait préparé son *Histoire* à partir des observations de Pellicier, je le récuse fermement pour ma part.'[10] On the contrary, he emphasises the important role played by the Montpellier prelate in supporting the work of the many naturalists in the region because of the key role of the university.[11]

Other people, famous or not, are also mentioned in comments bearing on specific details of the *Historia sui temporis*. Among the scientists of the first rank is Vesalius. On 18 January 1607, Clusius wrote de Thou a letter in which he stated that his sources on Vesalius' departure from Spain had misled him:

En lisant legerement ce qui s'est passé en l'an 1564, j'ai observé que vous avez esté mal informé de la façon de la mort de Wesalius, lequel partit d'Espaigne pour faire son voyage de Jérusalem, quasi en même temps comme j'y entray. Il en sortit par Perpignan et j'y entray par Guipuzcoa et Vittoria. Je vous advertiray avec plus de loisir comme son dit voyage s'est passé, l'ayant entendu partie en Madrid à la cour du roy d'Espaigne, partie l'année suivante à Bruxelles à mon retour d'Espaigne.[12]

A few pages later,[13] Clusius returns to the departure from Spain of the great physician, writing that Vesalius had never been able to get used to the Spanish way of life. An illness that he contracted in Spain[14] is the starting point

---

[9] On this point, see now Walter Hermann, 'Il commentario pliniano di Guillaume Pellicier (ca. 1490-1567) e la storia del codice parigino latino 6808', *Studi umanistici Piceni*, 17 (1997), 179-194.

[10] Ms. Dupuy 632, loc. cit.

[11] On this question, see for example Louis Dulieu, *La médecine à Montpellier*, vol. II, *La Renaissance* (Avignon, 1979).

[12] Ms. Dupuy 632, f. 77.

[13] Ms. Dupuy 632, f. 81: 'Is regem Philippum a Belgio excedentem est sequutus cum uxore. Nam Regius medicus erat. In hispania haerens capit taedio affici, ad Hispanorum ingenium et mores sese accommodare nequiens et libenter quidem in patriam rediisset, si Rex illum dimittere voluisset.'

[14] It is probably this text, preserved in the partial edition of Clusius's marginal notes to the *Historia sui temporis*, that the biographers of Vesalius followed in order to explain the context of his departure from Madrid. Jean Brocas, however, has shown that the situation of Vesalius in Spain was very complex and that there may have been several reasons for his sudden departure from the Spanish capital, ranging from the precarious position of the physician and the bad temper of his wife, his health problems, as to which the physician wonders whether they are physical or psychological, to

for the following anecdote. When he was seriously ill, Vesalius had vowed that he would go on a pilgrimage to Jerusalem if he recovered his health.[15] The Spanish king granted the physician's wish, allowing him to depart and providing him with all the necessary safe-conducts for the voyage as well as a large sum of money. It was only when Vesalius arrived in Perpignan that problems arose, not as the result of any difficulty on the road, but because of a difference of opinion between Vesalius and his wife. The physician wanted to take the eastbound route to Venice and then continue to the Holy Land, but his wife preferred to return immediately to the Southern Netherlands. All this took place, Clusius states, while he himself was travelling to Spain. The text then relates the well-known troubles that the great anatomist had to endure on his pilgrimage. The reasons which Clusius gives for the hasty departure of Vesalius are thus rather different from the established version based on the hostility of the Inquisition.

The identification of a person can be of considerable importance when it concerns someone who is little known. An example is the instructive case of a man named by Jacques Auguste de Thou as Jacobus Despota Sami. His story is so remarkable that Clusius relates it in full. A brief sketch of the main events will suffice here. Clusius begins by stating that he knows many facts concerning that well-known man in Montpellier:

De eo viro multa habeo quae dicam. Nam illum Mompelii noui. Nomen Iacobus de Marchetis tale certe dedisse, dum studiosorum matriculae inscriberetur, intelligebam. Sed quum Siculam se nunc diceret modo Despotam Sami sese appellaret, a plerisque nobilis Scorti Cortigianae di Messina in Sicilia filius est creditus. bene habito et robusto erat corpore, g[r]aecamque linguam vulgarem, Italicam, latinam gallicam callebat.
[I know quite a few details about that man that I will tell you, because I met him at Montpellier. I knew for sure that he had taken the adopted the name of Jacobus de Marchetis in order to enrol himself in the student matriculation register. But as he passed himself off as a Sicilian and called himself Despota Sami (Tyrant of Samos), he was believed by some to be the son of a famous prostitute from the court of

---

the most plausible reason, namely the opposition from the Inquisition. On this issue, see particularly Jean Brocas, *Contribution à l'étude de la vie et de l'oeuvre d'André Vésale* (Paris, 1958), 25-27; T.M.G. de Feyer, 'Biobibliographie de Vésale', *Janus*, 19 (1914); C.D. O'Malley, 'Andreas Vesalius' pilgrimage', *Isis*, July 1954, 138-144; R. Delavault, *André Vésale: Biographie* (Brussels, 1999); F.A. Sondervorst (ed.), *Histoire de la médecine belge* (St. Stevens-Woluwe, 1981); J.H. Talbott, *A biographical history of medicine: Excerpts and essays on the men and their work* (New York/London, 1970).
[15] Ms. Dupuy 632, f. 81: 'Itaque quum praeter voluntatem isthic haereret, in morbum incidit, quo cum quo cum difficultate curatus apud Regem denuo institit, ut abitum illi concederet.nam in morbo votum fecisse de adeundis Hierosolymis si sanitati restitueretur. Itaque si votum exoluere valde cupere, si per Regis voluntatem liceat.'

Messina. He was handsome and robustly built and excelled in vernacular Greek, Italian, Latin and French.][16]

This sad figure, Clusius continues, met the widow of a certain Patricius in Montpellier who had died rather opportunely in a fight with a Maltese knight, leaving her with a young son of two or three years of age. Jacobus de Marchetis did not let the grass grow under his feet and moved in with the young widow:

Elapsis vero paucis ab illius Patricii obitu septimanis, Iacobus ille (quem Mompelienses vulgo *le Grec* appellabant) in aedes viduae commigrauit, et familiariter cum ea versabatur.
[Barely a few weeks had passed since the demise of this Patricius when our Jacobus (whom the people of Montpellier commonly called 'the Greek') moved in with the widow and lived with her in all intimacy.][17]

Next the child was disposed of by the two lovers. Jacobus de Marchetis had noticed that the son of his mistress had the habit of playing hanging upside down in a large wardrobe. The movements of the boy caused the wardrobe to tilt, as a result of which it collapsed on top of him, killing him on the spot. At the time of the siege of Metz, Jacobus de Marchetis, who was in the following of a certain Saint-Hilari, departed to relieve the beleaguered army of François de Guise, leaving his mistress with Rondelet where she stayed barely three weeks. After the siege, the lover, having been informed of the infidelity of his companion, returned to her. Suspecting Sanravius[18] of having dishonoured him, he laid an ambush for him and killed him on a dusty road. In the face of royal anger and knowing the penalty for murder, he decided to flee to the Southern Netherlands.

The history of this sad individual clearly does not arouse much interest from the point of view of general history of the period, which is de Thou's focus. As can be judged from other comments made by Clusius, what mattered to him was to contrast the history of an individual with that of the world at large, and to describe in detail what de Thou had broadly sketched. For the modern historian, these comments have yet another interest for the insight they offer into the details of Clusius' own private life. The documents in Paris bearing on the last years of his life bear witness to the damage inflicted by the years. Clusius complains that he is not well able to walk, but congratulates himself that his visual ability is not impaired at the age of eighty-four.

---

[16] Ibid., f. 79, referring to vol. II of the *Historia sui temporis*, 127.
[17] Ibid.
[18] May this have been Joannes Sanravius?

The importance of these notes did not elude the successive editors of Jacques Auguste de Thou's work. Although the manuscript annotations were made after the publication of the work, they were reprinted in certain subsequent, late editions of the work. Clusius' notes did not therefore remain unpublished. They were incorporated, for example, in the edition published in The Hague by Scheurleer in 1740 and, two years later, in the pirate edition published by Brandmüller at Basel. The title clearly gives an indication of its contents: *Histoire universelle de Jacques Auguste de Thou, avec la suite par Nicolas Rigault; les mémoires de la Vie de l'Auteur, un recueil de pièces concernant sa personne et ses ouvrages y comprises les notes qui se trouvent dans les manuscrits de la bibliothèque du Roi de France, de Mrs Du Puy, Rigault et de Sainte-Marthe. Le tout traduit sur la Nouvelle édition latine de Londres. Et augmentée de remarques critiques de Casaubon, de Du Plessis Mornay, G. Laurent, Ch. De L'Ecluse, Guy Patin, P. Bayle, J. Le Duchat et autres.*[19]

Both editions publish the letter which opens the dossier of Clusius' remarks, acknowledging the receipt of Jacques Auguste de Thou's Latin history. Comparison of the manuscript and the printed edition of this letter does not reveal any notable variants apart from some minimal orthographic differences (the consonantal 'i' has been replaced by 'y' in the imperfects of the subjunctive and some spelling has been modernised). The accompanying note to this edition explains that the printed edition of these notes is taken from the Latin edition published in London, and that the French editor has moved them, together with some remarks by other scholars, to the corresponding points in the body of the text.[20] This dossier in Volume X of the French translation includes not only Clusius' despatch, but also the accompanying letter by Scaliger, and the response of Jacques Auguste de Thou. Bringing these documents together allows us to gauge the impact of the corrections proposed by Clusius and to see how the historian responded to the comments of these two famous correspondents. In the present case it concerns the date of the death of King Henry of Scotland.[21]

Ce que vous et Monsieur de L'Ecluse m'escrivez de l'année de la mort du Roy Henri D'Ecosse, me met en peine, d'autant plus que des lors que j'escrivis ce qui en est imprimé, ce scrupule me vient en l'esprit, que l'année n'estoit celle que j'ay mise, et que j'ay neantmoins trouvé telle en Buchanan[22]. J'en ay contesté fort avec des Ecossois qui

---

[19] H. Harrisse, *Les De Thou et leur célèbre bibliothèque* (Paris, 1905), 259.
[20] *Histoire universelle de Jacques Auguste de Thou [...]*, vol. X (Basle, 1742), 466, and note 1.
[21] It concerns the death of Henry Darnley (1546-1567), second husband of Mary Stuart, who was assassinated in 1567. The death of the queen's husband is interpreted differently by the historians of the period: François de Belleforest, Claude Haton, Paolo Giovio or George Buchanan.
[22] The allusion goes back to the historical writings of George Buchanan, published in Latin and frequently re-edited and translated until the end of the eighteenth century. According to the printed

estoyent lors au pays, lesquels toutesfois m'ont confirmé que la mort advint en l'année 1567 au mois de Febvrier que l'on compte encores en Angleterre 66; car, comme vous sçavez, l'année à eux (je ne sçay si aussi en Ecosse) commence au jour de l'Annonciation seulement, et lors mesmes nous ne comptions l'année qu'après Pasques[23] [...] Du surplus de vos bons advertissemens je feray fort bien mon profit comme de *Diuona Cadurcorum*.[24]

I shall here only present a few passages to show how Clusius' comments were used. The first concerns the physician of Montpellier, Guillaume Rondelet. The passage on the death of Rondelet and on the treatise on fishes is found in the pages dealing with the year 1566.[25]

La mort enleva la même année Guillaume Rondelet de Montpellier. Quoique François Rabelais en ait parlé avec mépris, dans cet ouvrage qu'il a composé avec une liberté satyrique, plus ingénieuse qu'irrépréhensible[26], on ne peut disconvenir qu'il n'ait été un habile médecin. A la vérité ses ouvrages ne répondent pas à la grande réputation qu'il s'étoit acquise, ni à l'opinion qu'on en avoit conçeue. Un des ses escris lui a fait plus d'honneur que les autres; c'est le traité des Poissons qu'il a fait imprimer, et qui lui auroit mérité plus de louanges, si on avoit l'attribuer à son industrie, et non à celle d'un autre. Car on prétend qu'il l'avoit tiré des Commentaires de Guillaume Pellicier, évêque de Montpelllier, homme d'une érudition peu commune et que cet ouvrage faisoit partie des scavantes observations que ce prélat avoit faites sur Pline et qui, pour le malheur de la République des lettres, ont été, ou perdues ou supprimées.

And here are Clusius' comments as they appear in the notes to the French edition of 1742:

Ce qu'on dit de Rondelet en cet autre endroit ne lui fait pas honneur. C'est sans doute à la persuasion de quelque jaloux de sa réputation. Le trait pourroit bien être parti des

---

catalogue of the library of Jacques-Auguste de Thou, he had two editions of Buchanan's historical work, one published at Geneva in 1583 and the other at Frankfurt in 1594. He had, by the way, access to other historical sources for these events in Scotland, for example the treatise by John Lesley, *Oratio pro libertate impetranda Mariae Scotorum reginae ad Angliae reginam* (Paris, 1574); he equally possessed two other small works, one for and the other against Mary Stuart: *Scotorum reginae Mariae Stuartae supplicium et mors pro fide catholica* (Cologne, 1587) and *Histoire de Marie reine d'Ecose touchant la conjuration faite contre le roy et l'adultère commis avec le comte Bothwel traduite du latin en français* (Edinburgh, 1572). For these references, see the *Catalogus bibliothecae Thuanae* (Paris, 1679), 430-431.

[23] It is thus simply a matter of defining the the style of dating, which depends on the date of Easter. The same problems could occur for events relating to France.

[24] This remars refers to a series of corrections of toponyms proposed by Scaliger. Jacques Auguste de Thou had given *Albiae Cadurcorum*, the French town of Cahors in the Département du Lot.

[25] In the edition of 1752, vol. III, 618-619.

[26] It is of course an allusion to Rondibilis in the *Third Book* of Rabelais, who is traditionally interpreted as a caricature of Guillaume Rondelet.

mains d'Honoré Castel[l]an, qui, de ma connoissance, pendant qu'il étoit professeur à Montpellier, avoit eu de grands différens avec Rondelet. A la vérité Castelan l'emportoit sur son adversaire par une grande facilité à s'exprimer sur le champ; mais celui-ci avoit plus d'érudition. Nous ne pouvions assez admirer la vivacité de son esprit. Je l'ai vu moi-même ordonner des remèdes dans un même instant pour deux maladies différentes, et dont la cure l'étoit aussi. Laurent Joubert écrivoit son ordonnance pour l'un des malades[27], tandis que j'écrivois d'un autre côté ce qu'il ordonnoit pour l'autre. Je ne sçaurois parler assez dignement de sa mémoire qui étoit merveilleuse.

This note by Clusius invites several comments. The first concerns its abbreviated character, for the manuscript in the Dupuy collection contains a good deal more.[28] This omission of what constitutes a body of very detailed commentary in the Paris manuscript is not an isolated case: Clusius' precision on the place of death of Sebastian Münster is equally omitted, thus showing that, although the remarks made to de Thou have had an incontestable influence on the correction of the successive editions, they have not all been considered as equally relevant nor subjected to a critique based on a detailed study of the differences between the various editions of the *Historia sui temporis*. Likewise, the toponomical precision with regard to the Hungarian city of Bins[29] is omitted. It also seems that the long note on the story of Jacobus de Marchetis is absent, as well as that on Count Egmont or the remarks on Clusius' stay in Granada. On the other hand, the onomastical note on the name Fassardo is taken over in its entirety,[30] as is the one on the Duke of Norfolk. The passage cited from the *Histoire universelle* is about the negotiations between the Duke of Norfolk and the Queen of Scotland, contrary to the interests of Queen Elizabeth.[31] The problem, as de Thou clearly explains, is the religious position of the duke.[32] The negotiation proceeds secretly through the intermediation of Ridolfi, who provides the link with the Holy See; he obtains the approval of the Pope. When the conspiracy is found out, Elizabeth has them all sent to the Tower:

On trouva une copie des lettres du pape avec le Mémoire de la reine d'Ecosse. Commte toutes ces circonstances chargeoient extrêmement le duc de Norfolk, et ses complices, Elizabeth l'envoya à la Tour avec le comte d'Arondel et le Lord Lumley son gendre.

---

[27] Laurent Joubert (1529-1582), physician of Montpellier, later chancellor of the the university and author of numerous medical works, such as the *Erreurs populaires et propos vulgaires touchant la médecine et le régime de santé, réfutés ou expliqués* (Bordeaux, 1579) and a French translation of Rondelet's treatise on fish.

[28] Ms. Dupuy 632, f. 78v.

[29] Book 13, vol. II, 236 in the French edition.

[30] *Histoire*, vol. IV, l 48, 352: 'One should read Fassardo and not Fajardo. Fassardo is the name of the family of Vélez.' The commentator continues: 'We shall write Fassardo from now on.'

[31] Vol. IV, 518; book 51.

[32] Ibid., 517.

The above passage draws on Clusius' comments:

Voyageant en Angleterre en 1571, dans le tems que le duc de Norfolk fut arrêté et mis à la Tour de Londres, je partis de Kingston, éloigné d'un mille d'Angleterre de Hampton cour, maison Royale pour aller à Nonsuch, afin de voir ce château, dont on me disoit de si belles choses. Il a été bâti par Henri VIII et ensuite vendu, après la mort du roi Edouard, par la Reine Marie au comte d'Arondel, qui l'a, dit-on, beaucoup embelli. Les domestiques de ce Seigneur me dirent, qu'ils ne pouvoient me laisser entrer dans le château, parce que le Comte y étoit. Ils me firent beaucoup de politesses et me montrèrent les jardins. Je retournai le même jour à Kingston où je passai la nuit. De là j'allai à Richemont, autre maison royale, où j'appris la prison du duc de Norfolk, et le commandement que le comte d'Arondel, gendre de ce Seigneur, avoit eu de ne point sortir de son château, sur le soupçon que la Reine eut, qu'il étoit entré dans le parti de son beau-père. Il y a toute apparence que ce fut la cause pour laquelle on ne me laissa pas voir le château de Nonsuch.

Other documents by Clusius

For the rest, even if the commentary on the *Historia sui temporis* is in fact the most important of all those preserved in Paris, a few remarks should be be made on the other letters, which form a second, more heterogeneous ensemble. These documents belong to two different groups. The first is composed of loose letters and pieces of verse without any coherence. The second, on the contrary, belongs to the body of the correspondence of Peiresc, a collection as important and coherent as the papers of de Thou.

Among the miscellaneous items, the first letter which attracts our attention is addressed by Clusius to the Blois naturalist Paul Reneaulme, whose presence among Clusius' correspondents is not surprising.[33] Although the context of this letter is not known[34], it argues the need for a division of labour to establish an ongoing discourse on the world of plants. Writing in an almost comforting manner, Clusius observes that it is impossible for a single person to have a complete knowledge of nature, and that the progress of scientific knowledge is based on the contributions of each of the naturalists involved:

[…] qui omnium plantarum cognitionem adquirere voluerit, frustra laborabit. Multorum igitur labore et diligentia hoc studium est extollendum. Tu aliquid conferes, ego aliquid, alii etiam aliquid, ut nostri posteri pleniorem plantarum cognitionem consequi

---

[33] He, too, belongs to the group of correspondents of Jacques Auguste de Thou. There is a letter in Latin by Paul Renaulme to Clusius in Leiden University Library, Vulcanius 101.

[34] The correspondence of Paul Reneaulme is preserved in the Bibliothèque Nationale de France. The letters which he wrote to de Thou are equally kept in the Dupuy collection.

possint. Propterea tametsi aliquid forte in hoc studio praestiterim, tu quam tua industria et diligentia adquisiuisti posteris inuidere non debes, nam non erit inutilis tua opera.

[He who wishes to know all plants will work in vain. This science has to be uplifted by the work and attention of many scholars. You contribute one thing, and I another, and again others do the same, in such a way that our descendants will dispose of a more complete knowledge of plants. Therefore you should not begrudge posterity of whatever I have been able to contribute with this study, and whatever you have acquired through your industry and diligence, for your work will be useful.][35]

The letters to Jacques Auguste de Thou also contain indications which bear directly on the literary genesis of the works and of the need for botanical precision.[36] Once again, Clusius' letter testifies to the fruitful and scholarly exchanges between Jacques Auguste de Thou, Scaliger and himself. In the field of botany, the starting point of the discussion is a passage in the *Observations* of Pierre Belon which, Clusius remarks in passing, will not be reprinted.[37] The first remark refers to the wild pine. Clusius demonstrates he has regularly translated as *picea*, corresponding to Theophrastus' Greek πεύκη ἄγριος[38]. The other botanical note concerns a problem which is ultimately as much a matter of taxonomy as of nomenclature. This time, it concerns the *ache*, or celery, translated by Clusius, as one would expect, as *apium*. The problem originates in the text of Belon, who distinguishes in French between the *ache majeur* and the *ache mineur*. Adopting this in Latin, Clusius writes in his *Observations* of the *apium maius* and the *apium minus*. Yet, he remarks, this does not really solve the problem, for there is only one plant with the name *apium*, unless it is not a question of the *ache*, but of the *lappa*, i.e. probably the *bardane* or burr. It is certainly useful, on the basis of these remarks by Clusius, to reconsider the problem as it presents itself in the botanical literature of the time, starting, it should be emphasised, with the direct source of Clusius, which is the Latin version of the *Observations* of Pierre Belon. The relevant passage is in Book I of the *Observationes*, the chapter on the flora of the island of Lemnos.[39] As

---

[35] Paris, BnF, Ms. 699, f. 139. Clusius' letter is dated May 1601.
[36] Paris, BnF, Ms.Dupuy 836, f. 122.
[37] Ibid.
[38] *v.* Theophrastus, περὶ φυθῶν ἱστορία. For pines, see particularly III, 9, 4.
[39] Pierre Belon, *Observations*, vol. I, 26: 'Nous avons vu le psilium croître par les champs, et le thlaspi et draba, le souchet tant rond que long. Les espèces de conizes le long des ruisseaux. La lampsane qui est une herbe qui ne croît ni en France, ni en Italie, et par ce nous est inconnue. L'on y trouve aussi de plusieurs espèces de joncs, du pouliot, de l'apparitoire, du cotylédon, de l'appe majeur et mineur que les Grecs nomment maintenant pattimendilla […]'; cited from the edition by Alexandra Merle (Paris, 2001, p. 117) which follows the first French edition published in 1553 by Gilles Corrozet.

the – slightly later – *Pinax* of Caspar Bauhin distinguishes numerous 'species' of *apium* on the basis of ancient and modern descriptions, it is certainly worth raising the issue anew.[40]

As in numerous other cases (in fact, it is always advisable to proceed in this manner), one has to begin with the information taken from ancient sources. The Latin word *apium*, in ancient botany, refers directly to the Greek σέλινον. The word is found in Latin texts, not only in the scientific literature, but also in texts lacking any connection with plants, nature or medicine. The following remarks on the sources available are intended to bring out the specificity of Clusius' method in the context of the botany of his day, and should not be taken to reflect on his work as a naturalist.

The dictionary of Latin plant names by Jacques André[41] clarifies the succinct information given by Bauhin: the author identifies some eight different plants corresponding to *apium*/σέλινον. The tradition of the classical homonymy persisted from late Antiquity to the end of the Middle Ages. There is no need here to enter into details, but Johannes Stirling has shown that quite a few plants designated by the Latin name *apium* are considered by botanists today to belong to different families.[42] Besides Pliny, it is mentioned in Pelagonius, Celsus, Apicius and even in the verses of Horace. Whether or not it concerns celery, *apium* thus belongs to the traditional corpus of plants used for both food and medicine. In his description of the plant, Isidore of Spain mentions explicitly that various *genera* exist, which are named successively *Petroselinon*, *Hipposelinon* and *Oleoselinon*.[43] Renaissance descriptions of the *apium* can be found in Brunfels as well as in the commentary of Mattioli. As for Clusius, Stirling refers to the *Nomenclator Pannonicus* of 1583.

Even though Clusius says that he knows only one type of *ache*, the vast majority of naturalists properly distinguish the plants named *apium* in order to justify the rich synthesis by Caspar Bauhin at the end of the sixteenth century. However, when – in the wake of many others before him – Bauhin distinguishes numerous plants called *apium*, he makes no mention whatsoever of a hypothetical opposition between *apium maius* and *apium minus*, and neither do his contemporaries. This appended remark, which does not figure in the

---

[40] Caspar Bauhin, *Pinax theatri botanici [...] siue index in Theophrasti Dioscoridis Plinii et botanicorum qui a seculo scripserunt opera plantarum* (Basle, 1623). In his introductory paragraph, Bauhin notes that the distinction between numerous plants in the category of *apium* / σέλινον goes back to antiquity. He states that Dioscorides proposes various categories (can one speak of species?) which can be distinguished essentially by the spot where the plant grows. But he also finds various species in Theophrastus and in Pliny.

[41] Jacques André, *Les noms de plantes dans la Rome antique* (Paris, 1985).

[42] Johannes Stirling, *Lexicon nominum herbarum, arborum fruticumque linguae latinae* (Budapest, 1995).

[43] Isidore, *Etymologiae*, XVII, 11.

translation itself, something of the order of the commentary, should therefore be interpreted as reflecting the project and the epistemological position of the naturalist. The remarks by Clusius concur with one of the most constant preoccupations of the naturalists of the Renaissance: the reattribution of names to plants in an attempt to harmonise the nomenclature of antiquity with their own experience-based nomenclature. Translation is one of the privileged instances of this ultimately futile exercise, and Clusius, who had published a resumé of Garcia da Orta from the Portuguese[44] and had translated Belon into Latin[45] and Dodoens into French, was one of the best specialists. Peiresc himself had not hesitated in 1605 to send Clusius a list of plant names including the Provençal variant of these phytonyms.[46] Clusius, like the majority of his contemporaries, compared plants with their names. This makes it easier to grasp the importance of the plant lists which figure in the despatches, as well as those which Rondelet included in the botanisation programme of the students at Montpellier and which were echoed by Rabelais in his *Gargantua*.[47]

Another group of letters concerns the international network which unified scholars all over Europe. Those relations, which are mostly epistolaries relationships, but sometimes direct as well, are well known. The documents preserved at the Bibliothèque Nationale de France provide information on the erudite circle of Pinelli[48], a circle which was revealed in its full coherence at the moment of the death of the man who had been a friend of Peiresc and Clusius, as well as having supplied books and manuscripts to Fulvio Orsini in Rome. The remarks concern the exchanges of books and news at the time of the death of Pinelli. It is not surprising to find such relations between three of the best experts in scientific and scholarly matters from the end of the sixteenth and beginning of the seventeenth centuries. To this group, of course,

---

[44] *Aliquot notae in Garciae aromatum historiam, ejusdem descriptiones nonnullarum stirpium et aliarum exoticarum rerum [...]* (Antwerp, 1582).
[45] *Petri Bellonii Cenomani, plurimarum singularium & memorabilium rerum in Graecia, Asia, Aegypto, Iudae, Arabia, aliisq. exteris provinciis ab ipso conspectarum observationes: Tribus libris expressae. Carolus Clusius Atrebas à gallicis latinas faciebat* (Antwerp, 1589).
[46] See Charles Joret, 'Liste des noms de plantes envoyées par Peiresc à Clusius', *Revue des langues romanes*, 4th ser., 7 (1893-94), 437-442.
[47] See Rabelais, *Gargantua*, chapter 21: 'Comment Gargantua feut institué par Ponocrates en telle discipline qu'il ne perdoit heure du jour.' 'Le temps ainsi employé luy frotté, nettoyé, & refraischy d'habillemens, tout doulcement s'en retournoyt & passans par quelques prez, ou aultres lieux herbuz visitoient les arbres & plantes, les conferens avec les livres des anciens qui en ont escript comme Theophraste, Dioscorides, Marinus, Pline, Nicander, Macer, & Galen. Et en emportoient leurs plenes mains au logis, desquelles avoyt la charge un ieune page nommé Rhizotome, ensemble des marrochons, des pioches, cerfouetes, beches, tranches, & aultres instrumens requis à bien arborizer.'
[48] On Pinelli and his correspondance, see Anna-Maria Raugei, *Une correspondence entre deux humanistes.*

one has to add Scaliger, who also was working and living in Leiden in close proximity to Clusius.

The letters written by Peiresc to Clusius are not the best known among the body of correspondence of the Provence magistrate. The first mentions Peiresc's role as intermediary in the dispatch of the letters of Clusius to Croatia and to Naples.

[…] je manquerois par trop à mon debvoir si je ne m'estudies de faire pour vous et pour le tres illustre Schaliger, l'honneur et la vertu duquel nous reverons uniquement en ces quartiers, ce que Monsieur Pinelli fairoit s'il estoit en vie. On a desjà envoyé votre lettre à Raguse à celuy auquel elle s'adresse, e[t] par la premiere commodité on envoyera à Naples le reste du pacquet au Duc d[e]lla Cerenza, neveu et heritier de feu M. Pinelli qui ayme fort semblables presens, e[t] qui sans doubte verrà de tres bon oeil ceux icy.[49]

The other letters to Clusius from Peiresc give an idea of the nature of the epistolary relationship between Pinelli and the Leiden naturalist. It comes as no surprise to see the letters accompanied by pieces of vegetation as well as books and drawings. In 1602 Clusius sent Pinelli via Peiresc four copies of his *Rariorum plantarum historia*, which he had published the previous year.[50] Writing to Clusius, Peiresc describes the circumstances under which this happened:

Monsieur, depuis que feu Mr Pinelli vous eut escrit du 18 avril de l'annee passee il receut les quatre exemplaires de votre histoire des plantes avec voz pourtraicts et la branche de clous de girophle qu'il estimoit beaucoup. Sur quoy il ne vous fit aucune responce, parce qu'il estoit couru un faulx bruit que vous estiès mort et dura iusqu'à ce qu'on vit voz dernieres lettres du 13 d'Aoust, lesquelles eussent fait survivre plusieurs jours dadvantaige le pauvre Sgr Pinelli s'il les eusse peu voir avant sa mort. car il en eusse receu un singulier contentement. Le duc de L'Acerenza, a ce que j'entends, a donné voz pourtraicts au Sr Imperato de Naples, avec la branche de Girophle et un de voz livres des herbes[51].

Peiresc's letter goes on to list other exchanges in which Pinelli was involved in one way or another. Thus the scene is set for the appearance of the great figure of Prosper Alpin, to whom Pinelli had promised a copy of an old manuscript then in the possession of one of his friends. The premature death of Pinelli dashed Prosper Alpin's hopes, the more so, Peiresc adds, since he did not even have the time to clarify from whom Pinelli had wanted to request

---

[49] Paris, BnF, n.a.f. 5172, f. 10.
[50] *Rariorum plantarum historia* (Antwerp, 1601).
[51] Paris, BnF, n.a.f. 5172, f. 10^v.

it: as is well known, Pinelli acted as intermediary between the Italian collections and the ancient manuscripts recently snatched from the old monasterial libraries.

The request by Prosper Alpin is probably anything but anecdotal. What matters to him is to shed light on the principles of methodical medical science within the framework of his own scientific activity. The requirement of philological precision which underlies the request for the ancient manuscript, makes particularly clear one of the most interesting aspects of the construction of knowledge among Renaissance scholars: for Prosper Alpin, as for the great majority of his contemporaries, that construction was predicated on a preliminary confrontation with the texts of antiquity and the conditions of their transmission.

To conclude, the specificity of the manuscripts of Carolus Clusius in Paris does not reside in their unpublished character.[52] On the contrary, many are copies of manuscripts of which the original is kept elsewhere, most notably in Leiden University Library, and the annotations to the *Historia* of de Thou are partially incorporated in later editons of the text. All the same, there are several reasons why they cannot be ignored.

The first is purely philological: even if the older date or the originality of a letter preserved at Leiden is completely attested, this does not mean that the copy could not be used for a comparison of forms, to resolve an ambiguity, or to decide on the orthography of a proper name. The notes on the *Historia sui temporis* are more complete in the manuscripts, although study of them also requires a preliminary and careful comparison of texts. These notes are the most interesting among the Clusius manuscripts preserved in Paris.

The second reason concerns the complementary information provided by the remains of Clusius' correspondence in Paris and by other collections of letters. The compilation of all this material enables a better understanding of the circulation of ideas and sciences at the end of the sixteenth century. What emerges is a network of professional and amateur naturalists, from Clusius to Peiresc and from Paul Reneaulme to Scaliger. As today, and perhaps more than today, the science of nature progressed in the seventeenth century from the exchange of objects to the exchange of knowledge. If one had to pinpoint one common feature in all this correspondence, it is without doubt the circulation of boxes with the latest books published by the humanist naturalists and the products of their fieldwork, the drawings made after life and the notes on ancient manuscripts. The knowledge of nature during the sixteenth and

---

[52] The description of the garden of Montpellier, recently opened by Pierre Richier de Belleval, is a copy the original of which is preserved in Leiden.

seventeenth centuries progressed by means of the accumulation of ancient knowledge and new observations. Clusius, who translated between both languages and disciplines, who realised that he had to translate Pierre Belon into Latin and Rembert Dodoens into French if their work was to be known, was thus the trait d'union between various works. His correspondence is the most immediate and tangible reflection of this.

Clusius in Montpellier, 1551-1554:
A humanist education completed?

Gillian Lewis

In his autobiographical memoir Clusius recalls that his father regarded the study of jurisprudence as 'a road which would lead to ample honours'.[1] Accordingly, after Latin and Greek letters in Louvain, he embarked, with some reluctance, on the study of civil law. In 1548 his father furnished him with the means to go to Marburg to pursue his legal studies further. Here for eight months he attended diligently the lectures of the celebrated jurist Joannes Oldendorp. When Oldendorp was summoned back to the Imperial Court, Clusius, with some alacrity, started attending instead the popular theology lectures of the Lutheran Andreas Hyperius, whose evangelical piety he admired, and in whose house he lodged. Hyperius for his part took a benevolent interest in this serious student, and suggested to him that he should go to Wittenberg, in order to meet Philip Melanchthon and to hear him teach.

Clusius took this advice. He abandoned the study of law, which he had found intricate and perplexing, and in which he was certain that he would never excel. On Melanchthon's advice he took up instead the study of 'philosophy' in the broad and generous sense in which Melanchthon (and indeed Hyperius) understood the word: not dialectic alone (although this had its place), but observation and analysis of all the works of God as manifested in Creation. Medicine and natural history had a prominent place in this comprehensive programme of religious and intellectual response to the world.[2] As a student in Marburg, Clusius had already enjoyed hunting for plants in the neighbouring woods.[3] Now he had a better reason to do so. At Wittenberg (where, according

---

[1] Clusius, autobiographical memoir, preserved in a letter written to him by Johan Posthius, dated 8 October 1588, Leiden University Library, VUL 101, ff. 176-177; a photograph of the memoir, together with a transcription was published by F.W.T. Hunger in Appendix II of the first volume of his *Charles de l'Escluse (Carolus Clusius), Nederlandsch kruidkundige 1526-1609*, 2 vols. (The Hague, 1927-43).

[2] Sachiko Kusukawa, *The transformation of natural philosophy. The case of Philip Melanchthon* (Cambridge, 1995), 201-210.

[3] Clusius, *Rariorum plantarum historia* (Antwerp, 1601), cxxi; Hunger, *Charles de l'Ecluse*, vol. I, 15, n. 1.

to Pierre Belon, Valerius Cordus in 1541 had given 'demonstrations and interpretations of the plants of Galen, Theophrastus and Dioscorides'),[4] he pursued his botanical interests further, recording his field trips and collecting specimens.[5]

When Clusius in the late 1540s discovered the powerful attraction of *res herbaria* he was encountering a phenomenon which had been gaining ground among scholars since the late fifteenth century.[6] The fashion, which owed a great deal to the rapid development of an international trade in printed books, was a dual one: in the first place it was a part of the demand for more accurate and critical versions of familiar classical texts, and for new editions of unfamiliar ones; in the second place it was part of a wave of curiosity about the sheer physical diversity of the natural world, the variety and complexity of the plants, birds, fishes, mammals and reptiles of land, sea and air. The fashion was, in origin, a bookish phenomenon, although one which increasingly took its adepts out into the fields and the mountains, the rivers and the coasts. By the 1530s it was beginning to attract followers among university teachers and students, cultivated apothecaries, noblemen, gentlemen, merchants and clerics all over Europe, their numbers being greatest in Italy, Germany, the Netherlands and France. Within the medical faculties of universities a keen interest in botany was closely associated with attempts to put the study of *materia medica* on a sounder basis through critical re-examination of classical texts.

The first phase in the fashion, from the 1470s to the 1530s, saw the publication of new printed editions and translations of Aristotle's books on animals, Pliny's Natural History, and Theophrastus and Dioscorides on plants. An important early contributor to the critical discussion of Pliny was the fifteenth-century Venetian nobleman Ermolao Barbaro. Aldus Manutius produced a Greek text of Dioscorides as early as 1499; by 1516 Latin translations and commentaries by Barbaro and by Jean Ruel had appeared. Dioscorides continued to attract attention: his book was translated into Italian and into French, and reissued many times. Pier Andrea Mattioli in his *Commentarii in libros sex Pedacii Dioscoridis Anazarbei*, first published in 1554, found a formula which was to enjoy spectacular success: after each short section of the Dioscoridean text he inserted his own comments, identifying the plants mentioned, giving them their Italian names, and including an illustration showing

---

[4] Karl H. Dannenfeldt, 'The University of Wittenberg during the period of transition from medieval herbalism to botany', in *The social history of the Reformation: Essays in honor of Harold J. Grimm*, eds. Lawrence P. Buck and Jonathan W. Zophy (Columbus [Ohio], 1972), 235.

[5] Clusius, *Historia stirpium per Pannoniam*, 327; Hunger, *Charles de l'Ecluse*, vol. I, 16, n. 5.

[6] An excellent account of these developments is given in Karen M. Reeds, *Botany in medieval and Renaissance universities* (New York/London, 1991), 14-24.

their appearance. Mattioli's book was frequently reprinted, with revisions and additions. It became a classic of its genre and a sixteenth-century best-seller. During the 1520s Galen's works were also re-issued in more critical Greek editions, and in new Latin translations from the Greek. The Hippocratic corpus underwent the same process. There was renewed interest within the universities in Galen's *De simplicium medicamentorum facultatibus*, and in Dioscorides. Lectures using new editions of classical texts first began to appear in the lecture-lists of medical faculties in the 1530s. By the 1540s they had come to dominate the scene, more or less supplanting Avicenna's *Canon* and the older digests and anthologies of Galenic material which had once been medical students' staple fare. Medicine as it was taught in Wittenberg by Casper Peucer in Clusius' day was in most respects similar to medicine as taught at that time in Basle, Paris, or Montpellier. Individual doctors had their own intellectual idiosyncrasies, but by and large a consensus was beginning to develop. It was a consensus that gave a central position to *materia medica*, thus opening a door, for those who wanted to go through it, into the wider world of *res herbaria*. Over the next fifty years – the span of Clusius' adult life – this unstructured enthusiasm for plants was to develop, on the one hand, into the new and more systematic discipline of morphological and taxonomic botany,[7] and on the other, into a new kind of learned horticulture closely related to the fashion for collecting natural curiosities of all kinds.[8]

The great herbals of the sixteenth century were ambitious attempts to identify, locate, name, describe and classify all the plants of a given area or indeed of the entire known world. One of the earliest, the *Kreutterbuch* of Hieronymus Tragus, did not have pictures (at least in its first edition of 1536), because its author believed that pictures would make lazy readers less likely to take the trouble to look at actual plants. Such objections were brushed aside by Otto Brunfels, whose *Herbarum vivae eicones* (Strasbourg, 1530), had magnificent illustrations of plants, far more accurate and lifelike than any seen in print before. Brunfels regarded the beauty and precision of the images (the work of Hans Weiditz) as an integral part of his discourse, illuminating and making sense of the text. Not surprisingly this attractive book sold widely. Its formula – text and image paired – was endlessly copied, and for Clusius and

---

[7] E.H.F. Meyer, *Geschichte der Botanik*, 4 vols. (Königsberg, 1857); E.L. Greene, *Landmarks of botanical history*, parts I and II, ed. Frank N. Egerton (Stanford, 1983).

[8] John Prest, *The Garden of Eden: The botanic garden and the recreation of Paradise* (New Haven, 1981); Krzysztof Pomian, *Collectors and curiosities: Paris and Venice 1500-1800* (London, 1990); Giuseppe Olmi, 'Ordine e fama: il museo naturalistico in Italia nei secoli XVI e XVII', *Annali dell'Istituto Storico Italo-Germanico in Trento*, 8 (1982), 225-274; Paula Findlen, *Possessing nature: Museums, collecting, and scientific culture in early modern Italy* (Berkeley, 1994).

his contemporaries this convention became the norm. Equally influential, and rather more heavyweight in its treatment of plants as ingredients for medicines, was the *De historia stirpium commentarii* (Basle, 1542) of Leonhard Fuchs, a medical professor at Tübingen. Fuchs employed three craftsmen – one to do the initial drawings from living specimens, one to copy these drawings on to boxwood blocks, and one to engrave the blocks for printing. In his book they are named and depicted to show just how they shared the task.

At Wittenberg Clusius would certainly have had access to this new literature, which would have been actively discussed. It would not have been surprising if he had decided to move on to Tübingen or to Basle. However, as he tells us in his autobiographical recollection, his thoughts were turning, in the summer of 1549, towards medical studies in France. Ravished by his studies in *res herbaria*, he longed to see for himself the 'exotic and diverse flora' of Gallia Narbonensis, of which he had heard tell.[9] It is not surprising, therefore, that he thought of going all the way to the ancient medical university of Montpellier, although in other circumstances, Paris would have been an obvious destination. By 1548 Montpellier had been for nearly twenty years the French university of choice for dozens of medical students from the parts of Germany and Swizerland where evangelical teaching was strong.[10] Possibly Melanchthon himself supplied Clusius with a letter of introduction to Rondelet, who was certainly already known to Conrad Gessner in Zürich, and whose reputation as a botanist and as an innovative teacher of medicine could perhaps have reached Wittenberg by this date (Ill. 6). But Clusius nowhere mentions anything of the kind; the conjecture may be without foundation. In any case, gossip among students and their teachers would have been enough to turn the attention of Clusius to Montpellier.

In July 1549 Clusius wrote to his old Marburg tutor Andreas Hyperius to tell him about his plan. Hyperius in his reply was encouraging. Medicine, he says, is a worthy study and a pious discipline, in many respects very similar to theology, since they both deal with the preservation and salvation of man, his body and his soul. He supplies Clusius with a comprehensive, if rather rough-and-ready, reading list, chosen from the medical books he himself has in his library. Hyperius is glad to hear that Clusius has decided to study medicine in France: this is a good choice, he says, since the theory and the practice of medicine are excellently taught there.[11]

---

[9] Clusius, autobiographical memoir, loc. cit.; Hunger, *Charles de l'Escluse*, vol. I.

[10] *Matricule de l'Université de Médecine de Montpellier (1503-1599)*, ed. Marcel Gouron (Geneva, 1957), 121.

[11] Letters from Hyperius to Clusius, December 1549; Hunger, *Charles de l'Escluse*, vol. I, Appendix III.

Ill. 6. Guillaume Rondelet. Woodcut by an anonymous artist, from G. Rondelet, *Libri de piscibus [...]* (Lyon, 1554).

The first stage of his journey took Clusius to Frankfurt, where he met a merchant who agreed to make him a loan to enable him to travel on into France. From Frankfurt he made his way to Lausanne and Geneva, descending the Rhone to Lyon, and to Avignon. He had arrived in Montpellier by early October 1551, just in time for the beginning of the new academic year.

On 13 October he enrolled as a student in the university. His entry stands out in the register not only for the meticulous legibility of his hand, but for the unusual amount of detail he provides.[12] He tells us that he has paid his fees, and that he has sworn to observe the statutes of the university. He also mentions that he had been 'examined in Dialectic and in natural philosophy'. Such an examination was regarded as indispensable, because the arts faculty in Montpellier was so small and so obscure as to be little more than a rather feeble grammar school. It did not award degrees. As a result home-grown Montpellier arts graduates were expected also to have studied elsewhere before the medical school would accept them. Most of the students at this time had studied in

---

[12] Archives de la Faculté de Médecine de Montpellier, S 19, f. 261v; Gouron, *Matricule*, 121.

other French universities, or in universities in the Low Countries, in the Swiss cities and in Germany. Even when they came armed with certificates, they had to convince the Montpellier university authorities that their earlier education in Latin, in dialectic and in philosophy was more or less equivalent to a mastership of arts from the University of Paris. Lastly, they had to persuade one or other of the four regius professors of the university to act as their sponsor or 'father', a tutelary relationship which was to last for the duration of their stay. There was no guarantee that these formal relationships would work out in practice in the way the statutes implied. Even the most assiduous teachers among the professors were frequently absent at court, or attending to patients elsewhere. But sometimes the arrangement worked admirably, with the professor taking an especially helpful interest in the progress of the more favoured of his protégés.

Sixteenth-century universities were practised in the promotion and transmission of the skills needed for competent performance in all the intellectual disciplines, from the arts of rhetoric and dialectic, mathematics and music, astronomy and astrology, natural and moral philosophy, to the 'sciences' of law, medicine and theology. But they did not monopolize the subsequent scholarly study of these things, nor did they invariably provide the most effective social context for innovation in these enquiries.

At least as productive were the informal relationships which sprang up, in universities, but also in cities and in noble and clerical households and at courts, between patrons and clients, masters and servants, colleagues and contemporaries, publishers and authors, authors and readers. The fortuitous character of such encounters should not be exaggerated: even when they came about accidentally in the first place, which in fact they rarely did, they were deliberately cultivated and strengthened by the protagonists, who understood how valuable – and how fragile – such networks could be.

At first sight, the experience of Clusius in Montpellier appears to furnish a clear example of the way in which informal contacts could prove more powerful and effective than the formal opportunities presented by an institution like a university. In his reminiscences and in his botanical works, Clusius refers often and affectionately to 'Mompelier' (as the locals, to this day, pronounce its name), but surprisingly he makes little mention of his studies in the University of Medicine itself. There is a striking contrast between the silence of the later letters and publications of Clusius on the matter of formal medical study at Montpellier, and the abundance of their reference to his experiences as a member of the household and circle of Rondelet.

Immediately after matriculation he should have started to attend lectures on the set texts prescribed by the statutes, but nowhere does he recall having done so. On the feast of St Luke (18 October) each year the doctors met in solemn congregation to assign among themselves the prescribed texts on which they would offer lectures during the coming academic year. Their decisions were recorded in the *Liber congregationum*, preserved as register S8 in the archives of the university. Unfortunately, there is a gap from 1547 to 1555 in the surviving run of this register, so that we lack this means of knowing what lectures were arranged, let alone actually given, during those years. The Journal of Felix Platter shows that in during the terms in 1552 and 1553 an assiduous student could have attended seven or eight lectures a day; he gives us the names of the doctors who lectured, but he does not tell us what texts these doctors had chosen, nor what topics they discussed.[13] Some conjectures are possible, using the evidence of the University records for the years immediately before 1547 and after 1555: in those years almost all the texts chosen for comment were Galenic treatises on practical medicine, de simplicium medicamentorum being an especial favourite; Rondelet lectured on Dioscorides in 1545, 1546 and 1558.[14] Almost certainly he also did so in 1552 and 1553. Clusius would also have had an opportunity to attend public anatomical dissections staged by the medical school; since 1550 it had been laid down that the University was to hold four public anatomies per year, to be performed by one each of the most 'suffisant et idoines' of the doctors and the surgeons.[15] Clusius could have attended one such occasion on 14 November 1552. It was a spectacle attended by students, noblemen, townsmen, young ladies, even monks. Dr. Guichard presided and a barber-surgeon operated. The autopsy was of a boy who was believed to have died of an abscess in the stomach; when they opened him up, however, no abscess was found, only a blueish stain, the lung being attached in that area by ligaments which they had to tear in order to get it out.[16] Exactly a year later, on 14 November 1553 Guichard again presided at a public anatomy, this time of an old man, whose lungs were in a bad state.[17] Platter records that the first of clandestine body-snatchings and secret dissections in which he personally was involved took place as late as November 1554, long after Clusius had left Montpellier.[18] It seems quite likely that had this kind of thing gone on in 1552 or 1553, Clusius would not have been involved. More surprisingly, there is no direct evidence in his later writings that he had become interested at

---

[13] Felix Platter, *Beloved son Felix: The journal of Felix Platter, a medical student in Montpellier in the sixteenth century*, transl. Sean Jennett (London, 1961), 47, 54-55.

[14] Archives de la Faculté de Médecine de Montpellier, S4, *Liber lectionum et clavium*, S8, *Liber congregationum*.

[15] Alexandre Charles Germain, *La Renaissance à Montpellier* (Montpellier, 1871), 136; Louis Dulieu, *La médecine à Montpellier*, vol. II: *La Renaissance* (Avignon, 1979).

[16] Platter, *Journal*, 47.

[17] Ibid., 70.

[18] Ibid., 89-90.

Montpellier in the philosophical problems in which the works of Galen abound, or that he had ever contemplated earning his living by teaching in a medical school, let alone by clinical practice. The university records (admittedly deficient for the years concerned) yield no evidence that Clusius proceeded to the bachelor's degree in medicine, or that he ever sought from Montpellier a licence to practise, although Vorstius was later to claim that he did.[19] Indeed, in a letter written in 1606 to Matteo Caccini, Clusius says 'I have never taken any degree at all. I have pursued my studies purely to indulge my own delight'.[20]

Perhaps Clusius did get down to work on the systematic reading of standard medical texts, especially the prescribed works of Hippocrates and Galen which he would have needed to master if he wanted to proceed to the bachelor's degree. There is one solid piece of evidence – and one only – that his obsession with plants, and his absorption in Rondelet's fish project did not distract him entirely from medicine. Vorstius, in his obituary of Clusius, after saying that in Montpellier he was 'an assiduous and sedulous listener to Rondelet, to whom, uniquely, he was beloved and accepted', adds that he 'took down in writing, word for word from Rondelet, the full-scale discussion of the treatment of diseases which Laurent Joubert was later to publish [under Rondelet's name] with the title *Methodus curandorum morborum particularum*'.[21] Such assiduous transcribing was by no means uncommon; for students it was the equivalent to today's photocopying or down-loading.

Rondelet offered him accommodation in his large town house. There was nothing unusual about this; to take in student lodgers was a common practice among schoolmasters and university teachers of the day. Here he joined several other young men, who included at that time Laurent Joubert of Valence and Jerome Betz of Constance. Almost immediately he was plunged into that household's busy social round: Rondelet's parties, with music and dancing, were well-known. More unusually, within a week or two Rondelet had enlisted his services as his Latin secretary to help him in preparing for the press the vast assemblage of material he had accumulated on *aquatilia*, that is to say fresh and sea-water fishes both bony and cartilaginous, aquatic mammals including cetaceans, seals and beavers, reptiles, amphibians, crustaceans and molluscs.[22]

At the same time Clusius began to explore the flora of the district, 'medica marina' being the very first plant he recorded having seen on the Mediterranean

---

[19] Everardus Vorstius, 'Caroli Clusii vita et obitu oratio', in Clusius, *Curae posteriores* (Leiden, 1611).
[20] Clusius, letter to Matteo Caccini, 29 September 1606.
[21] Vorstius, 'Caroli Clusii vita et obitu oratio'.
[22] See below, 84-85.

shore.[23] He mentions innumerable plant-hunting expeditions. Some of these may have been undertaken on his own, or in the informal company of a few friends, but some were almost certainly expeditions laid on by the university, in which groups of students led by bachelors and doctors of the faculty set out with spades and knapsacks and notebooks, and possibly also with dog-eared portable copies of botanical texts, to find and identify plants named by Dioscorides, and to bring specimens home, to plant out, or to press and dry between sheets of paper, or to take, for purposes of comparison, to the apothecaries' shops, where (it was hoped) the doctors would be able to 'demonstrate' the correct plant origin of the ingredients used in the pharmacists' remedies. Rondelet was certainly the leading light in these enterprises. He had been the first to lecture in Montpellier on Dioscorides, as early as 1544-45,[24] and he was well-known as an enthusiast for the improvement of pharmacy through the revival of a more meticulously learned *materia medica* based upon study of living plants in their natural habitat. His associates in this in 1551-52 included his colleague François Fontanon, and his son-in-law Jacques Salomon de Bonnail d'Assas, also a teacher in the university.

Most of the plant collecting was done quite near the town, in the nearby fields, or in the Bois de Grammont, or up on the aromatic south-facing slopes of the garrigues or down on the flat and waterlogged land near the salt-water lagoons between Montpellier and the sea, or out on the sand-spits of Maguelone. Years later, in the *Rariorum plantarum historia* Clusius carefully recorded the exact location where he had first seen a specimen of a particular plant. More than two dozen of these plants were given named locations near Montpellier.[25] Few of the plants named had a well-documented medicinal use. Plainly these expeditions were undertaken not only to instruct medical students in the plant materials used by apothecaries, but also to contribute to a long-term collaborative project to locate, identify, name and describe all the different flora of the various regions of Europe. Domestication of wild plants in gardens, and their propagation there for the purposes of conservation and exchange was also a part of the plan. On several occasions Clusius mentions digging up bulbs or plants and replanting them in Rondelet's garden. These included a narcissus which he found at Maguelone, and a hart's-tongue fern which he found on an expedition into the Cevennes.[26]

---

[23] Clusius, *Rariorum plantarum historia*, ccxlviii; Hunger, *Charles de l'Escluse*, vol. I, 23, n. 2.
[24] Archives de la Faculté de Médecine de Montpellier, S4, *Liber lectionum et clavium*, S8, *Liber congregationum*.
[25] Hunger, *Charles de l'Escluse*, vol. I, 26, nn. 5-19.
[26] Ibid., 154; Hunger, *Charles de l'Escluse*, vol. I, 26, n. 1; ibid., ccxiii; Hunger, *Charles de l'Escluse*, vol. I, 26, n. 2.

In the spring and early summer of 1552, Clusius and a group of friends set out for an ambitious expedition into Provence, proceeding via Nîmes to Arles and thence via Sâlons and the plain of Crau to Marseille.[27] At Marseille Clusius intended to take ship to Italy, the botanical delights of which were proverbial and where learned specialists (to whom Rondelet could no doubt have furnished him with useful letters of introduction) were to be found. He did get as far as boarding a vessel. However, at the very last minute – perhaps because he could not pay his fare – he had to disembark and return to Montpellier. Despite this disappointment, the expedition had been rich in botanical discoveries[28] and in observations of geographical and antiquarian interest.[29] No doubt he was also making notes on this journey for the map he made depicting Gallia Narbonensis between Carcassone and Marseille, which was to appear in print in 1570 in the *Theatrum orbis terrarum* of Abraham Ortelius.

This tour included the obligatory visit made by Montpellier students in Provence to the house of Francesco Valleriola, a visit made by Felix Platter, and by Jean Bauhin on other occasions. Valleriola was a learned physician and a prolific medical writer. He had a great library, a celebrated natural history collection and a garden which contained many unusual plants.[30]

On the way back to Montpellier, Clusius was taken ill with what he describes as a 'dropsy'. The great heat of the sun had led him, imprudently, to drink large quantities of cold water. When he arrived home, Rondelet dealt with the situation by giving him a decoction of wild chicory.[31] Rondelet's reputation as a physician was such that he was often called away to visit patients in distant parts of the entire region between Toulouse in the north west, and Marseilles in the south east. On at least one occasion, on a visit to Carcassone, he took Clusius with him, and they gathered plants on the way home.[32] Clusius remembered beach-combing on the shores of the Mediterranean for shells and fishes. He used to go down to the sea after a storm and see what creatures were to be found in the wrack along the shore.[33] On one occasion he found washed up on the beach an unusual sponge, shaped exactly like a broad-brimmed hat.[34] On another occasion he found what

---

[27] Ibid., 53; I Hunger, *Charles de l'Escluse*, vol. I, 27, n. 1; ibid., 54; Hunger, *Charles de l'Escluse*, vol. I, 27, n. 3.

[28] Ibid., 330; Hunger, *Charles de l'Escluse*, vol. I, 30, n. 2.

[29] The transcript he made of a Roman inscription he found near Saint-Chamas later found its way into print in Martinus Smetius, *Inscriptionum antiquarum [...] liber* (Leiden, 1588), f. cli.

[30] Platter, *Journal*, 102-103.

[31] Clusius, *Curae posteriores*, 35; Hunger, *Charles de l'Escluse*, vol. I, 30, n. 3.

[32] Clusius, *Rariorum plantarum historia*, 341; Hunger, *Charles de l'Escluse*, vol. I, 26, n. 4.

[33] *Notae Ephemerides Clusii*; Hunger, *Charles de l'Escluse*, vol. I, 24, n. 1.

[34] Clusius, *Exoticorum libri decem [...] Item Petri Bellonii observationes [...]* (Antwerp, 1605), lib. vi, cap. xi, 125-126.

Dioscorides called the 'third kind of sponge', an identification with which Rondelet agreed.[35]

Clusius tells us that throughout the two years and more during which he lived with Rondelet, he observed him dissecting a large numbers of fishes and marine creatures. Rondelet did this, according to Clusius, so that he could see what their internal parts were like, and so that he could make the descriptions in his fish-book as accurate and as perfect as possible (Ill. 7).[36] Presumably these dissections took place amid the aquaria and fish-ponds Rondelet had built at his country house. Clusius estimated that, in 1552 alone, Rondelet performed no less than one thousand five hundred such anatomies.[37]

He speaks also of Rondelet's 'museum' which had a spectacular example of the globe-fish hanging from its ceiling, and in which many other spiny and prickly specimens were displayed.[38] Elsewhere, Clusius recalls that local fishermen knew that the bishop, Guillaume Pellicier, would pay them for interesting specimens of fish, so they used to bring to him all kinds of odd things they had caught.[39]

Soon after his arrival Clusius seems to have become deeply involved in the highly specialized business of collecting, naming and preserving of fish specimens, presumably so that a draughtsman could make drawings of the creatures concerned, from which a wood engraver would later prepare the blocks for Rondelet's book. More surprisingly, it is possible to deduce from a letter dated 24 December 1551, from the Bavarian physician Lorenz Gryll, later professor of medicine at Ingolstadt, but at this stage resident in Paris, that as early as November 1551 Clusius had begun writing to distant friends, telling them enthusiastically – and evidently in some detail – about the botanical and ichthyological exploration and collecting he was doing, and even sending them unsolicited (and imperfectly preserved) specimens through the post. Gryll reports that most of the ones which had reached him had gone rotten on the way. Despite this, he turns out to be an eager customer for such goods, presumably intending to study them himself, alongside the texts of Pliny, Aristotle, Aelian and so on, or perhaps to sell on to some other scholar or gentleman who was building up a cabinet of curiosities. He is effusive in his thanks and

---

[35] Ibid., lib. vi, cap. xii, 127.
[36] Hunger, *Charles de l'Escluse*, vol. I, 33, n. 3.
[37] 'cum apud illum viverem anno quinquagesimo secundo supra millesimum & quingentesimum eum secantem scarum & illius anatomen facientem ut commodius describeret observabam, ut etiam plerosque alios, quorum in libro mentio'; ibid., 15; Hunger, *Charles de l'Escluse*, vol. II, 16.
[38] Clusius, *Exoticorum libri decem [...] Item Petri Bellonii observationes [...]*, lib. vi, cap.xxiii, 139; ibid., lib. vi, cap.xxiii, Hunger, *Charles de l'Escluse*, vol. I, 140.
[39] *Notae Ephemeridae Clusii*; Hunger, *Charles de l'Escluse*, vol. I, 32, n. 5.

LIBER VI. 175

Partibus internis ab aliis saxatilibus non multùm differt. Carne est tenerrima & friabili. Ob insignem teneritatem à multis negligitur. Optimus assus in craticula vel in sartagine coctus. Refrigeratus firmiore est carne. Ex aceto vel mali arantij succo edendus: priusquam coquatur sale conspergendus, vt gratioris sit saporis.

SECVNDI Generis turdorum is rectè censebitur, qui à nobis distinctionis gratia pauo nuncupatur. Hic priori similis est, maior tamen plerunque. Colore ex viridi cæruleo, vel indico, colorem colli pauonis referente in pinnis omnibus, & in cauda. Hinc pauonis nomen posuimus veteres imitati, qui merulas & turdos auium nomina piscibus dederunt, cùm liceat cuius rebus anonymis & nouis noua nomina ponere. Carne est molli, friabili, tenera. Viscidi aliquid habere videtur inter saxatiles.

TERTII Generis est is, qui *Minchia di Re* à nonnullis appellatur. Superioribus similis, colore verò dispar. Est enim varius admodùm, vixque vllus est, qui plures colorum differentias habeat. Viridis est magna ex parte, sed punctis purpureis, cæruleis, indicis,

Q 2

Ill. 7. Two woodcuts of fishes from G. Rondelet, *Libri de piscibus [...]* (Lyon, 1554), 175.

he declares himself willing to reimburse Clusius for expenses incurred in having such specimens despatched, properly dried, or, alternatively, gutted and stuffed with straw. He bombards Clusius with technical questions about the size, shape and variety of the specimens he can send, and describes to him what appears to be the egg case of a skate, which he has failed to identify. He asks Clusius to send him seeds, as well, through merchants in Lyon, and repeats that if Clusius sets out clearly how much it has cost him to comply with these requests, he will see to it that he is adequately repaid.[40]

Perhaps this letter arrived in Montpellier early in 1552 when Clusius had already set out on his expedition to Marseille. There is no further news of exchanges between the two men until September of that year, when, in a second surviving letter from Paris, Gryll expresses some anxiety about not having heard from Clusius recently.[41] Is he still in Montpellier, he asks? If so, please will he get in touch? Gryll longs to hear news about the plants and marine creatures Clusius has lately collected. He reminds him that he will pay for the expense of despatching specimens, and suggests that he use his discretion about the most appropriate way to prepare them, noting in passing that dried specimens quite often do not have the same form or shape as the one that creature had in life. He then appends a list of no less than thirty seven marine creatures he would like Clusius to send him. There is no obvious principle of selection at work in the list, which includes, for example, a pike, a sea-horse, a sea-urchin, a crab, a squid, a ray, a small tunny, a turbot, and a mackerel. Gryll does not seem to be specializing in any particular sort of fish, and none of the ones he names is especially rare or recognizable as the kind of bizarre-looking creature sought after for a cabinet of curiosities. All of the ones he listed were to figure (under similar names) in Rondelet's text, but in 1552 this was not yet published. Had Clusius given Gryll pre-publication access to its contents, or – more likely – was Gryll using Pierre Belon's fish-book, newly printed in Paris shortly before he wrote out this list of requests? All the creatures named are to be found there also.[42] Did Clusius ever send him any of the specimens he asked for? What would Gryll have done with them? Would he have packed them up again and taken them with him on his imminent return journey to Germany? Perhaps he had a purchaser in mind in Paris? There seems, at the moment, to be no means of

---

[40] Letter from Laurentius Gryllus to Clusius, dated 24 December 1551: Hunger, *Charles de l'Escluse*, vol. I, Appendix IVa, 399-400.
[41] Letter from Laurentius Gryllus to Clusius, dated September 1552: Hunger, *Charles de l'Escluse*, vol. I, Appendix IVb, 401-403.
[42] Pierre Belon, *De aquatilibus, libri duo cum eiconibus ad vivam ipsorum effigiem, quoad eius fieri potuit, expressis* (Paris, 1553).

finding out. The letter gives a glimpse, but a puzzling one, of the world of the mid-sixteenth-century collector.

In the light of all this evidence about plants and fishes it is tempting to conclude that the benefit which came to Clusius from enrolling himself on the books of the famous University of Medicine was negligible in comparison with the advantage he gained through living in Rondelet's household, and being treated not only as Rondelet's pupil, but as his assistant and his friend.

On reflection, however, to disparage the university entirely in favour of awarding praise to Montpellier's informal social networks, simply on the basis of what Clusius chose to recall, is to create a false dichotomy between public and private worlds. For Rondelet did, after all, operate inside the university, as well as outside it. The local Montpellier circles to which he belonged were profoundly affected by the presence of the medical university, an ancient and prestigious corporation, endowed with papal charters and presided over by the local bishop. The weight of the university in the town could be felt in the authority it claimed to license the practice of medicine in the diocese, and to regulate the local trade associations of the apothecaries and the surgeons. Its presence was emphasized by the grandeur and frequency of its public ceremonies, by the wealth and social standing of its leading doctors, and by the highly international character of its student body.[43]

Until 1534 the University's syllabus was still, in formal terms, based on the regulations of 1340, which prescribed Books I and IV (parts 1 and 2) of the *Canon* of Avicenna, the *Aphorisms*, the *De prognosticis* and the *De regimine acutorum* of Hippocrates, and some eleven of the treatises of Galen, either named individually or lumped together in the so-called *Tegni*. At the beginning of each academic year, each master had to choose which texts to lecture on, avoiding texts he had lectured on in the previous year. In this way it was hoped that every student would have an opportunity to hear lectures on the entire range of the syllabus during his four, or six, years of study. The archives of the Faculty do not contain any year-by-year records of which courses were actually taught between 1340 and 1488, but after that date the evidence improves. During the period 1488-1530 nearly all the lectures announced were on Avicenna. In a given year there would, typically, be three or four masters lecturing on various parts of the *Canon*, or on others works attributed to Avicenna. The *Aphorisms* of Hippocrates usually figured somewhere, and a very few Galenic works – usually the miscellany know as the *Tegni* or Galen's work on fevers. The *Liber*

---

[43] Archives de la Faculté de Médecine de Montpellier, S2, *Liber procuratoris*, S4; *Liber lectionum et clavium*; S8, *Liber congregationum* (1557-1598), passim; Gouron, *Matricule*, passim; Dulieu, *La médecine à Montpellier*, vol. II, passim.

nonus ad Almansorem of Rhazes made its appearance in 1512, and regularly thereafter. But Avicenna dominated the scene.

In the late 1520s, however, things began to change. Some of the students coming to Montpellier had enjoyed a good grounding in humane letters and in Greek. If they expected to find in Montpellier masters with a similar education, equipped to lecture from up-to-date editions of Hippocrates and Galen, by and large they would have been disappointed. At the University's request the King had endowed four paid regius professorships at Montpellier in 1498. This was intended to ensure the presence of at least four teaching masters, and to give the younger doctors some motive to stay in Montpellier in the hope of succeeding to a regius chair. Whenever one of the professors died, his successor was nominated by the three other survivors; this kept the posts in the hands of a tiny self-perpetuating oligarchy whose members were, by definition, successful practitioners who were likely to be absentees a good deal of the time, since their services were in demand among the great of the kingdom.

There was among the teaching doctors one man, however, who shared the aspirations of those students who had some Greek: he was called Jean Schyron, and he was an admirer and a follower of Jules-César Scaliger, an Italian humanist long resident in France. Schyron set out to transform the teaching offered in the university, by choosing to lecture on Hippocrates and Galen in preference to Avicenna, and by utilizing new critical editions of the Galenic texts. He did what he could to ensure the promotion within the Faculty of others who shared his views. In the academic year 1536-37 Schyron in a celebrated course of lectures on Galen's *De constitutione artis* ostentatiously used a Greek edition of the text.

Another protégé of Scaliger who turned up in Montpellier in 1530 was François Rabelais, whose reputation as a man learned in Greek and Latin letters and in medicine had preceded him. He was allowed to take his bachelor's degree only ten weeks after he matriculated with Schyron as his 'father'. He went straight on to give the short lectures bachelors had to deliver if they were intending to go on to the doctorate. He chose to comment on two central texts: the *Ars medicinalis* of Galen, and the *Aphorisms* of Hippocrates, allowing himself the theatrical and symbolic gesture of commenting on the Hippocrates from a manuscript of some antiquity, written in Greek.[44] This episode took place in a faculty where most of the doctors were still delivering traditional-style lectures on Avicenna or Rhazes, and where the few Galen texts chosen

---

[44] Roland Antonioli, *Rabelais et la médecine*, Etudes rabelaisiennes, vol. XII: *Travaux d'Humanisme et Renaissance*, no.143 (Geneva, 1976).

still had an old-fashioned look. Within three or four years, however, the picture was transformed.

The credit for this should go not only to the courageous Schyron, but to the bishop of Montpellier, a local man of truly outstanding intellect and erudition, Guillaume Pellicier by name. Accomplished in Greek and Latin letters, he was a scholar-courtier and diplomat who had served as ambassador in Venice for King Francis I. Here he consorted with the leading Venetian intellectuals and acted as an agent in the purchase of hundreds of Greek manuscripts for the King's new library at Fontainebleau; at the same time he bought as many books again, in manuscript and in print, for the private library he kept in his palace near Montpellier.[45] As bishop, he was, like his predecessors, *ex officio* head of the medical university, a duty he took very seriously. In 1534 he took an active part in drawing up with the doctors an important – and a tactful – reformulation of the ancient statutes of 1340. The changes, at first sight, did not look very far-reaching. Care was taken to preserve a rather conservative-looking framework. The names of the set books appeared to be more or less the same as in 1340, but the titles of the Galenic treatises are given in such a form that it is plain that an assumption is being made that the editions which will be used are the new ones, translated from the Greek. It was laid down that bachelors and doctors were permitted if they wished to lecture on Galen and Hippocrates alone. They were not compelled to do so. Avicenna's *Canon* was mentioned only in passing, but it was, as propriety demanded, still on the syllabus. Far more money than before was to be spent on the purchase of books for the library. This reformulation of the statutes was not so much a catalyst for change as a recognition of changes already under way. When Rabelais returned to take his doctorate in 1537 and to lecture once again on Hippocrates from the Greek, and when Schyron chose to lecture on Galen's *De simplicibus* in 1538 more than lip-service was being paid to Greek, and *materia medica* was beginning to move to the centre of the stage. Pellicier's preoccupation with Pliny, and his interest in everything animal, vegetable and mineral encouraged Rondelet to discuss with him similar interests which he too had been developing for twenty years. In the course of the 1540s the whole complexion and emphasis of the medical studies of the faculty, as reflected in the lecture-list, can be seen to have undergone a fundamental change.[46]

In the wake of the Lutheran upheavals in Germany a disquieting sense of religious division began to spread. Many of the students, especially the ones

---

[45] Henri Omont, 'Catalogue des manuscripts grecs de Guillaume Pellicier', *Bibliothèque de l'École des Chartes*, 46 (1885), 45-83, 594-624.

[46] Germain, *Renaissance*, passim; Charles Revillout, 'Les promoteurs de la Renaissance à Montpellier', *Mémoires de la société archéologique de Montpellier*, 2nd ser., 2 (1900), passim; Dulieu, *La médecine à Montpellier*, vol. II.

who had come there from Germany, showed sympathy with reformed religion. Inquisitors were sent with increasing frequency to Montpellier from Toulouse, and although foreigners among the students were left alone if they were discreet, known heretics who were also subjects of the King of France could no longer feel safe. A certain restlessness was growing in the university (and for that matter in the town) during these decades. Religious animosities were hardening between conservatives and reformers. It is too simple to suppose that the traditionalists and the innovators can tidily be equated with Catholics and Protestants – the ambiguities of Pellicier's own liberal evangelical position make this clear. But increasingly, individuals were being forced to make choices which revealed them to be in favour of, or against, the strict enforcement of ecclesiastical penalties against heresy.

In 1547 on the death of one of the four regius professors, a bitter quarrel broke out among the three survivors, who would normally have agreed in nominating a replacement. Honoré Castellan, an opponent of the evangelicals, refused to acquiesce in the nominee of the other two professors, Rondelet and Saporta. He was helped by the fact that there was in any case widespread resentment among some of the other teaching doctors of the existing practice in filling vacancies to the coveted chairs. Indignation was also expressed by spokesmen of the students. The malcontents appealed to Pellicier, and demanded that the post be filled not by nomination, but by competition, with the submission of theses to be disputed in public session. The case went to the Parlement of Toulouse, which ruled in favour of the petitioners: a competition was to be held, not only on this occasion, but whenever a vacancy for a regius chair occurred in the future. In the event, the candidate who was said to have done best in the competition was Jean Bocaud, a known evangelical. Some of the masters and the students refused to accept the outcome; bitter religious disputes occurred. First one party and then the other brought charges against their rivals in the royal courts. A tangle of litigation began. The matter was in the end resolved by arbitration, in which Pellicier seems to have played a constructive role, seizing the opportunity to clear up a number of obscurities in the governance of the institution. In 1550 an *Arrêt* was published by a royal judicial assembly known as the Grands-Jours of Béziers. Its aim (which by and large it achieved over the next few years) was to assert the authority of the regius professors over the student body, and over the younger masters, and to restate the regulations of the university, including the regulations concerning its teaching. The curriculum now for the first time took on an up-to-date humanist tinge, with the Greek medical authors given pride of place, and with provision made for regular public anatomies and regular public demonstrations of medicinal ingredients, notably but not exclusively botanical.

The *Arrêt* seems to have been welcomed, or at least acquiesced in, by almost all parties. However, when Clusius arrived in Montpellier in 1551, religious hostilities were just below the surface, and the interruptions suffered by the business of the university since 1547 had left scars. Record-keeping remained unreliable until after 1554. Matriculations were recorded regularly, but record of the congregations, of lecturing, examining and degree-giving was not kept; at least no such documents have survived. Some of the doctors, Rondelet included, had been fortunate enough to be away during the troubled 1540s, ensconced in the households of patrons and protectors elsewhere. By 1550-51 Rondelet was back from his travels. Clusius encountered him at an especially good moment in his career.

Guillaume Rondelet[47] came from a local family of prosperous spice-importers, suppliers of specialist goods to the apothecaries' trade. As an adolescent he had been sent to the University of Paris to be educated in the arts. On his return he had studied medicine in Montpellier. He was a prominent member of the student body in 1530 when Rabelais lectured there from a Greek text of the *Aphorisms* of Hippocrates, and he figured under the guise of 'Rondibilis' in the *Tiers livre*.[48] After taking his bachelor's degree in medicine in Montpellier he had returned to Paris, where he undertook specialized studies in anatomy with Jacobus Sylvius, and more general reading with the celebrated and eclectic medical scholar Jean Winther of Andernach.[49] He embarked on the observation of plants and fishes when he was tutor to the young Vicomte de Turenne in Auvergne. He then returned to Montpellier to take his doctor's degree, married, tried to make a success of clinical practice, and became a regular member of the university's teaching corps. Like Jean Schyron he was an innovator in the texts on which he chose to lecture.

Rondelet was no stranger to noble and prelatical patronage: in about 1540 he entered the service of the cardinal and diplomat François de Tournon, travelling in his small and learned entourage in the mid-1540s to Antwerp and other parts of the Low Countries, and then down to Bordeaux and Saintes where he learned about Atlantic fisheries and the whaling trade. Most important of all, he was fortunate enough to spend the year 1548-49 with Tournon in Italy, mostly in Rome, but with shorter visits to Padua, Bologna, Ferrara and

---

[47] The present writer is engaged upon a full-length study of the life and works of Rondelet in the context of sixteenth-century medicine and natural history.

[48] François Rabelais, *Le tiers livre*, ed. M.A. Screech (Geneva, 1964) [Textes littéraires français], xxx-xxxiv.

[49] Laurent Joubert, 'Gulielmi Rondeletii vita, mors et epitaphiae', in *Operum Latinor[um] tomus primus* (Lyon, 1579).

Pisa.⁵⁰ In these places he established acquaintance with the leading local botanists, spent time in botanical gardens, and probably learned from Luca Ghini the technique of making a herbarium or hortus siccus by sticking carefully dried and pressed plant specimens on to paper with glue.

Rondelet in 1551 was a prosperous and well established property owner in the town, a leading member of the University of Medicine, where he held one of the four coveted stipendiary professorships, and a successful medical practitioner with a wide reputation over the whole region between Toulouse and Lyon. He was also the owner of a mas, a fine country property between Montpellier and the sea, where he had a 'museum' of natural history specimens, and in the grounds of which he had constructed not only a garden, where he cultivated local and exotic plants, but a whole system of aquaria and ponds, some of fresh water and some of salt water, where he kept living specimens of fishes, crabs, turtles, and even a beaver.

Rondelet's friends in the town included the surgeons Michel Héroard and Barthélemy Cabrol and the apothecary Laurent Catelan, in whose house Felix Platter lodged. All three of these were leading figures in their respective trade-associations, active in promoting close co-operation between their trades, the local medical practitioners and the learned doctors of the medical university, several of whom (including Rondelet) offered vernacular lectures to apprentice apothecaries and surgeons, which medical students were allowed to attend.⁵¹

For twenty years Rondelet had been building up an encyclopaedic knowledge of freshwater and seawater fishes, mammals, amphibians, crustaceans, reptiles and molluscs. He had accumulated a mass of notes, drawings, dried and stuffed specimens, and living creatures, and by the autumn of 1551 he had written up part of the text of what was intended to be a comprehensive natural history of aquatilia of all kinds. The groundwork of Rondelet's knowledge was book-learning, but its originality lay in the extensiveness of his field knowledge of all kinds of fishes, and the sharp and discriminating eye with which he described them. His philosophical approach to their biology was Aristotelian, and much of his initial information came from Aristotle's own first-hand familiarity with marine creatures. After that he had used the fish-lore compiled by Pliny, with additional material one could extract from Aelian, Oppian, Galen and other ancient writers. It is clear from the way he refers to them in the text of the book, that Rondelet had read and re-read the classical sources. He may well have kept the references loosely in his head, relying on his memory

---

⁵⁰ Ibid.
⁵¹ Louis Dulieu, *La médecine à Montpellier*, vol. II: La Renaissance.

to recall them when needed. This casual attitude would tally with the picture Joubert gives of Rondelet's working methods. In this case it would have fallen to Clusius to locate, check, and transcribe the passages concerned. Rondelet's discussion of Aristotle's biological arguments, however, is always shrewd and critical; he pays attention to Aristotle's words. He is noticeably detached in his use of Pliny and the other classical sources, often expressing doubt about their reliability. It is clear that he is not relying on work done by a research assistant or a secretary. The arguments in the text could not have been put forward without using his own reflection on his own reading. The feverish pace of the dissections Rondelet performed throughout 1552 is clear evidence that he was asking new questions about the biology of fishes right up to the moment the pages were taken from him by the printer. Rondelet had also used the more recent work of Paolo Giovio and Oviedo, and above all he used his own eyes. His travels with the Cardinal de Tournon had furnished him with much of this material, built up from his own observations, from conversations with fishermen and whalers, and from studious (if slightly edgy) confabulation with three men in particular, namely his old friend Guillaume Pellicier, whose magnificent library he was allowed to use, and with whom he went fishing and beachcombing,[52] Pierre Belon, a rival in these matters, who managed to get his fishbook into print before Rondelet,[53] and Ippolito Salviani, an erudite ichthyologist in Rome, who did not. Salviani, in his own fish book, accused Rondelet of making off with some pictures of fishes which he lent him in Rome.[54]

Rondelet's pupil, friend and successor, Laurent Joubert, in his obituary of Rondelet, tells us that his wide and eclectic intellectual interests, his untidiness and disorganization, and his busy public life prevented him completing pieces of work, and made him postpone publication again and again. His botanical learning was acknowledged to be impressive, but none of it was ever brought to fruition in print, and even its outlines are now blurred for ever, concealed somewhere within the work of all those younger botanists who readily admitted that they owed him a debt. His ichthyological work might have gone the same way, had it not been for the coincidence, late in 1551, of two fortuitous events: one was the arrival of Clusius in Montpellier; the other was the sudden illness of Rondelet's patron François de Tournon, and Rondelet's success in restoring him to health. In October 1551, within a couple of weeks of the arrival of Clusius, Rondelet was called in haste to Lyon to attend Tournon who

---

[52] Joubert, 'Gulielmi Rondeletii vita, mors et epitaphiae'.
[53] Belon, 'Gulielmi Rondeletii vita, mors et epitaphiae'.
[54] Ippolito Salviani, *Aquatilium animalium historiae, liber primus, cum eorundem formis, aere excusis* (Rome, 1554).

had been taken ill. So rapid was the cardinal's recovery, apparently as a result of Rondelet's diagnosis and treatment, that the prelate awarded him in gratitude so handsome a reward, that he felt able, at last, to seek a royal privilege to protect his publication rights, and to go ahead without delay with the task of preparing the great fish work for the printer. For this enterprise he needed the help of a self-effacing assistant capable of undertaking the very considerable chore of transforming his accumulated notes and drawings of fishes into a systematically arranged narrative in decent Latin, furnished throughout with carefully checked references to classical texts. In Clusius, by great good fortune, he found just such a man. Clusius, with his neat, confident, elegant Latin, his meticulousness and assiduity, his bookish temperament, and his enthusiasm for everything to do with plants and animals and minerals, could not have arrived in Montpellier at a more opportune moment for Rondelet. It turned out to be propitious for Clusius as well.

It seems that Rondelet had not advanced very far in his work on the text of the fish book by October 1551. According to Clusius he had completed no more than the first four sections, which discussed the biology of fishes in general terms, and which did not need pictures.[55] The rest of the book probably existed in inchoate note form. Some of its material may have been locked up still in Rondelet's capacious memory, not written down at all. However that many written notes and miscellaneous collected drawings did already exist we can deduce from the anecdotal form in which many of the entries on individual fish species appeared even in the final text, and from the fact that the printed illustrations are so obviously based not on a single run of professionally executed drawings commissioned at one time, but on a miscellaneous compilation of sketches of widely different quality. Some are precise and accomplished, others are decisive but workaday, others are little more than amateurish scrawls. Among them one can distinguish at least five different hands. It must have cost the engraver Georges Reverdy (if it was he) an almighty effort to transform these assorted drawings into wood-blocks fit for the press in the course of 1554 and 1555. The unevenness of the quality of the drawings shows through. And yet individually the plates owe a great deal of their memorable impact to the bold muscularity of the drawings on which they are based. Nearly every one of these blocks is more than adequate for its intended purpose of giving the reader a striking visual image which tallies well with the appended text.

Hunger thought it strange that Rondelet made no acknowledgement anywhere in the book of the help Clusius had given him. However scholars quite

---

[55] *Notae Ephemerides Clusii*; Hunger, *Charles de l'Escluse*, vol. I, 23, n. 3.

commonly employed pupils as amanuenses and assistants, and there was no established convention that this demanded explicit acknowledgement in text or preface. Clusius gives no sign, in his reminiscences, that he felt slighted in any way. In the introductory material, which includes elegant Latin verses in Rondelet's honour, the contribution of Clusius is a lovely cornucopia of a poem, spilling over with slithering silvery scaly fish and crustaceans, all fins and eyes and tails and claws, like the deck of a fishing boat or the slabs and tanks of a fish market.[56] Within the decorous rhetorical conventions of such tributes, it is clearly the work of an affectionate friend.

Jacques-Auguste de Thou in his *Historia sui temporis* repeatedly used various tricks of rhetoric to rescue the reputation of some controversial figures of the recent past, and to disparage that of others. One of those who seemed to de Thou to be in need of posthumous rehabilitation was the diplomat and scholar Guillaume Pellicier, riding high in the reign of King Francis, but ruined and imprisoned under his successor. In a tragedy parallel to that of his political downfall, Pellicier's learned commentaries on Pliny had been denied to posterity, either by being accidentally lost or by being deliberately suppressed.[57] De Thou was prepared to utilize a story which blamed Rondelet for the disappearance of these celebrated observations on Pliny. 'It is claimed' ('on prétend'), says de Thou, that Rondelet stole these observations and used them to create his own book. De Thou does not name his informant, but he constructs a venomous passage, full of knowing innuendo, and with only the faintest pretence of impartiality to help carry conviction. He states, first, that in creating in the *Tiers livre* a comic character called 'Rondibilis', Rabelais was treating the real life Rondelet with scorn; the implication is that this was merited. Second, to show how fair-minded he is, he says that it has nevertheless to be admitted that Rondelet was an accomplished physician. Next, and most damagingly, he declares in his usual authoritative tone that 'Rondelet's writings do not live up to the great reputation he had acquired, nor are they worthy of the high opinion people had formed about him. His published book on fishes, which brought him more honour than his other works, would deservedly have brought him praise had it been genuinely the fruit of his own industry, and not of that of another'. No grounds are given for this sneer. One wonders whether de Thou had actually read the text.

The response of Clusius was sharp and clear. He was uncompromising in his rejection of the accusation of plagiarism, believing that the allegation dis-

---

[56] Appendix 1: *Gulielmi Rondeletii [...] Libri de piscibus marinis, in quibus verae piscium effigies expressae sunt* (Lyon, 1554), sig. a3v.
[57] Jacques Auguste de Thou, *Historiarum sui temporis libri XVIII* (Paris, 1604).

honoured Rondelet, and that it was almost certainly made by someone jealous of his reputation. He wrote: 'The gibe could well have been made in the first place by Honoré Castellan, who, to my knowledge, had great disagreements with Rondelet during the time he was a professor in Montpellier. It is true that Castellan often prevailed over his adversary by the great facility with which he could express himself on the spur of the moment; but Rondelet had much more erudition. We could not admire enough the liveliness of his mind. I myself have seen him simultaneously prescribing remedies for two different maladies, the cure of which was up to him as well. Laurent Joubert wrote down his instruction for one of the patients, while I wrote down what he prescribed for the other one. I cannot speak too highly of his memory, which I admire.'[58] In his *Notae ephemerides* he stated clearly: 'I am constant in my denial of the allegation that Rondelet cobbled together his *Historia piscium* out of the observations of Pellicier.'[59] He was entirely certain that Rondelet's knowledge of fishes was built up out of personal first-hand experience over many years, and that the 'more than one thousand five hundred dissections' Clusius had seen him perform were clear testimony to the genuineness, the independent enquiry and the almost obsessive perfectionism of Rondelet's absorption in ichthyology.

Pellicier was certainly a very learned man, known to other scholars for his erudite familiarity with a wide range of ancient texts of medicine and natural history, for his ongoing critical work on the natural history of Pliny, and for his first-hand knowledge of plants, animals, and minerals, fishes and other marine creatures in particular.[60]

However, Rondelet too had been collecting and recording information about fishes since his early days in Auvergne, and Pellicier and Rondelet had freely shared their knowledge, beach-combing together, and enjoying sea-trips in boats rowed by 'saracen' sailors. Pierre Pena and Mathieu de L'Obel, writing in the late 1560s, took the view that Pellicier had indeed taught Rondelet a lot of what he knew about plants and marine creatures, but that Rondelet himself had an independent and encyclopaedic knowledge in these matters, and that

---

[58] Clusius, manuscript annotations to Jacques Auguste de Thou's *Historia sui temporis*. Bibliothèque Nationale de France, MS Dupuy 632, f. 78v; cf. Marie-Elisabeth Boutroue, 'French manuscript sources on Carolus Clusius', in the present volume. She also points out that these annotations did not remain unpublished, being reprinted in, for example, the edition of de Thou's *Historia* published in The Hague in 1740. I am grateful to Marie-Elisabeth Boutroue for kindly drawing my attention to this passage.

[59] 'Sed quod Rondeletius suam Piscium historiam ex Pelliserii observationes concinnaverit, hoc ego constanter pernego'; *Notae ephemerides Clusii*; Hunger, *Charles de l'Escluse*, vol. I, 33, n. 4.

[60] Walter Hermann, 'Il commentario pliniano di Guillaume Pellicier (ca. 1490-1567) e la storia del codice parigino latino 6808', *Studi umanistici Piceni*, 17 (1997), 179-194.

the two men could best be described as friendly competitors and rivals.[61] In the Latin version of Part II of his book (in a passage omitted from the French translation) Rondelet refers to Pellicier as 'that most practised and most diligent reader of Pliny' and acknowledges that it was Pellicier who 'taught him' how to recognize that certain passages in Pliny were probably corrupt.[62]

Work on the *Historia piscium* must have proceeded at a great pace throughout 1552 and 1553. The book was still not safely through the press in the late autumn of 1553, and its second part, the unwieldy but even more pioneering *Universae aquatilium historia* was still being pulled into shape by another of Rondelet's young assistants, Jean Des Moulins.[63] Yet another assistant, Jacques des Bordes, 'a student of medicine learned in Greek, Latin and Hebrew' was engaged upon the task of polishing up Rondelet's *Librum de ponderibus & mensuris*, a treatise on pharmaceutical weights and measures, dedicated to Pellicier, and echoing Galen in its name.[64]

By the autumn of 1553 Clusius was clearly getting restless in Montpellier. There was probably little more he could do on the fish book at this stage. No doubt the printer, Macé Bonhomme, was anxious to get the manuscript once and for all out of Rondelet's tinkering hands. Were Clusius to stay on for another academic year, he would probably have to commit himself to straightforward medical study, and even to the unwelcome option of taking a medical degree. Religious tensions in Montpellier were worse than ever. It may have been difficult for Clusius to go on maintaining his Nicodemite discretion. The international news was bad. War had broken out once more between the Holy Roman Emperor and the King of France. The merchant on whose good offices Clusius relied for the transfer of funds from his father was compelled by the military situation to leave Lyon, and so the arrangement, which had worked well for over two years, now broke down. Clusius was deprived of all funding. For this reason, as well because of his father's insistent demands, in January 1554, he left Montpellier and set out on a circuitous journey home.

A trail of references to plants spotted en route from Montpellier to Lausanne enables us to trace the progress of Clusius in the early months of 1554. At Lausanne he paused, and wrote letters to his father asking him to

---

[61] Matthias Lobelius, *Stirpium adversaria nova* (London, 1570), 210; Hunger, *Charles de l'Escluse*, vol. I, 33, n. 1.
[62] Guillaume Rondelet, *Universae aquatilium historiae pars altera, cum veris ipsorum imaginibus* (Lyon, 1555), 'De piscibus fluviatilibus liber', cap. VIII, 178.
[63] Laurent Joubert, 'Gulielmi Rondeletii vita, mors et epitaphiae'.
[64] Jacques des Bordes (Jacobus Bordeus) matriculated in the Montpellier medical university with Rondelet as 'father' on 21 October 1559. Archives de la Faculté de Médecine de Montpellier, S 19, f. 332v; Gouron, *Matricule*, 147.

send funds to enable him to travel to Italy.⁶⁵ He waited there for some weeks, but when there was no response he reluctantly gave up his Italian plans and resolved instead to make his way home. He was, after all, an eldest son, and he had not been home since 1548.

He seems to have arrived at his father's house in the autumn of 1554 and to have spent the winter there. He took up his studies in Louvain once more, and then moved to Antwerp. After that he moved about the Belgic region for some time, staying in various places.⁶⁶ In 1560 he made his way to Paris. Here he worked on his plan to publish the Florentine *Ricettario*, a project very much in line with initiatives undertaken in Montpellier in the time of Laurent Joubert, and close to the heart also of the Paris doctor Jacques Goupyl, a friend of Rondelet, with whom Clusius was in touch at this time.⁶⁷

There is no clear evidence that he ever visited Montpellier again. Over the next ten years, all the news from 'Gallia Narbonensis' was bad. In a letter to Johannes Crato in 1563 Clusius says he has lost touch with Rondelet⁶⁸: because of the ferocity and chaos of the religious wars in France all communication with Languedoc is cut off. Even when he went to Spain in 1563-64 he travelled there by sea, avoiding the overland journey through France. He may even not have heard very quickly about Rondelet's unexpected death in 1566.

When he left Montpellier in January 1554, what did Clusius take away with him, in the way of physical and intellectual baggage? He was almost certainly advised by Rondelet to make his way to Leonhard Fuchs in Tübingen; Rondelet had written to Fuchs about him. The letter is lost, but at least part of its contents can be deduced from a belated reply written by Fuchs on 10 December 1556 to Rondelet: it seems that Rondelet had gone to the trouble of writing to Fuchs mentioning Clusius by name, and commending his talents. Fuchs complains that despite the fact that Rondelet's had suggested that Fuchs might find his protégé 'Carolus Lucius the Fleming' useful as an assistant on his current botanical projects, this paragon had never communicated with him, and had never turned up.⁶⁹ If Clusius did go away from Montpellier armed with letters of introduction to other botanists in Switzerland and Germany, he did not use them; if he had letters of introduction to scholars in italy, he did not have a chance to use them.

---

⁶⁵ Autobiographical memoir, ff. 177-178.
⁶⁶ Ibid.
⁶⁷ Edouard Morren, *Charles de L'Ecluse, sa vie et ses oeuvres 1526-1609* (Liège, 1875), 47.
⁶⁸ Johannes Crato von Crafftheim, *Consiliorum & epistolarum medicinalium liber* (Frankfurt, 1594), Epistola xxxi (1563); Morren, *Charles de L'Ecluse*, 12-13.
⁶⁹ 10 December 1556, from Fuchs at Tübingen to Rondelet at Montpellier, Universitätsarchiv Basel, Fr Gr II Sa, 44; Reeds, *Botany in medieval and Renaissance universities*, 67-68 and n. 94, 207-208.

It seems very likely that Clusius, like many other Montpellier students, departed with his saddle-bags stuffed with books, pressed plants, seeds, dried fishes, shells, and medical and botanical notes consisting in part of long copied-out passages from the medical teaching of the professors.[70] Did he also take away with him some of the drawings of fishes which had been used in the preparation of the plates for Rondelet's book? This was a conjecture made by Wegener in 1936 when he saw for the first time the magnificent water-colour depictions of fishes and other sea creatures in the Libri picturati.[71] It is a conclusion which, at first sight, is easy and tempting to reach. Almost all of the marine organisms depicted in lifelike colour and precise and subtle outline in this stunning assemblage of pictures are to be found in Rondelet's pages. There is a high degree of similarity in the positions from which they are drawn, and to some extent in the rough intimations of scale. The fact that the paper on which the fishes appear can be dated by watermark to Fabriano 1554 has been taken to rule out the possibility that the Libri picturati drawings were done for the book published in that same year.[72] But is the dating of the paper quite as conclusive as it might be? Watermark evidence is only as good as the data Briquet and others managed to collect: they found paper like this from Fabriano datable to 1554, not earlier. Is it possible that the Fabriano paper-mill was already using that watermark for that weight of paper several years earlier, and that no examples have yet come to light?

However, even were it to be the case that the highly professional botanical and zoological watercolours painted on the Fabriano paper, could be dated a little earlier than 1554, there would still be no evidence that they had come into the hands of Clusius, and in Montpellier, between 1552 and 1554. Furthermore, careful scrutiny of the zoological pictures shows up so many small differences in the precise representation of the cetaceans, cartilaginous and bony fishes, shell fish and amphibians that it becomes impossible to sustain the view that the illustrations in Rondelet's book were based on these particular paintings. The drawings in Rondelet's possession can be deduced from the plates to have been a rough and ready lot of uneven quality, the best of them vigorous representations which have a point to make, the weakest of them little more than formless scribbles. In contrast, most of the fish pictures in the Libri picturati are clearly the work of a one, or possibly two accom-

---

[70] Compare Platter, *Journal*, 128.

[71] Hans Wegener, 'Das grosse Bilderwerk des Carolus Clusius in der Preussischen Staatsbibliothek', *Forschungen und Fortschritte*, 12 (1936), 374-376.

[72] Florike Egmond, 'Clusius, Cluyt, Saint Omer. The origins of the sixteenth-century botanical and zoological watercolours in the Libri Picturati A 16-30', *Nuncius. Journal of the history of science*, 20 (2005), 11-67.

plished professional hands. They are beautiful to look at. In the way they combine form, colour, presentation on the page they look more like an album of natural history paintings to be kept alongside a collection than like preparatory drawings for an illustrated printed book. They surely belong to the same convention as the pictures Ligozzi did for the Grand Duke of Tuscany, or the ones Joris Hoefnagel did at the Imperial court. Charles de Saint Omer was a lucky man. If Clusius had anything to do with these pictures at all, it is likely to have been no more than advising, perhaps with the text of Rondelet in hand, on the identification of the specimens the painter had at his disposal.

What else had Clusius learned in Montpellier? While he was there he had consolidated his initial experience of practical botany in the field and developed his own codes for distinguishing one variety of the same species of plant from another. In the fields and hills and sea-shores around Montpellier he developed the authentic field-botanist's nose and eye for the habitats of different plants. It was almost certainly in Montpellier that he learned how to prepare a hortus siccus, where, among Rondelet's botanical students, it was all the rage.[73] In Montpellier Clusius learned to compare living specimens with the almost unrecognizable dried-up bits of leaves and roots to be found on the shelves of the apothecaries' shops. And in Montpellier he had the opportunity to pick up the crumbs of learning which fell from Pellicier's table – though only briefly, since by ill-luck the bishop, still active in the affairs of the Diocese and of the University in October 1551 was arrested by his political enemies on 12 November of that year, very soon after Clusius arrived.[74] He was incarcerated in Beaucaire, and in July 1553 taken, still as a prisoner, to the King's Court. He was absent during almost the whole of the time Clusius spent in Montpellier, awaiting trial on a wide range of charges, most of them concocted by his enemies with the aid of false witnesses.[75]

In the 1550s there were many private gardens in Montpellier, but, as yet, no official university botanical garden. There must already have been talk of the desirability of such a thing, to match the newly constructed anatomy theatre, and in emulation of the University botanical gardens Rondelet had visited in Italy. The turmoil of the religious wars was to delay the endowment and founding of the Montpellier garden for a generation,[76] but despite this it is clear that

---

[73] Platter, *Journal*, 88. The *herbarium* of Felix Platter was later seen and admired by Montaigne; it still exists, in Bern. The *herbarium* of Rauwolff has also survived. It is preserved in the University of Leiden.

[74] Louise Guiraud, *Le procès de Guillaume Pellicier* (Paris, 1907), 104.

[75] Guiraud, *Procès*, passim.

[76] Louise Guiraud, 'Le premier jardin des plantes français: création et restauration du Jardin du Roy à Montpellier', *Archives de la ville de Montpellier*, vol. IV (Montpellier, 1920), 265-396.

Clusius had learned in Montpellier how useful a botanical garden could be in a university, not only as a place to teach students about the basic ingredients of *materia medica*, but to serve as a 'Noah's Ark' where rare species could be nurtured and protected. Some years later he translated into Latin (under the title *De neglecta plantarum cultura*) Pierre Belon's remarkable book of 1558, the *Remonstrances sur le default du labour et culture des plantes et la cognoissance d'icelles* in which problem XX is an 'Admonitio ad Galliae respublica' on the desirability of founding botanical gardens in French university towns.

It is difficult to evaluate what Clusius had learned from the secretary's task of preparing Rondelet's rich (and probably chaotic) material for the press. All one can say, perhaps, is that the experience did not put him off, since repeatedly during the rest of his long life he voluntarily undertook and completed a succession of editorial tasks of similar complexity. The main difference seems to have been that after Montpellier he was no longer content to act as an anonymous assistant to a celebrated author, but preferred to make his own choices, to be his own master, and, in most instances, to have his own name on the title-page. Translating, re-branding, publicizing, publishing the important works of other authors continued to absorb his serious attention. He was very good at it, and he seems (mostly) to have enjoyed working with illustrators and with printers.

His earliest ventures into this world came soon after his return from Montpellier. In the work he completed with Rembert Dodoens on the translation of the *Cruydt-boeck* into French, Clusius adopted the role not of deferential assistant but of colleague on equal terms. In the 'letter to medical students' by Dodoens which appears in the French version, Dodoens was happy to acknowledge a debt to his translator and collaborator: he has derived profit, he says, from the industry of Clusius, from his universal knowledge of *materia medica*, from the utmost care and skill he has brought to the task of rendering the author's commentaries into idiomatic French, and from the benevolence and diligence he has shown towards the author while carrying out the work.[77]

The second publication Clusius prepared at this time, the *Petit receuil [...] d'aucunes gommes et liqueurs* also contained work by Dodoens, but evidence also that Clusius was concerned to de-mystify the exotic botanical substances he had seen in Montpellier during his visits to wholesale spice merchants' depots and apothecaries' shops.

---

[77] Rembert Dodoens, *Histoire des plantes, en laquelle est contenue la description entiere des herbes, c'est à dire, leurs espece forme, noms, temperament, vertus & operations: non seulement de celles qui croissent en ce pays, mais aussi des autres estrangeres qui viennent en usage de medecine [...] nouvellement traduite de bas aleman en françois par Charles de l'Escluse* (Antwerp, 1557).

The third publication of these years was a Latin translation of the Florentine *Ricettario* of 1550 and 1557. Clusius' part in preparing it for the press in Paris in 1561 was closely related to projects he had been witness to in Montpellier. Here he co-operated with Dr. Jacques Goupyl, another of Rondelet's friends.[78] The title he gives to his translation, *Antidotarium sive de exacta componendorum miscendorumque medicamentorum ratione*, is a direct echo, not only of more than one of Rondelet's published pieces, but of Galen's text on the same subject, which had formed the basis for several lecture-courses in Montpellier before his arrival, and right through until later in the century. This was universal stuff, of course: it was to be found in the teaching of most of the universities of Europe; but at Montpellier special efforts had been made to give the composition and preparation of medicines a central place in the training of medical students and apothecaries' apprentices alike. The vexed problem of Theriac preoccupied the more critical of learned physicians all over Europe; they debated whether or not it was a cure-all, precisely what its ingredients should be, and how to recognize and avoid fraud in its composition and superstition about its effects. Clusius attests Rondelet's participation in this matter.[79]

Clusius is so readily thought of as a botanist that it is easy to overlook the fact that long after Montpellier he also kept up his interest in animals, especially marine animals and fishes. There is abundant evidence of this in his published writings. Again and again in these matters he turns to the text of Rondelet for reference and guidance. Sometimes Clusius uses Rondelet's text to confirm or to deny a tentative identification. For instance, a mussel shell with a smaller shellfish attached to it, seen in Amsterdam in 1599, reminds him forcefully of the specimen which was described and illustrated in Rondelet's chapter 'De testaceis' in part II of the fish-book, the *Universae aquatilium historia*.[80] On the other hand a dried specimen of an East-Indian fish with a single protuberance like a horn on its head, seen in Leiden in 1601 is said by Clusius to be nothing like the scolopax or the capriscus in Rondelet, as anyone can easily see if they compare the description and the picture.[81] In a rare disagreement with the master he states categorically that the Pristis, or Serra (sawfish) should not be classified, as Rondelet classifies it, with the Cetaceans, which it does not at all resemble, but among the Dogfishes. He has based this judgement not merely on pictures, he says, but on his inspection of an entire dried specimen in Amsterdam.[82] Clusius also gives Rondelet credit for the fact that as early as the

---

[78] Joubert, 'Gulielmi Rondeletii vita, mors et epitaphiae', 153; Morren, *Charles de l'Écluse*, 48-49.
[79] Clusius, *Historia stirpium Hispaniae*, vol. II, cap.14; Hunger, *Charles de l'Escluse*, vol. I, 337, n. 2.
[80] *Exoticorum libri decem [...] item Petri Bellonii observationes*, lib. vi, cap. xv; Hunger, *Charles de l'Escluse*, vol. I, 130.
[81] Ibid., lib. xv, cap. v; lib. xxvi, cap. v; lib. vi, cap. xxviii; Hunger, *Charles de l'Escluse*, vol. I, 143.
[82] Ibid., lib. vi, cap. xix; Hunger, *Charles de l'Escluse*, vol. I, 135-136.

1550s he had understood that the diversity of fauna and flora in the world was far greater than the ancients had known, and that his generation and succeeding ones would come to learn more and more about this diversity. Confronted with an unfamiliar specimen, Rondelet first searched in the classics, consulting Aristotle in particular, as always, but then extending his search to recent writers. For the exotic mermaid-like manatee, and for the 'upside-down fish' Rondelet correctly followed Oviedo.[83] Rondelet's *Historia piscium* is full of anecdotal reference to fishes he has read about in travellers' accounts, but which he himself has not seen. He is scrupulous in distinguishing between different types of information, and also in indicating where he believes it to be the case that a particular the creature was unknown to the classical authorities. Rondelet's capacity for distinguishing between different kinds of evidence was one of the qualities of his mind which Gessner admired, and which explains his eagerness to secure Rondelet's permission to reproduce the text and pictures from the fish book more or less in their entirety in his own encyclopedia. The young Clusius, in working on Rondelet's manuscript, would have become familiar with this approach to the material; indeed it may have been part of his secretarial duty to check that it had been applied consistently throughout the text. It is not surprising that half a century later he still had enough confidence in the thoroughness with which classical sources had been utilized in Rondelet's book to go on using the test of presence or absence from Rondelet's text as a fair indication of whether or not a given fish had been known to the ancients. He does so, for example, in the case of an East India crab, a dried specimen of which he was shown in Leiden in 1603.[84] Clusius also recognized that sometimes the evidence did not allow a question of identification to be satisfactorily resolved. Was the large sea-creature caught in 1604 in the North Sea an example of the famous *Dux cetaceii* mentioned by Aelian in cap. xiii, lib. ii of his *De animalibus*? Clusius thinks this is unlikely. In any case the condition of the dried specimen was poor. Clusius was not able to examine its inner parts. He thinks the safest course would be to assume that it should be classed among Rondelet's dogfishes.[85]

How then are we to conclude? Clusius in later life recalled that it was Hyperius and Melanchthon, those two powerful advocates of a new-style evangelical humanist education, who had encouraged him as a young man in 1548 to

---

[83] Ibid., lib. vi, cap. xviii; Hunger, *Charles de l'Escluse*, vol. I, 135.
[84] 'Veteribus tamen prorsus ignotum fuisse arbitror'; ibid., lib vi, cap. xiv; Hunger, *Charles de l'Ecluse*, vol. I, 129.
[85] Ibid., lib. vi, cap.xx; Hunger, *Charles de l'Escluse*, vol. I, 137.

change his life's course. Any sense of obligation he may once have felt to his fee-paying father to continue his studies in the law evaporated in the warmth of their confidence in the moral and intellectual rectitude of their own view of the world. In this perspective he need not feel that his enthusiasm for *res herbaria* was mere self-indulgence, or an unworthy distraction from serious matters. On the contrary, it could be seen as part of an ampler vision of the purposes of learning, in which enlarging one's knowledge and understanding of God's creation in all its wonder and diversity would be at one and the same time an intellectual challenge and a lifelong act of religious devotion. It was in this frame of mind, apparently optimistic about the prospect of embarking on the study of medicine, that he had undertaken the long journey from Wittenberg to Montpellier, full of curiosity about the rich and exotic flora of Gallia Narbonensis. And once he had reached the Mediterranean, was not Italy within reach?

In the event, things turned out rather differently. If he ever had any inclination to qualify as a medical practitioner (which is doubtful) it seems to have weakened and died at the University of Medicine. He did not become an enthusiast for the study of human anatomy, despite the opportunities then available in Montpellier. More surprisingly, he seems not to have wanted to contribute actively to the new way of studying Hippocrates and Galen, despite his sound knowledge of Greek, which would have enabled him to take advantage of the readiness of some of the professors in Montpellier to tackle these ancient authorities using new and more critical versions of their texts. The planned voyage to Italy did not, in the end, take place. Furthermore he shows himself to have been jumpy about the volatile climate of international affairs, afraid of the threat and actuality of war between the kings of France and the rulers of the Empire. In Montpellier he had witnessed disturbing religious tensions and personal animosities with a confessional component. These enmities showed themselves in the politics of the University as well as in the town, and most dramatically in the arrest and imprisonment of no less a figure than the bishop himself. Petrus Lotichius Secundus, a good friend of Clusius since Wittenberg days, ran into trouble with ecclesiastical authority in Montpellier for openly disapproving of the prohibition of meat-eating in Lent, and had to be rescued through the good offices of Rondelet.[86] Rondelet himself became very worried after the arrest of Pellicier, and according to Joubert, secretly burned those books in his own private library which he thought might land him in trouble with religious inquisitors.[87]

---

[86] Hunger, *Charles de l'Escluse*, vol. I, 31-32.
[87] Joubert, 'Gulielmi Rondeletii vita, mors et epitaphiae', 154, lines 50-55.

The arrest was a crippling blow to the intimate and lively evangelical Catholic intellectual world which Pellicier had protected, presided over in Montpellier and in which he had been an active participant, with Rondelet as his coadjutor and friend. By 1552-54 it was Rondelet even more than the pioneering but now aged Schyron who upheld these values in the University, in bitter rivalry with Honoré Castellan, who hated evangelicals, and who took up a position ostensibly scornful of novelty, and articulate in defence of a traditional arts training in all its mathematical, logical and dialectical rigour.[88]

It is not surprising that, with his existing predispositions, Clusius became a Rondelet protégé and a Rondelet partisan, and that he spent so much of his time on field botany, and on the great fish book. Like Rondelet's other protégés at this time and later, Clusius also showed concern about the need to improve the practice of pharmacy by making the latest knowledge of materia medica available to physicians and apothecaries alike. Dealing with printers and illustrators, and working as secretary to another scholar must have been useful experiences. He was willing to repeat them, with Dodoens, only just over a year later, but with some significant differences in the terms of professional engagement: Clusius was no longer the servant, the employee, the maid-of-all-work as he had been for Rondelet. Now he was in charge: he was the initiator of the project, the translator and presenter of an existing work to a new audience. At the same time he was anxious to work in co-operation with the creator of the book, and to show willingness to welcome that author's desire to include new material in the Clusius publication. This was half-way towards the practice Clusius adopted in later publishing projects, where, by and large, he selected modern works, but ones whose author was dead; in this way Clusius had plenty of scope to plan and present the translation as he wished, and even, in some cases, to add to its text notes and comments of his own. He became more practised, also, in his dealings with publishers.

In Montpellier Clusius had not become immersed in an ordinary medical student's grind. He had no need to do so, since he did not intend to become a bachelor of medicine, let alone a doctor. Instead he enjoyed apprentice membership of a very particular humanist evangelical milieu, that of Pellicier and of Rondelet, just before it was destroyed by the forces of bigotry and reprisal. Perhaps it was in Montpellier, also, that Clusius learned that in matters of religious commitment, a very low profile might well prove necessary for survival.

It may be that Clusius decided to leave Montpellier at the beginning of 1554 not merely on account of the rumours of war, and because his money had run out, but also because he felt that in some way he had come to the end of what

---

[88] Honoré Castellan, *Oratio Lutetiae habita, qua futuro medico necessaria explicantur* (Paris, 1555).

the town and the university had to offer him. In any case he needed to go home and to discuss matters of property and of family obligation with his father. Even when that was done, he remained rudderless for some years, accepting temporary employment and temporary hospitality as opportunity offered, hovering around on the edges of the printing trade, frustrated by the civil wars in France, hampered above all (after the premature death of Saint Omer) by his failure to find a permanent patron whose intellectual interests dovetailed neatly with his own knowledge, skills and talent.

And yet the two years in Montpellier had certainly furnished the young Clusius with a variety of useful experiences and, as it turned out later, with a lasting enlargement of his existing network of scholarly contacts and correspondents. It had furnished him also with a number of intellectual projects and ambitions. These included the drawing up of regional botanical surveys of a kind only patchily undertaken before, the utilization of the printing trade and of translation to make available to a Latin-reading public pioneering vernacular works on flora and fauna of the world beyond the seas, and the conservation and propagation of rare plants of all kinds in the botanical gardens and the pleasure grounds of Europe. These projects, conceived and in part gestated during his time in Montpellier, were to absorb him and to occupy his considerable energies until his dying day.

Appendix 1

CAROLI CLUSII ELEGIA
Quisquis squamigeros pisces, genus omne natantum,
Nosse cupis, praesens perlege Lector opus.
Quod tibi florenti contexuit integer aevi
Rondeletus medici gloria rara chori.
Invenies hîc qui saxis stabulentur opacis,
Incurvísque colant proxima litoribus.
Hîc lupus immitis, soleae, triglae, melanuri,
TrachurΩnque greges, sepia, mugil, elops,
Anthias hîc, scarus & percae, turdique vírentes,
Hippuri celeres, Iulides & lamiae:
Et muraena ferox, auríque imitata colorem
Chrysophrys, sargus, scorpius, atque faber:
Et congri, rapidíque canes, & glaucus utroque
Solstitio nunquam conspiciendus, adest:
Et qui praeduro clauduntur tegmine pisces,
Cancer & obliquis gressibus aequor agens:
Missilibus spinis horrens spectatur echinus,
Et variis gaudens polypus insidiis:
Fallax limosas piscatrix rana paludes
Incolit, & rigido conficit ore cibos,

Difficilísque trygon, quo cernitur aequore toto
Nil mage pestiferum, sed medicamen habet:
Callida torpedo mira se fraude tuetur,
Dum piscatorum membra rigere facit:
Nautilus aequoreis cautus spatiatur in undis,
Et gaudet celeres assimilare rates:
Succantem puppim fluctus echeneïs, & altis
Flatibus impulsam, sola tenere potest:
Invenies blandos curui delphinis amores,
Ut duri xiphiae in tristia fata ruant:
Ut Phocae catulos miro sectentur amore,
Ut stolidus scomber retia sponte subit:
Quas pelagi sedes teneant immania cete,
Ut passim rapiant squamigerum omne genus.
Denique quicquid habent & pontus, & aspera saxa,
Codice in angusto picta videre licet.
Ergò cum vario celebrentur carmine vates,
Et quisquis medica nomen ab arte tulit,
Et qui solertis naturae arcana recludunt.
Cur laudem praesens non mereatur opus?
Ex hoc nanque potes rerum cognoscere causas,
Orta vel immenso monstra stupenda mari.
Quinetiam morbis medicam superantibus artem,
Vnus saepe tulit piscis, & alter, opem.

Clusius' exchange of botanical information with Spanish scholars

Josep L. Barona

The scientific and personal relations of Carolus Clusius with the Spanish scientific community began as a result of a trip made during 1564 and 1565 across the Iberian Peninsula. The Flemish botanist spent one and a half years crossing Spain and Portugal in the company of his pupil Jakob Fugger. The journey had a commercial purpose, as Jakob Fugger, the son of the German banker Antoni Fugger, was an outstanding trader in American products. In fact, he owned the commercial monopoly of *guayaco,* a natural product from America used for the treatment of gout and rheumatism.[1] During that long trip through Spain, Clusius became aware of the important natural wealth of the New World and discovered the original works of some Spanish naturalists and doctors. In the years immediately following his return, he was to publish and translate into Latin the works of a selected group of Spanish naturalists on exotic flora.

Clusius made the acquaintance of the *Coloquios dos simples [...]* by the Portuguese naturalist Garcia da Orta in Spain, and translated it immediately into Latin as *Aromatum et simplicium aliquot medicamentorum apud indos nascentium historia* (Antwerp, 1567). The work of Garcia da Orta was the best natural historical description of the so-called East Indies. Clusius later added to this book *Aliquot notae in Garciae aromatum historiam* (Antwerp, 1582) in order to complement the work of Garcia da Orta with more recent observations deriving from the Pacific expedition led by Francis Drake.[2]

From the mid-1560s on, the scientific communication of Clusius with the Spanish scholars was close and frequent. In the same year in which it was published in Spanish, Clusius translated into Latin the work of Nicolás Monardes (ca. 1507-1588) entitled *Primera y Segunda y Tercera Partes de la Historia Medicinal de las Cosas que se traen de nuestras Indias Occidentales que sirven en Medicina [...]*

---

[1] Cf. J.L. Barona and X. Gómez Font, *La correspondencia de Carolus Clusius con los científicos españoles* (Valencia, 1998), 29-30.

[2] Ibid., 31-32. Clusius refers often to this trip in his *Rariorum aliquot stirpium per Hispanias observatarum historia* (Antwerp, 1576).

(Seville, 1574), which was a crucial work for the renewal of the traditional Galenic pharmacopoeia with new American products.[3] Monardes was a prestigious physician linked to the trade in medicinal products and slaves with America.[4] Clusius translated his work as *De simplicibus medicamentis ex occidentali India delatis quorum in medicina usus est* (Antwerp, 1574).

During his stay in England Clusius obtained a copy of the *Tratado de las drogas medicinas y plantas de las Indias orientales* (Burgos, 1578) by Cristóbal de Acosta, a doctor of Jewish Portuguese origin who practised in Burgos. The work of Acosta was a continuation of Garcia da Orta's work. Clusius produced a Latin version of it under the title of *Aromatum et medicamentorum in orientali India nascentium liber* (Antwerp, 1582).

All these examples indicate that Clusius developed a wide interest in Spanish botany, especially in all the publications relating to the American and East Indian flora. The works of Garcia da Orta, Cristóbal Acosta and Nicolás Monardes were widely disseminated among European botanists through Clusius' Latin translations. Moreover, it is necessary to add his close scientific relation with the university professor of medical botany at the University of Valencia, Juan Plaza, and his exchange of information and seeds with a group of correspondents with whom he regularly corresponded. Clusius maintained a personal, intellectual, friendly relationship with some of them, as in the case of Benito Arias Montano. Others, like Pedro and Hipólito Martín, wrote to him on purely occasional or even bureaucratic matters.

The exchange of botanical species and scientific information was concentrated in a small group of scholars, mostly naturalists in Seville: Simón de Tovar, Juan de Castañeda, and Rodrigo Zamorano. This network of collaboration on botany was forged during Clusius' Spanish journey, an occasion that allowed him to collect plants in some parts of Castile, Andalusia and Valencia, and put him in touch with certain aspects of the Spanish natural historical tradition and its relation with the New World and the East Indies. Clusius discovered not only the printed works of scientific scholars, but also the commercial relevance of the newly discovered plants.

Clusius' Spanish journey began in Castile: Valladolid, Salamanca and Alcalá were some of the towns he visited first. He then went on to Madrid, Toledo and Extremadura, later continuing to Portugal. After visiting Lisbon, Clusius left his pupil Jakob Fugger and went to Seville, which was the centre of commercial and scientific relations with the colonies of the New World. He

---

[3] F. Guerra, *Nicolás Bautista Monardes. Su vida y su obra (ca. 1493-1588)* (Monterrey [México], 1961); J.M. López Piñero and M.L. López Terrada, *La influencia española en la introducción en Europa de las plantas americanas (1493-1623)* (Valencia, 1998); J. Pardo Tomás, this volume.

[4] Cf. J.M. López Piñero, *Diccionario histórico de la ciencia moderna en España*, vol. II (Barcelona, 1983).

continued on his way to Cadiz, Gibraltar, Málaga, and Granada, and reached Valencia at the end of 1564.⁵ According to his own testimony, he remained in Valencia for about three months, where he established a solid scientific and personal relationship with Juan Plaza. This friendship is reflected in the numerous references to the Valencian university professor in Clusius' botanical works.

In mid-April 1565 Clusius returned to Madrid as the first stage of a long journey that led him back to Antwerp, which he reached at the beginning of June. During his stay in Spain he learnt Spanish and Portuguese, and he integrated the importance of the American and Asian flora in his scientific perspective. The confluence of these two aspects made Clusius a key figure in the diffusion of Spanish botany in Europe. Helped by Juan Plaza, Simón de Tovar and other Spanish naturalists, Clusius also had the opportunity to collect plants in several territories of the Iberian Peninsula. Two hundred new plants growing in Spain that had not been described before are included in Clusius' works. As a result of all these activities, he published *Rariorum aliquot stirpium per Hispanias observatarum historia* (Antwerp, 1576), and *Rariorum plantarum historia* (Antwerp, 1601), containing more than one hundred new botanical species. At the end of his life, the *Exoticorum libri decem* (Leiden, 1605) presented much of his previous work, adding information about new exotica. In 1611 a posthumous edition of this work appeared under the title *Curae posteriores*. Plantin himself was dead by then; the edition was a joint publication by Franciscus Raphelengius Jr in Leiden and Jan Moretus in Antwerp, the successors to Plantin.

The Spanish flora and the friendship with Benito Arias Montano

Between 1568 and 1573 Clusius lived in Malines in the house of his friend Jean de Brancion. He spent most of his time there on the preparation of a book on the flora hispanica.⁶ The design of a map of the Iberian Peninsula was a second project on which he was engaged.⁷ But some financial difficulties on the part of his friend the printer Christopher Plantin created problems for

---

⁵ There are plenty of details about this trip in his autobiography. Ch. De Backer and L.J. Vandewiele collected them in 'Le botaniste flamand Carolus Clusius (1526-1609) et ses relations avec l'Espagne', *Medicamento, historia y sociedad. Estudios en memoria del profesor D. Rafael Folch Andreu* (Madrid, 1982), 183-186.
⁶ A Spanish versión of this book has been published under the title: *Charles de l'Écluse de Arras, Descripción de algunas plantas raras encontradas en España y Portugal,* transl. Avelino Domínguez García and Florentino Fernández González, ed. Luis Ramón-Laca Menéndez de Luarca and Ramón Morales Valverde (s.l., 2005).
⁷ De Backer and Vandewiele, 'Le botaniste'.

these ambitious projects, and in the end the Spanish flora could not be printed. On the other hand, two printed copies of his geographical maps of Spain have been preserved, which are therefore of great scientific and bibliographical value.[8] The period in Malines was distinguished by a close relation with Benito Arias Montano (1527-1598), one of the most representative figures of Spanish humanism in the sixteenth century. He was a key figure in the intellectual network of the court of Philip II. Arias Montano was responsible for the publication of the famous *Biblia Regia* [*Biblia políglota Complutense*], and for a time he was in charge of the Library of El Escorial, the most important intellectual institution and project of the King of Spain.[9]

Arias lived in Antwerp between 1568 and 1575. It was during this stage that he became integrated into the circle of intellectuals, artists and scientists around the powerful printer Christopher Plantin. From then on he played an important role in connecting the important nucleus of scientists in the Netherlands (Ortelius, Mercator, Gemma Frisius, Dodoens, Clusius) with some Spanish scientific circles, who were becoming more and more isolated from the new European trends of science and culture as a consequence of the isolationist policy promoted by Philip II and his Counter-reformation.

Several historical testimonies indicate that Arias Montano frequently sent scientific books and instruments to Spain, introduced Spanish physicians, naturalists, cosmographers and other scientists in Flanders, and helped them with the publication of scientific books in the great printing workshops of the Netherlands.[10] Simón de Tovar and the surgeon Francisco de Arceo were among them. During the final stage of his life, Arias Montano conceived an unfinished work devoted to a general exposition of natural theology based on the Bible. He only managed to publish one volume in Antwerp, under the title of *Historia Naturae,* and this was three years after his death.[11] Faced with this ambitious project, he tried to appoint his friend Carolus Clusius, who was already old. A great number of Spanish naturalists were expected to collaborate with Clusius. Arias was in fact the main link between Clusius and the circle of Sevillian naturalists with whom he maintained a regular correspondence and a fruitful scientific exchange of information, seeds and plants.

---

[8] Ibid.

[9] On the historical significance of Benito Arias Montano, cf. T. González Carvajal, 'Elogio histórico del Dr. Benito Arias Montano', *Memorias de la Real Academia de la Historia*, 7 (1832); B. Rekers, *Arias Montano* (Madrid, 1973). Rekers also analysed his correspondence in 'Epistolario de Benito Arias Montano (1527-1598)', *Hispanofila*, 9 (1960), 25-37.

[10] Some of the letters exchanged with Clusius refer to this commitment, and Rekers' research on Arias Montano's *Epistolario* also gives evidence on this.

[11] B. Arias Montano, *Naturae historia, prima in magni operis pars* (Antwerp, 1601).

The period in Valencia: The scientific relation with Juan Plaza

The main testimonies of Clusius' scientific relations with the university professor of botany and *materia medica* of the University of Valencia, Juan Plaza, are to be found in his own published works. Nevertheless, in spite of the close scientific collaboration between them, Plaza was not one of Clusius' group of correspondents, according to the manuscript documents and letters preserved in Leiden University Library.[12] Other testimonies suggest that Plaza maintained a frequent exchange of information with Clusius, but no historical testimony of it has been found.[13] However, according to his own testimony expressed in his *Rariorum aliquot stirpium* (1576), the months that Clusius remained in Valencia and the teaching of Plaza had a decisive influence on his scientific project of studying and printing a book on the Spanish flora.

Throughout the pages of *Aromatum et simplicium* (1567), the references that testify to this close relation are frequent. For instance, in chapter II of the first book, when talking about the *persea*, Clusius states that 'its fruit matures in autumn, according to what was explained to me by the Illustrious Don Juan Plaza, physician and Valencian professor, who showed it to me in the mentioned place and assured me that the villagers called it *Mamay*, although the Spaniards who described America designate by this name a very different tree.'[14] When Clusius talks about the fruits of the *persea* called 'always green', he records the comment of Plaza that the fruits of this plant are not similar to the colour of the grass, but are black.[15]

In chapter XLI he refers to the *casia*, a plant that grows abundantly in the kingdoms of Granada and in the Valencia region. It was initially identified by the name *osiris*, although the name *casia* was later generally accepted. This is the name it was given in some parts of Spain. For its identification according to classical botany, Clusius appealed to the authority of 'the wise Plaza', who identified it as the *polygonus* of Pliny.[16] Plaza's opinion also influenced the identification of the so-called *Rubus Idaeus Valentinus*, or hawthorn of Ida from Valencia. The observed variety lacked thorns, and that contradicted the classical description of the plant. In that case, the Valencian botanist gave Clusius a possible solution: a Latin codex of Dioscorides stated that there was a variety of the hawthorn of Ida without any thorns. Such an observation did not appear in the Greek versions of Dioscorides. Clusius comments in his book on

---

[12] Leiden University Library, Vulcanius correspondence.
[13] On Juan Plaza, cf. López Piñero, *Diccionario histórico de la ciencia moderna en España*, vol. II.
[14] C. Clusius, *Rariorum aliquot stirpium per Hispanias observatarum historia* (Antwerp, 1576), book I, chapter 2. I am grateful to Luis Laca for his kind information about Plaza in this work.
[15] *Rariorum aliquot stirpium*, book I, chapter 2.
[16] Ibid., book I, chapter 41.

the Spanish flora that Plaza had shown him a specimen of the Hawthorn of Ida in the same convent garden in which they observed the *persea*, a convent located scarcely a mile from the city of Valencia.

In a reference to the variety of *Narcissus* that he calls 'Esparganio of Plaza', Clusius affirms:

In the northern part of Valencia,[17] in certain rough and stony places, a plant grows. It has two leaves that extend close to the ground, long, thick leaves of a dark green colour, very similar to those of the narcissus described before, but in whose internal face a quite wide and white line is drawn of longitudinal form. The root is bulbous and is formed of many layers, like the onion, of which the outside is black and the inside white and full of a viscous and dense juice, and it tastes disagreeably bitter. The illustrious doctor Don Juan Plaza said that this plant had a light stem a foot long (I only managed to unearth the roots with the leaves thanks to his indications). At the top of it there are some flowers which are similar to those of the iris, white, and the subsequent capsules of it were round and a small seed was hidden in them. He called it *esparganio*.

Hemerocallis Valentina is another one of the plants mentioned by Clusius as native to the Valencian coast. It had been named *Hemerocallis* by Plaza 'because he said that it had flowers like those of the iris, but yellow, slightly dark and of a pale tone.'[18] Clusius also mentions the authority of Plaza when he describes the *ruda silvester*, a variety mentioned by Dioscorides that was rejected by Ruelle:

The first to show this plant to me was the illustrious gentleman Don Juan Plaza, physician and professor of Valencia. We saw it one mile from the city of Valencia, in the monastery consecrated to the Holy Virgin also called of Jesus, in the same orchard where the avocado grows. I saw it later in other places. I saw it pulled up, with the biggest roots of all the plants, in the suburban property of His Grace Don Pedro Alemán, who had offered me his hospitality. The plant was about two elbows long and in the surroundings thirty dwarf plants had come up. I pulled up two of them and took them to Belgium with me.[19]

When discussing the identification of some plants or sometimes when expressing his doubts, Clusius often appeals to the judgement of Juan Plaza. However, sometimes Clusius expresses an opposite opinion, as in the case of the species called *onobrichis*, identified by Plaza as *astragalus*; Clusius considers that the detailed description contradicts that identification.[20]

---

[17] Ibid., book II, chapter 1, *Narcissus*.
[18] Ibid., chapter 14.
[19] Clusius is referring to the Franciscan monastery of Santa María de Jesús, a quarter of a league from the city of Valencia, founded in 1420 by King Alfonso the Magnanimous.
[20] *Rariorum aliquot stirpium,* book II, chapter 86.

The frequent references to Plaza in Clusius' work indicate a close relationship and shared experiences of collecting plants in the Valencian region during Clusius' stay there. He states: 'I have only seen this plant [*Anthyllis Valentina*] next to the drains of the city of Valencia, in Spain, next to the city gate oriented towards the *Castillo Real*, and I was able to see it because Mr Plaza showed it to me.'[21] The appreciation Clusius expressed of the Valencian botanist was great: 'While we gathered several plants on the coast of Valencia, the *doctissimus* Plaza thought that *Alipon* was the *Alypum*.'[22]

The circle of physicians and naturalists in Seville

Through the connection established by Benito Arias Montano, down to the end of his life Clusius maintained an epistolary relation with a close circle of physicians and naturalists in Seville based upon the exchange of plants, scientific information and seeds. Seville was then the nucleus of Spanish trade and commercial activities with the New World. As stated before, one of the main figures in this field was Nicolás Monardes. Nevertheless, we have not found documentary evidence for the existence of a direct relation between Clusius and Monardes, a specialist in the study of and trade in American medicinal products, the commerce of slaves, and the collections of tropical and exotic products. Neither the correspondence of Clusius preserved in the University Library of Leiden, nor that preserved in Spanish libraries, nor the autobiographical data on the two men point to direct contact between them.

The Sevillian scientific circle of Clusius' correspondents had three main protagonists: Simón de Tovar, Juan de Castañeda, and Rodrigo Zamorano. Simón de Tovar (died in 1596) was a well-known Sevillian doctor, founder and director of an important botanical garden in the city, where he carried out experiments in acclimatisation and exchanged seeds. He made periodic catalogues and lists of plants, mainly of the new botanical species from America. He maintained a wide relation with other European naturalists, especially with those in the Netherlands through Benito Arias Montano. Among his main correspondents we find Bernardus Paludanus and Carolus Clusius, to whom he sent his catalogues of plants and seeds, which were later published in some of Clusius' works. Moreover, his main scientific activity was focused on therapeutics, the most suitable use of healing plants, and the composition of medicines.[23] Due to his close relation with the *Escuela de Náutica* [School

---

[21] Ibid., book II, chapter 92, *anthyllis valentina*.
[22] Ibid., book II, chapter 94, *álipon*.
[23] His two printed medical books were devoted to these topics: *De compositorum medicamentorum examine nova methodus [...]* (Antwerp, 1586); *Hispaniensium pharmacopoliorum recognitio* (Seville, 1587).

for Navigation], he also published a book on the techniques and instruments of the art of navigation.[24]

Rodrigo Zamorano (died 1620) was a mathematical cosmographer who worked in the *Casa de Contratación* of Seville from 1575 on, where he was in charge of teaching the art of navigation and of the manufacture of scientific instruments, especially those connected with the art of sailing. Zamorano had a wide range of scientific interests that went beyond the simple art of navigation to the New World. He also devoted his efforts to cartography and astronomical observation. Zamorano had a great scientific reputation and was appointed as *piloto mayor* in the *Casa de Contratación*, the highest position with responsibility for the scientific and technical activities in the Sevillian institution.[25] During the period in which he maintained a close epistolary relation with Clusius, Zamorano created a botanical garden and a rich collection of curiosities and exotic animals and plants from the colonies.

We have less information about the other correspondent of Clusius, Juan de Castañeda, who is scarcely mentioned by Spanish bio-bibliographical scholars. The *Enciclopedia universal Española e Hispanoamericana* refers to him, not without mistakes, as 'a learned Spanish botanist of the seventeenth century born in Seville. He wrote to the famous Clusius many scientific letters, fourteen of which have been printed. He also sent seeds and other elements of study to Clusius and composed a speech and *Quaedam carmina* in honour of the work of this famous naturalist.'[26] Through some of the references that appear in the letters to Clusius, we can deduce that Castañeda probably belonged to the scientific circle of Rodrigo Zamorano, and it is possibly through Zamorano's mediation that he initiated a long epistolary relation with Clusius.

Exchanging botanical information

Clusius carried out a fundamental work of assimilation, translation and dissemination of the Spanish flora and the American and Asian exotic new plants through the printed works of García da Orta, Nicolás Monardes and Cristóbal de Acosta. His experience in collecting plants in Spain and exchanging scientific information with Juan Plaza was another aspect of his knowledge about the Spanish flora and the exotic American plants. Moreover, once Clusius moved to Malines, Vienna or Leiden, he maintained a constant botanical exchange of information with the Sevillian physicians and botanists from afar.

---

[24] S. Tovar, *Examen y censura [...] del modo de averiguar las alturas de las tierras, por la altura de la Estrella del Norte, tomada con la Ballestilla* (Seville, 1595).
[25] U. Lamb, 'Zamorano, Rodrigo de', in *Diccionario histórico de la ciencia moderna en España*, vol. II, 443-444.
[26] *Enciclopedia universal española e hispanoamericana* (Barcelona, 1919-21), vol. XII, 211.

That correspondence was compiled to a large extent by Ignacio Jordán de Asso at the end of the eighteenth century, who published a Latin transcription of most of the letters in a small book entitled *Clariorum Hispaniensium atque exterorum epistolae, cum prefatione et notis. [...]* (Zaragoza, 1793), written in defence of the Spanish scientific tradition, a topic widely discussed during the Enlightenment. This work is not easily available to the present researcher.[27] Jordán de Asso was an important jurist and historian of his time who wrote some books on jurisprudence. He also represented the Spanish government abroad and was an enthusiast of botany and natural sciences, devoting part of his time to collecting plants in his native Aragón, as well as showing an interest in geology and palaeontology.

In the mid-1770s, Jordán de Asso was the Spanish consul in Amsterdam. This stay in Holland allowed him to get in touch with the Dutch academic and scientific community. During a journey to Leiden he discovered the manuscript correspondence of Clusius with the Spanish scholars and decided to include it in his *Clariorum Hispaniensium*. The *Praefatio* summarised the key moments of the Spanish scientific tradition, mainly from the sixteenth to the eighteenth century. Nevertheless, some of the letters we know nowadays passed unnoticed by the Aragonese jurist.[28] In addition, he wrote a manuscript dated 1788 entitled *De claris Hispaniis historiae naturalis cultoribus* and prepared it for publication, but it was never published. This manuscript included a chronological survey of the Spanish naturalists and made reference to the letters to Clusius from Spanish scientists that he had discovered in Leiden.

The compilation and transcription in its Latin original of the correspondence of Clusius with the Spanish scholars made by Asso at the end of the eighteenth century has been useful to make historians of science aware of the existence of a close scientific relation of Clusius with Spain. Nevertheless, the corpus collected by Jordán de Asso was incomplete, contained important mistakes of transcription, avoided some parts of letters that were damaged or difficult to understand, and lacked any analysis of the content. Nevertheless, a positive conclusion could be deduced from Jordán de Asso's printed work and manuscript: at the time when Jordán de Asso saw the letters, the number of them preserved in Leiden was no larger than their present number, which means that no significant materials have been lost. Neither considerable losses

---

[27] An edition of this book is currently available in the Biblioteca Nacional de Madrid.

[28] The whole correspóndence between Clusius and the Spanish scholars has been analysed in Barona and Gómez Font, *La correspondencia de Carolus Clusius con los científicos españoles*. Jordán de Asso did not know the letter by Benito Arias Montano dated 22 April 1569. Perhaps it was not in the Archives when he saw the Clusius correspondance (Rekers, 'Epistolario de Benito Arias Montano (1527-1598)', no. 26).

nor impairments have taken place, although it has to be said that some of the manuscripts are very difficult to read and transcribe. Our research work has also made it possible to augment the correspondence with letters that Jordán de Asso did not know, both from the Leiden University Library and from the Boerhaave Museum in the same city.

Taken as a whole, the corpus of letters constituting the correspondence of Clusius with Spanish scholars is made up of six letters written by Benito Arias Montano dated between 1568 and 1596; fourteen letters written by Juan de Castañeda, dated between 1600 and 1604; one letter by Pedro Martín (1569); one letter by Hipólito Martín (1570); two letters by Simón de Tovar (1596); and one by Rodrigo Zamorano (1603). This constitutes a total of twenty-five letters. Jordán de Asso transcribed nineteen of them in Latin.

Although Clusius travelled in the Iberian Peninsula between 1564 and 1565, the majority of letters from the Sevillian scholars preserved in Leiden date from the end of the sixteenth century and the first years of the seventeenth, that is to say, the final period of Clusius' life in Holland. The most personal relation with Benito Arias Montano is the only one that covers the whole period after the visit to Spain down to his death. On the other hand, the letters of Pedro Martín and Hipólito Martín refer to personal questions relating to the Spanish military service in Flanders and lack any relation with scientific matters. It is quite possible that previous scientific correspondence with Juan Plaza or Simón de Tovar has been lost. The first letter by Tovar is sent immediately after the publication by Clusius in 1593 of his compilation containing the three works of Nicolás Monardes, Cristóbal Acosta and García da Orta.

The correspondence between Benito Arias Montano and Clusius started during the Spaniard's long stay in the Netherlands. The first letters of Arias Montano were written in Antwerp and were addressed to Clusius in Malines, where he resided under the protection of their common friend Jean de Brancion. Even at that early date – 1569 –, the exchange of seeds and the lists of plants and botanical catalogues of species constituted one of the commonplaces usually present in the letters they exchanged, in addition to others of more a personal character. The role of Christopher Plantin as an intermediary is evident, since Plantin frequently participated in the exchanges, sending books, packages and other objects. Some small incidents can be mentioned and a certain tension arose due to the non-payment of the cost of postage. References to Spanish naturalists like Bernardino de Burgos or Duranus, who were Montano's collaborators in his projected publication of a Spanish flora, frequently appear in the correspondence with Arias Montano, but Montano's collaborators were not relevant in the context of the Spanish naturalists nor in the academic context, and it is difficult to find historical testimonies that allow their identification.

In the last letter by Arias Montano kept in Leiden, he makes explicit reference to the exchange of seeds and the participation of Simón de Tovar as the main protagonist in those exchanges.[29] Arias Montano mentioned Tovar as the recipient of the seeds that Clusius had sent him. In this last letter a problem was mentioned that was probably not infrequent in the exchange of scientific material at the time: the loss of the labels that identified the seeds as a consequence of the opening of the packages during the marine passage. Loss of plants and seeds and even the theft of scientific materials in the boats were often reported in the letters. In 1596 the correspondence between Clusius and Arias Montano was no longer frequent, and the Spaniard talked about the past with a nostalgic feeling, appealing especially to the pleasant memory of the eight years spent in Flanders, that he qualified as 'the happiest time of his life'. Those last letters attest the memory of the past and the friendship that united him with Clusius. Simón de Tovar, a close friend of Arias Montano in Seville, kept Arias informed. Through him Arias knew the delicate state of health of Clusius and expressed his solidarity in the difficult days. The last letters of Arias Montano were written from his final retirement in *Campo de Flores*, next to Seville.

The letter by Pedro Martín dated 15 December 1569 refers to personal matters relating to the Spanish army in Flanders, but it reflects the good relations between Clusius and Plaza. Martín explains some details to Clusius about Plaza's journeys to the Pyrenees to collect plants, and provides a testimony of the reception of the Latin version of the book of Garcia da Orta sent by Clusius to Plaza at the time. It is a testimony of the cordial relationship they still maintained, although any documentation is unfortunately lost.

The two letters by Simón de Tovar are dated in 1596 and have a completely different purpose, since they are long and refer strictly to botanical matters. The correspondence of Tovar included continuous references to Benito Arias Montano, who was close to him in Seville and was an intermediary in the exchange of seeds between the two botanists. In those days the health of Clusius and Arias was delicate, according to the letters. Some practical difficulties that they had to confront are indicated. These are usually connected with delays in the departures of boats, problems with the passage, and thefts of plants and seeds in the course of the trips. Tovar sent Clusius periodic catalogues of the plants he cultivated in his botanical garden in Seville. From this correspondence it can be deduced that Clusius used to send Tovar plants and seeds for

---

[29] The last letter from Arias Montano is dated 1596, when he was almost seventy years old. He died in 1598. Simón de Tovar was introduced by Arias to continue the botanical exchange of information with Clusius, starting in 1596.

acclimatisation purposes. Sometimes Tovar complained about the difficulties involved in the process:

> With regard to the trees and shrubs from the Indies, they produce very few fruits in our soil. Those which do produce rarely bring their fruit to fruition, for they produce it so late and out of season that they lose their fruit before it is ripe under the first autumnal colds.[30]

The correspondence between Clusius and Tovar also contains references to and reflections on the vernacular and Latin names of many new botanical species, as well as discussions of the difficulties raised by their cultivation and acclimatisation. The existence of a frequent exchange of letters between Simón de Tovar and Bernardus Paludanus can be deduced from those letters. Tovar announced that he was prepared to send his catalogue of plants to them both. He also describes certain American species with plenty of details, as in the case of the *narcissus jacobeus*, which received the indigenous name *azcal xochitl*. The letters contain and discuss a wide and valuable botanical terminology, which is reproduced in the appendix to this chapter. The scientific relation and exchange was reciprocal, as is explicit when Tovar enumerates bulbs and seeds that he has received from Clusius, or when he asks Clusius to ship him certain botanical species. Even the discussion of some cartographical matters and news on navigation charts are present in their correspondence.

When the exchange of letters with Simón de Tovar ended,[31] a new correspondence with Juan de Castañeda began, in which Rodrigo Zamorano occasionally took part. From the last letter of Zamorano, dated 3 June 1603, it is easy to understand the importance and abundance of the exchange of species and seeds at the beginning of the seventeenth century. Until then it constituted a truly international framework susceptible to piracy and theft.

The fourteen letters of Juan de Castañeda cover the period from 17 September 1600 to 13 February 1604. They are written in Spanish, which shows the knowledge that Clusius had of the Spanish language and the non-academic status of some of Clusius' correspondents and cultivators of the Hispanic botany. They did not share the common use of Latin among scientists. The letters by Juan de Castañeda often answered specific queries by Clusius about American seeds to which he did not have access and that he wished to incorporate in the botanical garden in Leiden. Most of those plants came from the New World and are included in the botanical appendix to this chapter. The complaints about the damage to plants and seeds during

---

[30] Letter of Simón de Tovar to Clusius, 13 and 19 March 1596.
[31] The last letter conserved in Leiden UL, VUL 101 by Simón de Tovar is dated June 1596.

the trip and their arrival in a bad condition and state of conservation were frequent at the time. On the other hand, the relationship of friendship and scientific collaboration between Juan de Castañeda and Rodrigo Zamorano is present in many letters. On some occasions one refers to the other, or to certain materials sent by Zamorano to Clusius, including exotic animals that Castañeda considered interesting enough to be reproduced in print. He wrote to Clusius:

I promise Your Grace that for your taste and for men as cultivated as Your Grace, there are here, if we had someone able to engrave and paint them, the greatest curiosities one can desire: of all the animals and fishes that have natural shells and defences like land tortoises and sea turtles, others like snails or armadillos, nacre mother of pearl, and other very different animals, which cannot be sent because of the lack of a duplicate. When I have time, I will send you their names; but if Your Grace has among your learned books something on animals or fishes or has the intention of writing one, you would be delighted to have news of all these new and peculiar things so that they could be brought to light through your very learned and excellent style, for until now nobody had information about them and they are being discovered now. If you had someone curious enough to print it, it would be a very new and peculiar thing, and of great benefit. They have been assembled by the Licenciado Zamorano. As he is an examiner of masters of the fleet to the Indies, each master who travels to the New World brings some new or extraordinary things to him, and thus the walls of the vestibules of his house are all filled with these shells, fishes and animals on show. Please let me know if you could assume that task, because I will help you in every way I can. In this city of Seville, 20 October of the year 1600.[32]

The interest of Zamorano in establishing a relationship with Clusius becomes evident from some of the letters sent by Castañeda. Zamorano held the most important post in the Seville School of Navigation, he was 'piloto mayor de Indias', and was very interested in botany and zoology. He tried to attract the interest of Clusius to his work and in several letters he kept insisting on the good properties of the rosemary oil he had obtained and applied for therapeutic purposes. But it seems that Clusius did not pay very much attention to this new remedy.

In the letter he sent to Clusius on 29 April 1601, Castañeda answered a request of the Flemish naturalist: 'I will try to find out what happened to that book entrusted to Nardo Recchi of Naples and I will inform you of what I can find out.'[33] The reference is to Clusius' interest in what had happened to the excellent materials that the royal physician Francisco Hernández had brought

---

[32] Letter of Juan de Castañeda to Clusius, 20 October 1600.
[33] Letter of Juan de Castañeda to Clusius, 29 April 1601.

from the Vice-Royalty of New Spain, as a consequence of the scientific expedition instigated with the backing of the Spanish monarch Philip II.[34] Recchi was in charge of publishing a summarised version of the materials collected during the expedition with the purpose of introducing them to the Accademia Lincei. This is a further testimony of the exceptional interest that the American flora aroused in Clusius.

Another remarkable aspect of these epistolary exchanges is the description and the drawing of new botanical species. In his letter of 21 August 1601, Castañeda stated: 'I would like to send you the drawings of so many new plants before the completion of the printing of your book. If you had sent me one assistant from your country, where people are more keen to work than those here and do their work better, it would have been a great help'.[35] It is quite possible that some of the drawings and engravings that Christopher Plantin bought from Clusius came originally from the Spanish botanists.

Some letters document the interest that the pharmacopoeia and the therapeutic application of the plants aroused among the naturalists. Besides Zamorano's references to the properties of rosemary oil, Juan de Castañeda refers to the collaboration of a Flemish specialist who was in charge of the preparation of medicines. According to Castañeda's testimony, he was very knowledgeable about his profession and had a very successful practice in Seville.[36]

The botanical flora and the botanical terms in the correspondence

In a previous work published in 1998, the correspondence of Carolus Clusius with the Spanish scholars was analysed, taking as point of departure the existing letters in the Vulcanius collection of the catalogue of letters of the Leiden University Library and others kept at the Boerhaave Museum in the same Dutch city.[37] In making a critical edition of the correspondence, an attempt was made to solve important mistakes and omissions present in the previous work of Jordán de Asso, to analyse the scientific content of the letters, and to enlarge the corpus of documents. In addition, a translation was made into Spanish of all the Latin letters, which had never been translated before. The main problem lay in producing a reliable translation, while the critical edition had to contend with the extremely chaotic scientific terminology, which was deficient because of the lack of a universally accepted system of denomination and classification

---

[34] See I. Baldriga in this volume. The Vice-Royalty of New Spain included a wide region of Central America, mainly coinciding with the present state of Mexico.
[35] Letter of Juan de Castañeda to Clusius, 21 August 1601.
[36] Letter of Juan de Castañeda to Clusius, 15 January 1602.
[37] Barona and Gómez Font, *La correspondencia de Carolus Clusius con los científicos españoles*.

for the botanical species. To understand the botanical terms of the time, one has to apply a complicated and sometimes dangerous process of interpretation, mainly drawing on the previous works of Nicolás Monardes, Cristóbal Acosta and García da Orta.

The naturalists of the time normally limited themselves to review catalogues of plants, indices and long lists, although on some occasions a precise description of the medical properties and the therapeutic applications was included. The identification of each plant became somewhat easier when it referred to easily traceable species or indigenous plants, but it proved difficult to specify the plants exactly in the case of generic terms like 'bledo', 'haba', 'cardo', 'mate', 'name', to mention only a few.

Appendix *Catalogue of plants mentioned in the correspondence between Clusius and the Spanish scholars* (as a general rule, the original spelling has been kept)

Abas coloradas, faseolos mates colorados
Abitas de todos los colores
Absynthium inodorum
Acacia
Achiole
Adonidis flos
Aguacate
Alcactunge
Alisum
Alsina repens
Amir
Anamu
Ananas
Anchusa neapolitana
Anemones
Anones
Anthyllis altera seu Chamaepitis spuria
Antirrhinum
Añil
Apios fuchsii o Terrae glandes
Aquilexia cerulea
Aracus o Chicaros de Lisboa
Aranhera planta
Árbol coral o Árbol del coral
Árbol Judd
Arbutus
Arisarum
Aro Egiptio
Arpella Valentinorum

Arreboleras
Arrhenogonum
Arum aegyptium
Asarum
Asplenum
Aster atticus
Aster montanus o Montana Hisp. Vulgo dictus
Astragallus
Astrágalo crético
Astrantia nigra
Azcal xochitl
Bella dona
Bisnaga
Bledo especioso purpúreo
Bledo espinoso
Bledo que mira el sol
Boniato o batata blanca
Buenas noches
Bulbus muscari
Buphthalmus
Cabuya
Calamintha
Camesice indica
Campanillas azules
Campanula
Canime
Cantueso
Caña fistola
Capinos

Çapote
Çapotillo de la Abana
Cardillo de la Abana
Cardo santo del Perú
Cardus benedictus del Perú
Cariophyllati
Caucalis
Cebolla de Caracas
Centaurium folio Lapathi
Cantaurum magnum
Centella
Cepae rotundae
Cerefolii magni
Cerinthe seu Maru
Cerulea Indica
Ceybilla de Yndias
Chamaeleon albis et niger
Chamaepitis spuria
Chamaesyce
Chelidonium maius
Chere folium
Chicama
Chicaros de Lisboa
Chiote
Cisto o cistus
Colocapsia
Colocasiam
Colutea arbor
Convolvulo
Coral árbol
Coralii arboris indici
Coronae imperiales
Coronaria purpurea
Crocus vernus
Cunde amor
Curú
Cytissus
Daucus creticus
De Mexía (aceite de)
Delfinum bucinum
Delfinum bucinum flore album
Delfinum flore speciosso
Ditamo de Onduras
Doradilla o Asplenum
Draba
Draconum arbor
Elaphoboscum
Elleborum nigrum

Epipactis
Ericae omnes
Eriophoros
Escobas de Abana
Faseolos mates colorados
Ferrum equinum
Ferula galbanifera
Frisoles
Frisoles blancos
Fumaria adulterina
Gallii
Gayomba
Gengibre
Genista hortensis
Gladioli
Guanábana
Guanábana ex Santa Marta
Guayabo
Habas del Perú
Hemerocallidis valentinae
Hicaco
Hyacinthum quem orientalem Indicum
Hyacinthus
Hyacinthus anglicus
Hyacinthus orientales
Hypecoi
Hyppoglosum valentinorum
Indicum narcissum
Irides
Iris bulbosa bethica
Iris bulbosa latifolia
Jacinto blanco
Jacobaeus
Jecomaustle
Kreeft oogen
Lacrima Jobis
Lactaria indica
Laurus tinus
Ledum
Lentisco o lentiscus
Leucoio bulboso vel lilio rubro chalcedonico
Lilia persica
Lilio rubro chalcedonico
Lilium persicum o lilia persica
Lilium susianum
Lúpulo
Lycium
Lycium quorundam melosilla

Madre del cacao
Madreselba
Mafasa
Maguey
Maíz cariaco
Malanga
Mamei
Mamones
Maro hispaniae, vulgo Sclarea
Mates cenicientos
Mates de Indias
Mates de Yndias pardos
Mechoacan o mechuacan
Mechuacan
Melanthium
Meleagris
Melissa
Millefolium
Millium solis
Moly homericum
Montana Hisp. Vulgo dictus
Myrhidis maioris
Myrti variae
Narciso de Alger o narcissum de Argel
Narciso índico jacobeo o narcissum indicum iacobeum o narcissus indicus iacobaeus
Narciso que trae las flores blancas
Narcisos
Narcissi marini
Narcissum de Argel
Narcissum indicum iacobeum
Narcissus albus multiflorus
Narcissus iacobaeus
Narcissus indicus iacobaeus
Narcisus maximus
Nasturtium indicum
Nopal
Ñames
Onobrychis
Oroçuz
Pancracium valentinum
Papaver curniculatum
Papaver poliantos
Parra de calabazas largas leonadas y coloradas
Pendejera
Pepinos del Abana
Petroselinum macedonicum
Phyllon Thelygonum
Pimientos dulces
Piñas o ananas
Piñones de purgar
Pita
Plantago mayor
Platano o musa
Plumbago
Poleomontano
Polipodio
Polygonum montanum
Polyphyllorum
Pseudohyssopus
Pulsatilis
Punçela
Pyrethrum
Quaecececpathi
Radix rhodia
Raponticum aliorum verum
Rabarbarum
Rhannunculi genus hisp. Dictum centella
Rhannunculus lusitanus
Romero
Ruta harmel
Ruta montana seu sylvestris
Ruta sylvestris
Sandalidae creticae
Sanicula foemina
Scabiosa tenuifolia
Sclarea
Scorçonera
Scrophularia maior
Securidaca
Sedi maioris
Sedum maius o majus
Smilace
Smilax aspera
Solanum hortense vulgo ital. bella dona
Stramonia magni secreti
Superba
Sylibus
Tamariz
Tanacetum album
Tanacetum pannonicum
Terebintho o terebinthus
Terebinthus
Terrae glandes
Tetrahil
Thymelea

Thymus capitanus
Tithymalus characias
Tomate
Tomillo
Tordilium dodonei
Toronjil
Torvisco
Tragacantha
Trifolium bitubinosum
Trifolium engibar
Tudesca
Tulipae
Tuna
Ubero utraque Elaphoboscum
Vadea de agua
Verbascum lychnite

Verbenaca
Verónica
Viola matronalis lutea
Viola matronalis purpurea
Virginis Matris Palla
Viva
Xiphia
Yerba ala
Yerba cidrera
Yerba de las quentas
Yerba mimosa
Yerba tua-tua
Yerba viva
Yerba Y
Yuca

PART II

Clusius and individual correspondents: Two case studies

'Qui me unice amabat.' Carolus Clusius and Boldizsár Batthyány

Dóra Bobory

Introduction

There is a well-known anecdote that Clusius tells in his *Fungorum in Pannoniis observatorum brevis historia*[1] about a certain soup served at the table of Boldizsár (Balthasar) Batthyány, one of the most powerful Hungarian aristocrats of his time. The soup was made of mushrooms,[2] and it had a peculiar yellow colour which Clusius, who enthusiastically studied but rarely ate mushrooms, attributed to saffron rather than to the mushrooms themselves. Such ignorance in matters of cuisine, displayed by the pioneering and most thorough researcher of the dish being served made the other guests and the count himself laugh heartily. The warmth of the anecdote gives us a glimpse of the relationship between Clusius and his Hungarian host, who continued to invite the famous botanist both to his castle(s) in Western Hungary and on the field trips that he made in neighbouring lands. It seems, however, that too little has been said about Batthyány, while, fortunately, more and more is known about Clusius.

---

[1] C. Clusius, *Fungorum in Pannoniis observatorum brevis historia* (published as an attachment to his *Rariorum plantarum historia* [Antwerp, 1601]), cclxxiij. 'I recall the year 1584 when we gathered for the harvest (he used to summon me by sending a carriage two or three times each year) in the stronghold of Németújvár, and we retired to the inn, and once, while we were having our lunch, a dish of mushrooms boiled in their sauce was put on the table: then I, who very rarely eat mushrooms, and not knowing that the yellow colour was the mushroom's juice, asked him in French (since that hero besides his native Hungarian tongue, spoke perfectly foreign languages: Latin, Italian, French, Spanish and Vandalic, that is, Croatian) and asked whether it was saffron that coloured the soup yellow. He, heartily laughing, addressed the other noblemen, since usually some eight or ten of them normally sat at his table at lunch: Clusius Uram, that is, My Lord Clusius (and he said more in Hungarian to them) thinks that this soup is yellow because saffron was added to it: then they all started to laugh excessively, and wonder at my ignorance concerning the nature of mushrooms, especially those who knew that in that same year and already some years before I had been diligently studying – as well as many other kinds of plants – the varieties of mushrooms which grew on his lands.'

[2] According to József Csaba, the mushroom was *amanita caesarea*. See J. Csaba, 'Magyar ethnobotanikai adatok Clusius műveiben' (Hungarian ethno-botanical data in the works of Clusius), *Vasi Szemle*, 27 (1973/4), 598.

Who was this generous count, who (as Clusius reports) spoke seven languages, was a devoted collector of botanical curiosities, and – while successfully fighting the Turks – found time and energy to organise a great library and pursue alchemical experiments? How did Batthyány become such an intriguing combination of amateur scientist and professional patron? How did his co-operation, even friendship, with Clusius begin and develop? What role did Clusius (and indirectly also Batthyány) play in the introduction of new species of plants to Hungary? What sort of garden did Clusius plan for Batthyány in Szalónak?

On the basis of both the abundant and mostly unpublished[3] correspondence of Batthyány with various humanists and scholars of the age and the related secondary literature, I will attempt to answer these and other questions. All this will be done with the conviction that the introduction of Boldizsár Batthyány, as someone who recognised, appreciated and supported talent, will give us more insight into Carolus Clusius as well.

Boldizsár Batthyány's Humanist circle

The Batthyány family had large possessions in western Hungary from the early sixteenth century onwards, and, due to the position of their lands, the male members were also warriors who had to face the disturbing proximity of the Ottoman Empire (Ill. 8). The first prominent member of the family was Ferenc Batthyány, one of the few who not only participated in but also survived the Battle of Mohács in 1526. He was a real courtier, one who grew up in Vienna with the young King Louis II, who himself died in the aforementioned battle. Later he was to be on good terms with King Ferdinand I. He married twice but remained without children of his own, which moved him to 'adopt' the children of others: he and his wife transformed their court in Németújvár (Güssing) into a school for young nobles, a tradition also kept and cherished by Boldizsár. He sponsored the studies of Boldizsár and his brother Gáspár (who died young). It is through his connections and influence that Boldizsár spent almost two years in France.

---

[3] Only a small number of letters and other documents were published by B. Iványi, 'A körmendi levéltár memorabiliái/Acta Memorabilia in tabulario gentis principum de Batthyány reperibilia' (The memorabilia of Körmend Archives), *Körmendi Füzetek*, 2 (1942), and idem, *A körmendi Batthyány-levéltár reformációra vonatkozó oklevelei I: 1526–1625* (Charters of the Batthyány Archives of Körmend related to the Reformation I), ed. L. Szilasi, Adattár 29/1 (Szeged, 1990). I am working on the annotation and publication of approximately two hundred of the letters written by various members of Batthyány's informal humanist circle.

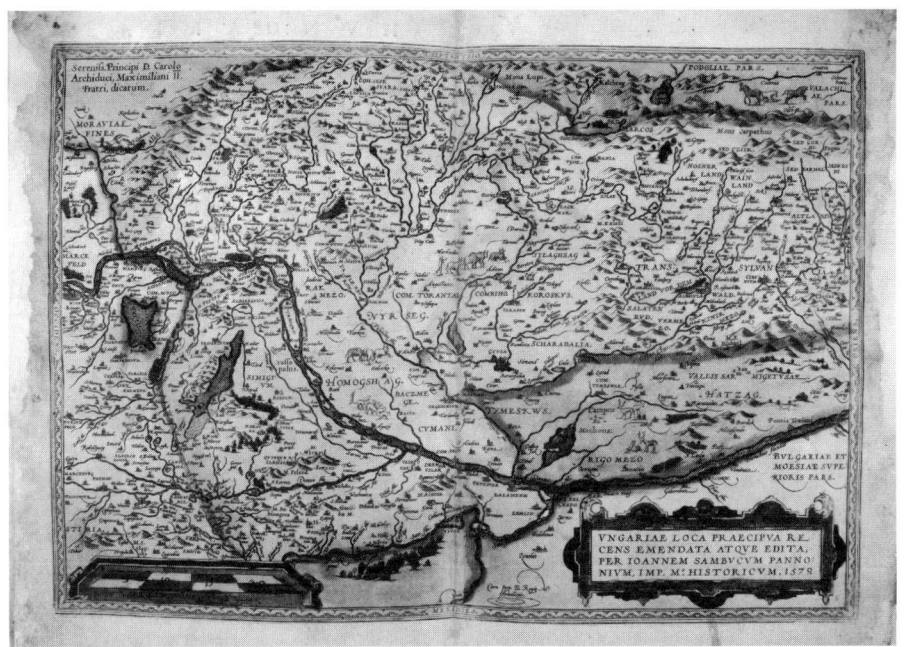

Ill. 8. Map of Hungary from Abraham Ortelius, *Theatrum orbis terrarum* (Antwerp, 1595), no. 96. (See also colour plate 16).

Boldizsár (Ill. 9) was born in 1537. He pursued his elementary studies in Croatia, Slavonia (Zagreb, Trakoščan and Vinica) and Austria (Graz and the Viennese court), and in the Batthyány court of Németújvár. From 1559 possibly until 1561-1562, he was in France serving the young king Francis II and Mary Stuart collecting the experiences that were to have a lifelong influence on him.[4] These years deeply affected his conduct in matters of religion and confessional debates. There are suggestions as to his standpoint, but most of them remain unsatisfactory and, indeed, highly speculative.[5] The idea that Batthyány

---

[4] S. Eckhardt, 'Batthyány Boldizsár a francia udvarban' (Boldizsár Batthyány at the French court), *Magyarságtudomány*, 2 (1943), 36-44.

[5] In general, see the following: the articles of I. Katona, 'Clusius és kora' (Clusius and his age), *Vasi Szemle*, 27 (1973), 398-407; 'Sárvár és a Nádasdyak a XVI. században és a XVII. század elején' (Sárvár and the Nádasdys in the sixteenth and early seventeenth century), *Savaria*, 1 (1963), 239-255; *Brueghel és a Batthyányak* (Brueghel and the Batthyánys) (Budapest, 1979); 'A Báthoryak, Batthyányak, és Zrínyiek Habsburg ellenes mozgalma' (The anti-Habsburg movement of the Báthorys, Batthyánys and Zrínyis), *Savaria*, 2 (1964), 159-174; 'A Batthyányak és a reformáció' (The Batthyánys and Reformation), *Savaria*, 5-6 (1971-72), 435-466.

Ill. 9. Portrait of Boldizsár Batthyány by an unknown painter, seventeenth century.

had an inclination towards Anabaptism[6] because the painting of Pieter Brueghel the Elder depicting the preaching Saint John the Baptist (an outstanding example of Anabaptist iconography) was on display in the main hall of Németújvár castle, is not convincing. Neither is the theory according to which he might have sympathised with Calvinism because he knew a couple of people who later openly declared themselves Calvinists[7] (such as István and András Beythe or István Pathay). What we know for sure is that he had a Protestant upbringing,[8] and that the witnessing of the massacre of the Huguenots had a deep impact on the young Protestant nobleman. In his adult years we can see that he acted with tolerance towards different religious groups and confessions. He, for instance, allowed Antitrinitarians to settle on his lands,[9] while the printer Joannes Manlius during his stay in Batthyány's castle published mostly Lutheran works.[10] Furthermore, the count corresponded with one of the propagators of Calvinism in Hungary, István Szegedi Kis,[11] and also with both the bishop of Győr, János Liszthi[12] and the bishop of Eger, István Radéczy.[13] Thus, whatever his inclinations really were, they remained an intimate form of faith rather than an active (let alone aggressive) propagation of any of the religious trends of the time.

After 1575 Boldizsár inherited the entire Batthyány domain, with its centre at Németújvár castle, where he not only followed in the footsteps of his granduncle, keeping the informal courtly school running, but also made it a focal point of humanist culture. He created a scientific circle not so much through

---

[6] I. Katona, *Brueghel és a Batthyányak*, 93: 'The 'Baptist' of Brueghel suggests the Anabaptist alternative. Obviously, Batthyány wanted to demonstrate through that painting that he accepted neither the Lutheran, nor the Catholic alternative.'

[7] Ibid., 80.

[8] A. Koltai, *Batthyány Ádám és könyvtára* (Ádám Batthyány and his library) (Budapest/Szeged, 2002) [A Kárpát-medence kora újkori könyvtárai (Early modern libraries of the Carpathian basin), 8]: 'To sum up, Boldizsár Batthyány publicly followed and supported Lutheranism on his lands, while it seems very probable that he also had Calvinist ideas, and the influence of other currents is also not to be excluded.'

[9] Letter of István Bátai to Boldizsár Batthyány, 21 December 1570, Veszprém. Published in Iványi, 'A körmendi Batthyány-levéltár', 66-67. The pastor Bátai reproaches the count for letting the heretical teachers stay on his lands.

[10] I. Monok, P. Ötvös and E. Zvara, 'Balthasar Batthyány und seine Bibliothek', in *Bibliotheken in Güssing im 16. und 17. Jahrhundert*, vol. II (Eisenstadt, 2004) [Burgenländische Forschungen, Sonderband, 26], 221-222.

[11] See the introduction of Ötvös in Monok, Ötvös and Zvara, *Balthasar Batthyány und seine Bibliothek*, 9-10. He allegedly had a copy of the work of Szegedi Kis on the Holy Trinity, ibid., 26.

[12] See the letters published in Iványi, 'A körmendi Batthyány-levéltár.'

[13] There are eight letters to Boldizsár Batthyány in the Hungarian National Archives (Magyar Országos Levéltár, abbreviated as MOL), P 1314: correspondence of the Batthyány family (microfilm no. 4886).

settling a permanent community of intellectuals in his court, but rather through extensive correspondence with humanist scholars dwelling in various parts of Europe, in the tradition of *respublica litteraria*. His actual geographical itinerary as a young man is significant (some assume that he even visited the Netherlands, more precisely the court of Queen Mary of Habsburg in Brussels[14] – again through the connections of Ferenc Batthyány – during his apprenticeship in France). Even more important, however, is the intellectual journey throughout the rest of his life through different fields of science, with the help and co-operation of both acknowledged scholars and other enthusiastic amateurs like himself, with whom he intensively corresponded. Thus the primary sources for our research are the letters, written in Latin, German, French, Italian, Croatian and Hungarian to Batthyány (and only a few by him), preserved in the Hungarian National Archives.[15]

An extraordinary library and hundreds of letters

It is important to note here that apparently Boldizsár Batthyány had never gone to university, and most probably had not even been to Italy, although it was long assumed that he had studied for some time in Padua.[16] He was thus a 'self-made scholar' who, without special education, cherished a serious interest in primarily natural sciences. Apart from the letters, another expression of this lifelong passion for learning was his library, one of the greatest book collections in Hungary of his time. The research pursued by Béla Iványi in the 1940s, as well as that recently conducted by István Monok and Péter Ötvös,[17] has shown that Boldizsár Batthyány had approximately one thousand volumes in his library.[18] However, only a part of it has survived in Güssing, where Ádám

---

[14] E. Thury, referring to documents which have since been lost, claimed that Boldizsár did indeed pay a visit to Queen Mary of Habsburg in Brussels. See E. Thury, *A Dunántúli Református Egyházkerület története* (History of the Transdanubian diocese) (Pápa, 1908), 30.

[15] MOL, P 1314.

[16] Some researchers, Sz.Ö. Barlay, T. Klaniczay and Gy.E. Szőnyi, for instance, support the idea that Batthyány spent some time in Padua, while A. Koltai suggests that this supposition is due to the mistranslation of a place in a letter that a 'familiaris' of the Batthyánys wrote from Italy. This refers, however, to Gáspár, his younger brother, who studied in the Jesuit College in the Italian city. Even if Boldizsár also visited Italy, he only spent some months there, without enrolling officially at the university. However, his knowledge of Italian, his acquaintance with some of the so-called 'Padovans' (a group of intellectuals from Hungary who spent time at the university) and many small details does indeed lend support to the previous alternative.

[17] Monok, Ötvös and Zvara, *Balthasar Batthyány und seine Bibliothek*, passim.

[18] The greatest library in Hungary in Boldizsár's time, with 6,500 volumes, was that of János Zsámboky (better known as Joannes Sambucus). After him comes Joannes Dernschwam, procurator of the Fuggers in Hungary, whose collection consisted of approximately 651 books and

Batthyány, the grandson of Boldizsár, deposited them in the newly founded Franciscan monastery in 1641 to 'get rid of'[19] the unwanted Protestant books of his ancestors.[20] Other books with Boldizsár's owner's mark can be found in various libraries today, including Győr, Körmend, Sopron, the University Library of Budapest, and also the National Széchényi Library,[21] and probably also in libraries outside Hungary (Vienna, for instance).

The present-day book collection in Güssing numbers 334 titles, which does not reflect the true dimensions of Boldizsár Batthyány's formerly extant library. In the absence of a contemporary catalogue or inventory, we have to rely also on the data recovered from the book bills issued by various book dealers (Aubry, Hiller, Widmar) between 1571 and 1589, to be found in the Hungarian National Archives. Yet another source of information – since Batthyány acquired books through some of his humanist friends, and primarily through the poet Corvinus – is his correspondence.[22] Summing up all the titles mentioned in the bills and the letters, adding to it the books today kept in Güssing, we reach a total of more than 670 volumes. Thus the estimate by Monok and Ötvös, which ranks his library as the fourth or fifth largest among contemporary collections in Hungary, seems correct. In an inventory[23] from 1780, we find an interesting classification of Catholic books (1571 volumes) and non-Catholic, 'heretical', books (1281 volumes). This latter category refers primarily to the books used by the Protestant school, although it is possible that some of the works on the natural sciences were also considered 'heretical'.[24]

1,162 various printed materials. Boldizsár Batthyány occupies roughly the fourth or fifth place among the ranks of contemporary book collectors. According to the evidence from the early seventeenth-century catalogues of the Nádasdys and Thurzós, their libraries contained some 400 volumes. For more information, see the following: *Magángyűjtemények Magyarországon 1551–1721* (Private collections in Hungary 1551–1721), ed. I. Monok, Könyvtártörténeti Füzetek 1 (Szeged, 1981); *Magyar könyvtártörténet* (The history of Hungarian libraries), eds. Cs. Csapodi et al. (Budapest, 1987); G. Kelecsényi, *Múltunk neves könyvgyűjtői* (Famous book collectors from the past) (Budapest, 1988); B. Iványi, 'Batthyány Boldizsár a könyvbarát' (Boldizsár Batthyány the bibliophile), in *A magyar könyvkultúra múltjából* (Records from the past of Hungarian book culture), ed. B. Keserű (Szeged, 1983), 389-433; and R.W. Evans, 'The Wechel presses. Humanism and Calvinism in Central Europe 1572–1627', *Past and present* (1975), Supplement no. 2, 1-74.

[19] M. Horváth, 'Egy növényjegyzék hátteréből. Adalékok a németújvári (güssingi) könyvtár alapításának körülményeihez' (The background to a nomenclature. Information concerning the circumstances of the foundation of the Güssing library), *Magyar Nyelv*, 78 (1982), 197.

[20] Koltai, *Batthyány Ádám és könyvtára* and Horváth, 'Egy növényjegyzék hátteréből,' 191-203.

[21] Monok, Ötvös and Zvara, *Balthasar Batthyány und seine Bibliothek*, 17.

[22] See my contribution in Monok, Ötvös and Zvara, *Balthasar Batthyány und seine Bibliothek*, 223-235.

[23] T. Tabernigg, 'Die Bibliothek des Franziskanerklosters in Güssing', *Biblos*, 21 (1972), 173.

[24] Horváth, 'Egy növényjegyzék hátteréből', 197.

A large part of the once-extant book collection mirrors Batthyány's passion for alchemy, and a significant portion of letters testify to this same interest. On the basis of the surviving letters it has been possible to delineate the following group,[25] which could be indicated as Batthyány's alchemical-medical circle: the first and most important member is the Viennese poet laureate, Elias Corvinus,[26] the second is the Styrian Count Felizian von Herberstein,[27] who had interests in various mines (supposedly in Transylvania as well) and who was also devoted to the exploration of nature, the third is the Italian physician, Nicolaus Pistalotius,[28] also living in Vienna, while the fourth is another doctor from Pettau (today Ptuj, Slovenia), Joannes Homelius.[29] Corvinus is the most important element in this minor network, since he also functioned as a book agent to Batthyány. Unfortunately the letters that Batthyány wrote to him, or to any other member of this circle, have not yet been found. A catalogue made before the 1940[30] (consequently before the two great tragedies in the history of Hungarian archives[31]) lists many other intriguing names, humanists and doctors, such as a certain Cesare Franco from Padua, who allegedly was also involved in an exchange of alchemical recipes, or the famous architect Pietro Ferrabosco, with whom Batthyány corresponded in Italian.

---

[25] Here I must mention articles by Sz.Ö. Barlay, the only scholar who has dealt with Batthyány's scientific circle. See the most important summary of his findings in German: Sz.Ö. Barlay, 'Boldizsár Batthyány und sein Humanistenkreis', *Magyar Könyvszemle*, 95 (1979), 231-251.

[26] The 'freshest' short biography was written by O. Sárkány in his introduction to the epic poem of Corvinus, in the early twentieth century. See E. Corvinus, *Joannis Hunnadiae res bellicae contra Turcas. Carmen epicum*, ed. O. Sárkány (Leipzig, 1937) [Bibliotheca scriptorum medii recentisque aevorum] and also Sz.Ö. Barlay, 'Elias Corvinus és magyarországi barátai' (Elias Corvinus and his Hungarian friends), *Magyar Könyvszemle*, 93 (1977), 345-353.

[27] Little is known about Felizian. There is a funeral speech written to commemorate him. See D. Reuss, *Zwo Leich und Trostpredigten ober dem seligen Abschied und Begrebnis des [...] Herrn Feliciani Freyherrn zu Herberstein [...] und Raymundi auch Freyherrn zu Herberstein* (Leipzig, 1595).

[28] Pistalotius was the doctor of Ferenc Nádasdy. He was considered an excellent physician, and consequently he was invited to cure the most various diseases in different parts of the country. See Gy. Magyary-Kossa, *Magyar orvosi emlékek* (Hungarian medical records) (Budapest, 1931; rpt. 1995), 201.

[29] Homelius is the least well-known figure of all. He wrote his letters from Marburg and Pettau where he worked as a doctor. See, *Zbornik splošne bolnišnice dr. Jožeta Potrča Ptuj 1874-2004* (Miscellany of the general hospital 'Dr Jože Potrč' in Ptuj 1874-2004) (Ptuj, 2004), 120-121. There lived another Doctor Homelius in Pettau who hosted Paracelsus for the year the latter spent in the city but this was in the 1520s or 1530s, while 'our' Homelius composed his letters to Boldizsár Batthyány in the 1580s. See A. Poznik, 'Osnovne Paracelsusove teze in njegovo bivanje v Ptuju' (Paracelsus' basic theses and his stay at Ptuj), *Zbornik za zgodovino naravoslovja in tehnike*, 8 (1985), 115-126.

[30] MOL, microfilm no. 4770.

[31] On the history of the Batthyány archives, see A. Koltai, 'A Batthyány család körmendi központi levéltárának kutatástörténete' (History of the research pursued in the Central Körmend Directory of the Batthyány family archives), *Levéltári Közlemények*, 71 (2000), 207-231.

However, those letters are no longer available. We, furthermore, have some information about an alchemical notebook that Batthyány probably used to write down his ideas, findings, or the problems that he wanted to discuss with others.[32]

As a patron and book collector, Batthyány's interest in the classical learning is also remarkable, with a large part of his library consisting of classical authors such as Cicero, Homer, Tertullian, and so on. Yet it would be too far fetched to say that he himself inclined towards textual studies. Rather, he acquired those books for the school in Németújvár for didactic purposes. As a well-known patron, he was asked to support the publication of various manuscripts: Simon Forgách (brother of the historiographer Ferenc) called his attention to the work of the historiographer Giovanni Michele Bruto (Joannes Michael Brutus) in 1587,[33] and when Joannes Sambucus (János Zsámboky) recommended a rare and precious Greek manuscript to him in 1582,[34] he was ready and very much willing to finance its translation and publication, supposedly because it was related to alchemy.[35]

As an influential aristocrat, diplomat and warrior, he was connected to the intellectual elite of the entire kingdom of Hungary, from the neighbouring landlords, the Zrínyis and Nádasdys, through the literary circle of Bishop Radéczy in Pozsony (Pressburg, Bratislava), to Miklós Istvánffy, politician and historiographer, and even the prince of Transylvania and king of Poland, István Báthory. He was acquainted with most of the courtly intellectuals in Vienna, such as Ogier de Busbecque, who addresses Batthyány as 'mein lieber herr und freundt',[36] and, at least indirectly, he knew Sambucus, Crato von Craftheim, Rembert Dodoens and also the imperial librarian, Hugo Blotius. This impressive list of names shows that Batthyány did not isolate himself

---

[32] It is briefly described in *Horváth Tibor Antal hagyatéka. Regesták a körmendi levéltár missiliseiből vegyes tárgyakra vonatkozóan. XVI-XVIII. sz.* (The heritage of Tibor Antal Horváth. Excerpts from the 'Missiles' of the archives of Körmend concerning various issues. Sixteenth and seventeenth centuries). Ms. 5264/1, Manuscript collection of the Hungarian Academy of Sciences, Budapest. Since Horváth, in the 1950s, no one has seen this intriguing notebook.

[33] S. Takáts, 'A magyar és török íródeákok' (Hungarian and Turkish scribes), *Rajzok a török világból* (Sketches from the Turkish period), 3 vols. (Budapest, 1915-17), vol. I, 37.

[34] See the study by Á. Ritoók-Szalay, 'Zsámboki János levelei Batthyány Boldizsárhoz' (Letters of János Zsámboky to Boldizsár Batthyány), in 'Nympha super ripam Danubii', *Tanulmányok a XV-XVI. századi magyarországi művelődés köréből* (Studies on Hungarian cultural history of the fifteenth and sixteenth centuries) (Budapest, 2002) [Humanizmus és Reformáció], 213-217.

[35] Even Clusius became involved in this affair. After the death of Sambucus, Batthyány still wanted to finance the translation of the manuscript and asked Clusius to try and find it for him but it could no longer be found. See Ritoók-Szalay, 'Zsámboki János levelei Batthyány Boldizsárhoz', 217.

[36] Augerius de Busbecke to Boldizsár Batthyány, 2 February 1570, Vienna. MOL, letter no. 7811 (microfilm no. 4794).

from cultural influences, even though he preferred to stay on his lands and invite scholars to come to him rather than become a courtier in Vienna. He created a pleasant ambiance, enjoyed by both people and plants, as it will be demonstrated in the following section.

Horticulture in sixteenth-century Hungary

Before discussing the garden of Boldizsár Batthyány in particular, it seems necessary to make some points in general about horticulture in Hungary[37] in the second half of the sixteenth century. Although we are lacking in systematic descriptions or inventories from this period, by relying on the data available from the abundant private correspondence, we can see that in the territory of the Kingdom of Hungary fruit production was the most widespread. During the reign of King Ferdinand an interesting competition was even cultivated among the Hungarian aristocrats, who tried to please his majesty by presenting him with the nicest and most delicious fruits – as early as possible. It was considered to be a triumph to be the first in the season to provide the ruler with the sweetest melon, peach or plum, and the noble ladies – traditionally responsible for the gardens – were encouraged by their husbands from the court to try and give of their best.[38] The value of good fruit was almost as great as a well-written political speech in the Hungarian parliament, as is testified by Boldizsár Batthyány's request to his wife to send a wagon of melons to Pozsony so that he could distribute them among the members of the Diet.[39] It was definitely also a good move to bribe state officials with certain delicacies before submitting a plea, as is evident in the letter from a servant whom Batthyány had sent to Vienna. In it, the servant named Pál, reports that before he handed the plea to Joannes Listhius, he gave him the fruits, which the bishop of Győr accepted with much pleasure. Only then did Pál present the official papers to him.[40]

In this regard the role of the Nádasdy family is, beyond doubt, of primary importance: the fame of the high quality of their fruits reached further than Vienna. Even Queen Mary of Hungary asked them for grafts of pear, apple,

---

[37] On the general impact of Italian botany on Hungary see A. Ubrizsy-Savoia, *Rapporti italo-ungheresi nella nascita della botanica in Ungheria* (Pécs, 2002).
[38] S. Takáts, 'Kertészkedés a török világban' (Gardening in the Turkish period), *Rajzok a török világból*, vol. III, 366.
[39] S. Takáts, 'Dinnyeszüret a hódoltság korában' (Melon vintage in the Ottoman period), *Rajzok a török világból*, vol. III, 392-393.
[40] Letter from Pál to Boldizsár Batthyány, 9 January 1574, Vienna. Quoted in Takáts, 'Kertészkedés a török világban', 371.

peach and plum.[41] They produced various types of pears and plums, and they also grew melon and apparently had orange and lemon trees in their gardens.[42] From a letter we have exact data about a certain 'big garden' that they built in 1546, with 104 orange trees and 50 lemon trees, which – taking the climate into account and supposing that they did not have glasshouses – seems to have bordered on true exhibitionism. However, it apparently did not produce enough fruits, since Palatine (in Latin *palatinus*, the highest authority in Hungary after the king) Tamás Nádasdy ordered new trees over the course of years to come.[43]

In Hungary, medicinal plants – similarly to exotic and rare spices – were often the most precious and were even planted in the same segment with them and they were also surrounded by fences. Thus, the difference between medicinal and botanical gardens was frequently vague.[44] The design and establishment of a *hortus medicinalis*[45] required specialised education which very few doctors in Hungary possessed. Gáspár Szegedi Kőrös (Fraxinus) was one of the few who studied at the University of Padua, and later became the doctor of Palatine Tamás Nádasdy: Takáts[46] cannot say with certainty whether Fraxinus created such a professional garden during his stay at Sárvár, while Fazekas[47] claims that the erudite physician not only founded a great herbal garden with the help of the gardener, István Kerti, but also commissioned someone to paint these valuable plants. According to Fazekas, the Nádasdys produced herbs in commercial quantities in their huge garden and had some true exotic plants, as well, such as almond and fig.[48]

Interestingly enough, the only professional *hortus medicinalis* we are sure of to have existed in this period[49] belonged to a good friend of Clusius, the doctor,

---

[41] Gy. Komoróczy, *Nádasdi Tamás és a XVI. századi magyar nagybirtok gazdálkodása* (Tamás Nádasdi and the agriculture of the Hungarian dominium in the sixteenth century) (Budapest, 1932), 84, and Takáts, 'Kertészkedés a török világban', 359-360.

[42] Ibid., 84-85.

[43] J. Stirling, *Magyar reneszánsz kertművészet a XVI-XVII. században* (Hungarian Renaissance horticulture in the sixteenth and seventeenth centuries) (Budapest, 1996), 28.

[44] J. Stirling, 'Orvosi kertek Magyarországon a XVI. században' (Medical gardens in sixteenth-century Hungary), *Orvostörténeti Közlemények*, 109-112 (1985), 112-114.

[45] The classification system of Conrad Gessner is quoted in R. Rapaics, *A magyarság virágai. A virágkultusz története* (Flowers of the Hungarians. History of the flower-cult) (Budapest, 1932), 220-221.

[46] S. Takáts, 'Orvosságtudakozás és orvoslás a hódoltság korában' (Exchange of medical information and healthcare in the Ottoman period), *Rajzok a török világból*, vol. III, 123, 129.

[47] Á. Fazekas, 'A magyar nyelvű herbárium-irodalomról' (On the Hungarian herbals), *Orvostörténeti Közlemények*, 97-99 (1982), 52. Unfortunately, he provides no further references or footnotes. Thus it must remain an intriguing piece of unsubstantiated information.

[48] Ibid., 47.

[49] Stirling, 'Orvosi kertek', 112.

poet and botanist Georg Purkircher[50] from Pozsony. His was the first example of small, urban botanical gardens, a phenomenon which, due to the delay in the urbanisation process in Hungary, became widespread only in the seventeenth century.[51]

Boldizsár Batthyány's garden

After this brief overview of Hungarian horticulture in the second half of the sixteenth century, we shall attempt to reconstruct the garden and gardening activity of Boldizsár Batthyány from the sources, the trends that he followed, and what role Clusius played in this whole enterprise. Relying on the data provided by Béla Iványi, who could work on sources that have since been lost, we find a surprisingly early encounter that Boldizsár Batthyány had with botany. In the year 1553, when he was only 16 years old and living in the Slavonian town of Vinica with his tutor, Mihály Pomagaics, he writes asking his father Kristóf for garden plants:

> would you please send us floriferous plants, smaller cypresses, marjoram, lavender and various other types of plants, which we are willing to serve to Your Magnificent Lord. We would like to ornament the garden where sometimes we could relieve our hearts from the fatigue of studying.[52]

Similarly to the Nádasdy archives, a great number of data concerning the exchange of grafts and various fruits can be found in the Batthyány letters as well. Thus it can be concluded that the Batthyánys were also involved in fruit production on a relatively large scale (probably second in the region). These letters also testify to the fact that Batthyány not only collected plants but was also a well-known source of rare plants, one to whom others frequently turned. Elias Corvinus thanks him in a letter for the peach tree that the count had sent to him and to the bishop of Würzburg,[53] while Joannes Homelius, for instance, asks for various flowers, such as tulips, daffodils and violets.[54]

---

[50] Stirling, *Magyar reneszánsz kertművészet*, 75. See also, I. Weszprémi, *Succinta medicorum Hungariae et Transsilvaniae biographia*, 4 vols. (Leipzig, 1787); E. Gombocz, *A magyar botanika története* (History of Hungarian botany) (Budapest, 1936), 81-82.
[51] Stirling, *Magyar reneszánsz kertművészet*, 75.
[52] Boldizsár Batthyány to Kristóf Batthyány, 12 April 1553, Vinica. In Iványi, 'Batthyány Boldizsár a könyvbarát', 396.
[53] 'I have received the two baskets of 'Duránc' peaches, for which I am infinitely grateful and I have given some also to my brethren, the bishop of Würzburg; he was pleased by that gift from Hungary'. Elias Corvinus to Boldizsár Batthyány, 12 September 1577, Vienna, MOL, letter no. 8099 (microfilm no. 4795).
[54] 'I beg you, Your Magnificence, to give me some of your bulbs, daffodils, Jericho rose, seeds of double violets and other more handsome ones.' Joannes Homelius to Boldizsár Batthyány, 3 April 1587, Marchburg, MOL, letter no. 19 604 (microfilm no. 4830).

When Boldizsár married Dorica Zrínyi (the daughter of the defender of Sziget, Miklós Zrínyi) in 1566, he laid down the foundations of his independent life. Unfortunately, we have no letters written by Boldizsár, and only a few to him, preserved from the period between 1558 and 1571. Thus it is not known whether anything important happened in the establishment of his humanist network, or at least the events of his life must be reconstructed from other sources. What is relevant from the point of view of our research is the year 1570, when Boldizsár inherited Szalónak (Schlaining, today Stadtschlaining, in Austria) and Rohonc (Rechnitz, today in Austria) since this is the beginning of his large-scale book collecting activity. A book bill, issued by the Paris book dealer Jean Aubry in 1571,[55] already reveals Batthyány's passion for the natural sciences. Among other things it contains the following titles: *Idea Medicinae* (Basle, 1571), a treatise in the spirit of Paracelsus by Petrus Severinus; the *Mercuriorum liber* (Cologne, 1567) and *De quinta essentia* (Cologne, 1567) of Raimundus Lullus; *De ratione conficiendi lapidis philosophici* (Basle, 1571) by Laurentius Ventura and Joannes Garlandius; Gherhardus Dorn's *Artificii chymistici physici* (Basle, 1568).

Batthyány's library may also prove useful to help us to see whether he possessed books related to botany. In a bill written by the Viennese bookseller Erhardt Hiller (after 1588),[56] for instance, we can find Pliny's *Historia naturalis* (Venice, 1507), in another bill (after 1586)[57] we find a certain *Pflantzbüchlin*,[58] perhaps the one written by Johann Domitzer. He also had a copy of Rembert Dodoens' *Frumentorum, leguminum, palustrium et aquatilium herbarum* (Antwerp, 1566) and the above-mentioned *Florum, et coronarium odoratumque nonnullarum herbarum historia* (Antwerp, 1568).[59] Last but not least, Boldizsár had at least one work by Clusius, the *Aromatum et simplicium aliquot medicamentorum apud Indos nascentiva historia*;[60] others, which he definitely also owned are today no longer in this collection. The great discovery, however, remains the *Stirpium*

---

[55] Monok, Ötvös and Zvara, *Balthasar Batthyány und seine Bibliothek*, 21-24.
[56] Ibid., 83-93.
[57] Ibid., 93-99.
[58] Ibid., 94.
[59] Ibid., 133-134: Rembert Dodoens, *Frumentorum, leguminum, palustrium et aquatilium herbarum, ac eorum, quae eo pertinent, historia* (Antwerp, 1566) (Coll. 1); *Florum, et coronariarum odoratarumqve nonnullarum herbarum historia* (Antwerp, 1568) (Coll. 2).
[60] Ibid., 125-126: Carolus Clusius, *Aromatum, et simplicium aliquot medicamentorum apud Indos nascentiua historia. Primum quidem Lusitanica lingua per dialogos conscripta, a D. Garcia ab Horto prosegis Indiae Medico. Deinde Latino sermone in epitomen contracta, et iconibus ad vivum expressis, locupletioribusque, annotatiunculis illustrata a Carolo Clusio Atrebate* (Antwerp, 1579^3) (Coll. 1); Nicolaus Monardes, *Simplicium medicamentorum ex Novo Orbe delatorum, quorum in medicina usus est, historia. Hispanico sermone descripta a D. Nicolao Monardis [...] Latino deinde donata et annotationibus iconibusque affabre depictis illustrata a Carolo Clusio Atrebate* (Antwerp, 1579^2) (Coll. 2).

nomenclator Pannonicus, which László Fejérpataky[61] found among the books of the Franciscan monastery in Güssing, and which first appeared in print in 1583 by Manlius in Németújvár. This index has received much attention in scholarship,[62] since it represents the first glossary of Hungarian plant-names, and most of them are still valid today. The book was lost during the 1940s so that the only copy we know of today is in Göttingen.[63]

A very interesting detail is the fact that among the books of the Güssing library there are six volumes which have the name of Clusius in them, and which, allegedly, were donations[64] by the botanist to the Protestant school, since they mostly are Latin grammar books and works by classical authors in Latin.[65]

How did the co-operation, or even, friendship begin between this young and ambitious aristocrat and the globe-trotting botanist who would never enjoy wealth or stability very long in his life? According to Andrea Ubrizsy-Savoia[66] and Ladislaus Batthyány-Strattmann,[67] Batthyány met Clusius in the Viennese court; Jeanplong and Katona[68] claim to know that Batthyány was even a member of an informal Viennese imperial academy, together with courtly physicians, historians (such as Miklós Istvánffy, for instance) which

---

[61] L. Fejérpataky, 'A németújvári ferences zárda könyvtára' (Library of the Franciscan monastery of Güssing), *Magyar Könyvszemle*, 8 (1883), 101.

[62] See, for instance, its new editions in *The beginnings of Pannonian ethnobotany: Stirpium nomenclator Pannonicus*, ed. S. Beythe (1583), C. Clusius (1584), D. Czvittinger (1711), etc., (Szombathely, 1992) [Ethnobotany and ethnobiodiversity, Bio Tár, Collecta Clusiana, 2, ed. A T. Szabó]; J. Jeanplong and I. Katona, 'Clusius in Westpannonien. Beziehungen zu Boldizsár Batthyány und István Beythe', in *Carolus Clusius Fungorum in Pannoniis observatorum brevis historia et Codex Clusii. Mit Beiträgen von einer internationalen Autorengemeinschaft*, eds. S.A. Aumüller and J. Jeanplong (facsimile, Budapest/Graz, 1983), 34-39; V. Petkovšek, 'Clusius' *Nomenclator Pannonicus* und seine Zusammenarbeit mit Joannes Manlius', in *Carolus Clusius und seine Zeit. Symposion in Güssing 1973 (Vorträge)* (Eisenstadt, 1974) [Wissenschaftliche Arbeiten aus dem Burgenland, 54, Kulturwissenschaften, 19], 24-32; Csaba, 'Magyar ethnobotanikai adatok', 595-599; Sz.Ö. Barlay, 'A Clusiusnál található magyar növénynevek kérdése' (The question of Hungarian plant names in Clusius), *Magyar Nyelv*, 44 (1949), 69-72; M. Szlatky, 'A magyar nyelvű természettudományos és orvosi irodalom a XVI. században' (Natural scientific and medical literature in Hungarian in the sixteenth century), *Orvostörténeti Közlemények*, 109–112 (1985), 91-97.

[63] *Régi Magyarországi Nyomtatványok 1473-1600* (Old prints from Hungary, 1473–1600), eds. G. Borsa, F. Hervay, B. Holl, I. Käfer and Á. Kelecsényi (Budapest, 1971), no. 536.

[64] Horváth, 'Egy növényjegyzék hátteréből', 198.

[65] Monok, Ötvös and Zvara, *Balthasar Batthyány und seine Bibliothek*, 90, 146, 164, 192, 207, 213.

[66] A. Ubrizsy-Savoia, *Die Beziehungen des Lebenswerkes von Carolus Clusius zu Italien und Ungarn* (Vienna, 1977), 12.

[67] L. Batthyány-Strattmann, 'Güssing und die Batthyány zur Zeit des Clusius', in *Festschrift anläßlich der 400jährigen Wiederkehr der Wissenschaftlichen Tätigkeit von Carolus Clusius (Charles de l'Escluse) im pannonischen Raum* (Eisenstadt, 1973) [Burgenländische Forschungen, Sonderheft, 5], 112.

[68] Jeanplong and Katona, 'Clusius in Westpannonien', 34.

Clusius also found very interesting. Csapody[69] says that Batthyány could have met Clusius in the Low Countries during his travels. When and how did Clusius and Batthyány really meet? Unfortunately we do not know exactly. However, Vienna seems to be the most plausible place, since both of them went there often.

Again, there are many suggestions concerning the earliest stay of the botanist in Hungary: Stirling[70] relies on Csapody[71] when indicating the year 1579 as the first time Clusius visited Szalónak, while Aumüller[72] suggests a date between 1578 and 1579, and Jeanplong and Katona claim that Clusius stayed there as early as 1575.[73] The letters (Ill. 10) do not help us out here since they are fragmentary. There is no correspondence in the strict sense: the following summary will show how enormous the chronological gaps are between letters. Istvánffi[74] published all the letters he knew of (twelve of them), but somehow he did not find half of the first, all of the second and also letter no. 8021 of Clusius to Batthyány, which today are in the Hungarian National Archives (MOL).

Letters of Clusius to Batthyány

21/10/1577	Vienna	French	MOL	8014	second page unpublished
30/11/1577	Vienna	French	MOL	8015	unpublished
04/05/1578	Vienna	French	MOL	8016	Istvánffi
02/06/1578	Vienna	French	MOL	8017	Istvánffi
05/07/1578	Vienna	French	MOL	8018	Istvánffi
23/07/1578	Vienna	Latin	MOL	8019	Istvánffi
19/12/1579	Vienna	Latin	MOL	8020	Istvánffi
30/09/1587	Vienna	Latin	MOL	8021	unpublished

---

[69] I. Csapody, 'Clusius magyar mecénása és munkatársai' (The Hungarian patron and colleagues of Clusius), *Vasi Szemle*, 27 (1973), 408; he, however, adds that if not then, they could have met at the Viennese court at the latest (409).
[70] Stirling, *Magyar reneszánsz kertművészet*, 25.
[71] Csapody, 'Clusius magyar mecénása és munkatársai', 412.
[72] S.A. Aumüller, 'Wissenschaftliche Tätigkeit in Wien', in *Carolus Clusius' Fungorum in Pannoniis observatorum brevis historia et Codex Clusii*, 31.
[73] Jeanplong and Katona, 'Clusius in Westpannonien', 35.
[74] Gy. Istvánffi, *A Clusius-Codex mykologiai méltatása adatokkal Clusius életrajzához* (Mycological evaluation of the Clusius codex with references to the biography of Clusius) (Budapest, 1900).

10/12/1587	Vienna	Latin	VUL 101	11	Istvánffi
Description of *Tabuco latifolium*		Latin	MOL	no number	unpublished
Notes		Latin	MOL	no number	unpublished

Letters of Batthyány to Clusius

13/11/1584	Németújvár	Latin	VUL 101	no. 7	Istvánffi
22/02/1585	Németújvár	Latin	VUL 101	no. 8	Istvánffi
16/08/1585	Németújvár	Latin	VUL 101	no. 9	Istvánffi
26/01/1586	Vinica	Latin	VUL 101	no. 10	Istvánffi
11/11/1588	Németújvár	German	VUL 101	no. 12	Istvánffi

The first element in the correspondence is thus Clusius' letter from November 1577, which, judging by its content and tone, is not the very first one during the course of his acquaintance with Batthyány. In this letter the botanist gives a short report on his situation in Vienna, because of which he is unable to accept Batthyány's invitation to visit his lands. What is clear from this, chronologically first, surviving letter, is that Clusius had already visited Batthyány's domain – 'I have decided to concern myself with the description of plants which I have observed on my way to you, and others which I found while travelling through the mountains in Austria'[75] – and he even contemplated commissioning someone to paint the plants that he had collected and described, and publishing the work in his homeland. He promises Batthyány that as soon as his position is clarified at the Viennese court and he receives the payments due to him, he will visit the count and prepare his garden for spring: 'Since I would not leave this town before kissing your hands, and before I thank you for the many benefices which you have given me, I would stay at your place for 8 or 10 days to dress your garden up for spring.'[76] In the second, unpublished, part of the letter Clusius explains that Hubertus Languetus could not satisfy Batthyány's request to acquire a certain 'Docteur Quercetanus' for him, because he was ill. There is a work written by Josephus Quercetanus – not surprisingly, dealing with metals – among those that were in the library of the Güssing Franciscan monastery; thus it seems that Batthyány used all his

---

[75] Carolus Clusius to Boldizsár Batthyány, 21 October 1577, Vienna. MOL, letter no. 8014 (microfilm no. 4794). Istvánffi, *A Clusius-Codex*, 205.

[76] Carolus Clusius to Boldizsár Batthyány, 21 October 1577, Vienna. MOL, letter no. 8014. Istvánffi, *A Clusius-Codex*, 205.

Ill. 10. A letter of Carolus Clusius written to Boldizsár Batthyány.

connections to buy books that really interested him.⁷⁷ For the reconstruction of Batthyány's garden, it may be interesting that Clusius also promises to send a 'cytronnier' (lemon tree) and an 'arbrisseau de pseudocapsicum' (Jerusalem cherry), which he recommends should not be exposed to winter cold – together with the *canna indica* (Indian-shot) he sent earlier that year – and put them under the window inside the house.⁷⁸

An unknown painter and a mysterious job

Letter no. 8015, completely unpublished, is rich in historical data (such as news about the religious wars) but does not contain anything concerning botanical issues, while letter no. 8016 is of primary importance in this regard. Stirling⁷⁹ does not use the information from Takáts, according to which Batthyány had paprika in his garden, because the latter does not provide the source for his statement. On the contrary, everything is clear from the otherwise short letter which runs as follows:

A *Palingenius*, and the recipe for the small tree. Daniel the gardener told me that he had just bought the daisies which have to be planted in beds similarly to balsam and paprika since they have nice flowers.⁸⁰

From this quotation it is evident that Clusius had produced an exact plan for the garden of Batthyány in Szalónak (a sketch that unfortunately did not survive). The botanist explained in detail where to put the various plants, and how.⁸¹ Apparently, Batthyány at this time was already growing *balsamina* (garden balsam) and *capsicum* (paprika), since Clusius tells him to plant daisies in the same manner. Apart from the plants, Clusius also sends him a book,⁸² which today is still to be found in the collection of the Franciscan monastery in Güssing.

---

⁷⁷ Josephus Quercetanus, *Ad Iacobi Auberti Vindonis De ortu et causis metallorum contra cymicos explicationem [...] breuis responsio. Eiusdem de exquisita mineralium, animalium, et vegetabilium medicamentorum spagyrica praeparatione et vsu* (Lyon, 1575) in Monok, Ötvös and Zvara, *Balthasar Batthyány und seine Bibliothek*, 34-35.
⁷⁸ Carolus Clusius to Boldizsár Batthyány, 21 October 1577, Vienna. MOL, letter no. 8014. Unpublished.
⁷⁹ Stirling, *Magyar reneszánsz kertművészet*, 24.
⁸⁰ Carolus Clusius to Boldizsár Batthyány, 4 May 1578, Vienna. MOL, letter no. 8016. Istvánffi, *A Clusius-Codex*, 205.
⁸¹ 'I am sending you a wooden casket full of herbs, as I have written down for you on a piece of paper: they should be distributed in the order I have indicated there, similarly to the grains. The garden mallow can be planted in a circle following the line of the walls. I am also sending you a pattern for the trees which have to be placed around the segments or beds.' Ibid.
⁸² Marcellus Palingenius, *Zodiacus vitae, hoc est, de hominis vita, studio ac moribus [...]* (Basle, 1566) in Monok, Ötvös and Zvara, *Balthasar Batthyány und seine Bibliothek*, 66.

In the next letter Clusius says, 'I have changed the plan of your small garden – which I know was there in Szalónak – for a bigger one.'[83] However, this letter contains more intriguing information which has given grounds to a series of inaccurate interpretations. Here, Clusius talks about a certain painter for whom Batthyány has a task, some sort of job to accomplish.

I have spoken to one of the better painters in this town who is from our country, and who is very good at the art of wall painting and so on [...] I assure you that it is very hard to find a good painter here who is able and who knows his art well enough: mostly they are nothing more than apprentices. Thus, I spoke to one who is among the better masters and one of the main ones. He is good at painting natural subjects and certain historical subjects, and any similar thing that you may desire.[84]

This is in the letter which follows, as well: 'The painter I wrote you about, My Lord, each day promises me to offer you his services in the job you have for him in Szalónak.'[85] And also in letter no. 8019:

The painter that I spoke first with said that as soon as he is dismissed from his position, he will go to Your Magnificence, since he has accomplished the better part of the work which His Majesty the Archduke entrusted him with. That job will take another 8 – 10 days, then he will be completely free. In the meantime, he wanted to see what he will have to do in the castle of Your Magnificence, and to agree with Your Magnificence about the price, then he would like to come back here [...] and he could have 8 to 10 days of free time[86]

According to a letter which has since been lost, the poet Corvinus was also involved in the search for a painter, as mentioned in Istvánffi: 'Elias Corvinus was equally engaged in the search for a painter.'[87]

In the earlier literature, these places were unanimously interpreted as referring to the skilful painter of the watercolours of mushrooms, today in the collection of Leiden University Library known as the *Codex Clusii*.[88] However,

---

[83] Carolus Clusius to Boldizsár Batthyány, 2 June 1578, Vienna. MOL, letter no. 8017. Istvánffi, *A Clusius-Codex*, 205-206.

[84] Ibid.

[85] Carolus Clusius to Boldizsár Batthyány, 5 July 1578, Vienna. MOL, letter no. 8018. Istvánffi, *A Clusius-Codex*, 206-207.

[86] Carolus Clusius to Boldizsár Batthyány, 23 July 1578, Vienna. MOL, letter no. 8018. Istvánffi, *A Clusius-Codex*, 207.

[87] Elias Corvinus to Boldizsár Batthyány, 21 August 1578, Vienna. Quoted in Istvánffi, *A Clusius-Codex*, 185. There is no letter by Corvinus among the private letters of the MOL with this content and date.

[88] The only exception is Hunger's great work which seems to have been neglected on this issue by later scholarship. He also finds it hard to accept that this, earlier, correspondence would regard the depiction of mushrooms. F.W.T. Hunger, *Charles de l'Escluse (Carolus Clusius), Nederlandsch kruidkundige 1526–1609*, 2 vols. (The Hague, 1927-43), vol. I, 160.

if Clusius started to work on mushrooms only in the first half of the 1580s, or more precisely, in 1584,[89] then we can exclude the idea that the hunt for a good painter in 1578, involving both Clusius and the poet Corvinus, was for the sake of the depiction of the mushrooms.[90] In these years Clusius worked more on the Pannonian and Austrian flora.[91] In 1579 he met Plantin in Antwerp to discuss the details of its publication. The final manuscript was ready by 1580, and the work appeared in print in 1583. The first letter in which Batthyány mentions the mushrooms and their depiction is from the year 1584: 'I was really pleased that you visited me last summer, the time when I desired to have depictions of the various species of mushrooms before they were sent to print.'[92] The letter which follows chronologically[93] from Batthyány does not mention mushrooms again. Therefore we can conclude that the work had been carried out and indeed, the watercolours had been produced by the painter.

Due to the gap between the letter written in the summer of 1578 and the 'next' one, from December 1579, we cannot know with certainty what exactly Batthyány wanted the painter to paint for him, and whether in the meantime they had managed to organise the artist's travel to Szalónak, and whether he had finished the job. Since Clusius says that the painter, his compatriot, could find eight days in his schedule to do what Batthyány required, it can be assumed that the task was not a large one.

If these conclusions are right, then one more question remains: what was the job that Batthyány had for the painter in the year 1578? It could equally well have been either something for the decoration of his castle or the depiction of plants. Taking into account the expressions that Clusius uses to describe the painter's strengths, such as 'depicting natural subjects' ('contrefaire au naturel') or his saying that summer is 'the right season for that work' ('la saison propice à ce faire'), we might suspect that the job involved some plants. If so, is it possible that the watercolours painted in 1584 were not the first that Batthyány sponsored for Clusius? If this job was ever carried out, it could even have been related to an earlier work of Clusius'.

---

[89] Aumüller, 'Wissenschaftliche Tätigkeit in Wien'. Also, the anecdote I referred to in my introduction is from 1584, when Clusius obviously stayed with Batthyány. At that time it was already well known that he was concerned with mushrooms.

[90] Both Istvánffi and Ubrizsy-Savoia assume that this earlier correspondence concerned the depiction of the mushrooms, although Istvánffi has doubts that the painter mentioned here actually executed the painting of the mushrooms in 1584. See Istvánffi, *A Clusius-Codex*, 186.

[91] C. Clusius, *Rariorum aliquot stirpium per Pannoniam, Austriam et vicinas quasdam provincias observatarum historia quatuor libri expressa* (Anwerp, 1583; facsimile edn. Graz, 1965).

[92] Boldizsár Batthány to Carolus Clusius, 13 November 1584, Némtújvár. Leiden University Library, VUL 101, no. 7. Istvánffi, *A Clusius-Codex*, 208.

[93] Boldizsár Batthány to Carolus Clusius, 22 February 1585, Némtújvár. Leiden University Library, VUL 101, no. 8. Istvánffi, *A Clusius-Codex*, 208.

If for a moment we recall the words of Clusius from his first remaining letter to Batthyány, 'I have decided to engage myself into the description of plants which I have observed on my way towards you, and others which I found while going for the mountains in Austria, and a part of next summer (if God gives me long life) to have the plants painted, so that on my return to the country I can have them published',[94] it is possible to connect Clusius' intention of having the plants depicted and Batthyány's intention of finding a painter for a job exactly in the same period. Indeed, the *Rariorum aliquod stirpium per Pannoniam* is richly illustrated, containing 358 woodcut images, although that is far too much work to do in the eight days that Clusius mentions at one point in a letter. Furthermore, in this case we know who the artist was: either Gerard van Kampen who made the illustrations on the basis of the drawings of Clusius and Peeter van der Borcht, or the son of Virgil Solis in Frankfurt. Van Kampen and van der Borcht were compatriots of Clusius but it is questionable whether any of them worked in the Imperial court of Vienna, and whether any of them might be the painter for whom we are looking.

In conclusion, the 'hunt' for a skilful painter in 1578 cannot be connected to the depiction of mushrooms in 1584, because in that period Clusius was involved in other projects and publications and had not started his research on mushrooms yet. Furthermore, the formulations in the letters are so unclear that we can take for granted only the simple fact that Batthyány had the intention of hiring a painter in the summer of 1578 for an unknown job.

The next letter which contains some botanical information is from 1578, and apart from mentioning the painter for the last time, Clusius promises to bring some bulbs to plant in the garden of Szalónak, as they had earlier agreed: 'I will go with him (if Lord Althan comes back) to you, bringing along bulbs which I will arrange in your garden at Szalónak, as I promised last time.'[95]

More than a year later, Clusius sent many plants to Batthyány through the latter's 'familiaris', Farkas (Wolff) Schaller, who obviously functioned as a courier between the count and Vienna:

I have given him some seeds, two sorts of nice nasturtium, one which has to be sown together with thyme in early spring. The lupine silvestris, which has extremely fragrant and very elegant yellow flowers, has to be planted with legumes around March, if it is

---

[94] Carolus Clusius to Boldizsár Batthyány, 21 October 1577, Vienna. MOL, letter no. 8014. Istvánffi, *A Clusius-codex*, 205.
[95] Carolus Clusius to Boldizsár Batthyány, 23 July 1578, Vienna. MOL, letter no. 8019. Istvánffi, *A Clusius-Codex*, 207.

not too cold, or in April. Instead of *pisum silvestris Grebensis* I send you *pisum sativum elegans* which has flowers on the edge of its stem. It should be sown with the other peas.[96]

In the same letter Clusius thanks Batthyány for his invitation to Szalónak, although due to his obligations in Vienna he cannot accept it. He is, however, most welcoming about the idea of an excursion to the Styrian Alps in the spring. In a letter dated as being from more than nine years later, the botanist is still busying himself around the enrichment of Batthyány's garden: he sends two Indian-shot plants, which have to be planted in small baskets or wooden boxes, and kept in the room: 'I send you two *canna indica* plants, one older and smaller, germinated from seed this summer. They have to be kept in wooden baskets or boxes so that you put soil on them and keep them inside the house.'[97] He also forecasts a particularly cold winter, because of which neither grapes nor mushrooms will grow in great quantities. In the last surviving letter we have no remarkable information concerning gardening issues. Rather, it is a nice example of how private letters served the purpose of spreading news, especially if the correspondents were in contact with different parts of Europe.

In connection with the book on mushrooms, I have already mentioned the letter by Batthyány, written in 1584, which is the first of the few letters that have come down to us. In it Batthyány asks Clusius for various things: 'I very much beg you, My Lord, to be kind and help me with the correct arrangement of seeds: especially with the thyme seeds, and also, tell me which flowers the bees most prefer.'[98]

Batthyány's last surviving letter to Clusius from 1588 has often been referred to because it reveals the enthusiasm and passionate collecting euphoria of this 'amateur scientist', who set his valuable Turkish prisoner Ali Bey free in the hope that he would bring him bulbs that could only be found in the gardens of the Turkish Sultan.

I have recently set my prisoner, Ali Bey, free to go to Turkey from here. He claims that he will bring me nice flowers from there. He says that the hyacinths which we have here are not the same as those in the garden of the Turkish Sultan, since those have 36 petals[99]

---

[96] Carolus Clusius to Boldizsár Batthyány, 19 December 1579, Vienna. MOL, letter no. 8020. Istvánffi, *A Clusius-Codex*, 207.
[97] Carolus Clusius to Boldizsár Batthyány, 30 September 1587, Vienna. MOL, letter no. 8021. Unpublished.
[98] Boldizsár Batthyány to Carolus Clusius, 22 February 1585, Némétújvár. Leiden University Library, VUL 101, no. 8. Istvánffi, *A Clusius-Codex*, 208.
[99] Boldizsár Batthyány to Carolus Clusius, 11 November 1588, Némétújvár. Leiden University Library, VUL 101, no. 12. Istvánffi, *A Clusius-Codex*, 209-210.

In the same letter, Batthyány sends some antique pieces (perhaps coins) and gold to Clusius in Frankfurt, asking him to bring him more bulbs and buy him new books:

Hereby, I send you, My Lord, 24 antiquities together with a golden florin; My Lord may be satisfied with them until I will be able to get something more, which I will then send to My Lord. And I also beg you, My Lord, to look around for new books and send them to me along with some new flowers which we do not have here.[100]

Newness and exoticism are thus of primary importance. The plants, flowers and herbs were doubtlessly a pleasure to look at, and were useful in thousands of other ways, but to have something in one's garden that no one else had: that was a challenge and a triumph.

Let us summarise what plants and what sort of garden Boldizsár Batthyány had between 1570s and 1590 in Szalónak. On the basis of the testimony from the few surviving letters, he definitely received a 'cytronnier' (a lemon tree), *pseudocapsicum* (Jerusalem cherry), *malva hortensis* (garden mallow), *balsamina* (garden balsam), *capsicum* (paprika), various herbs, grains, bulbs, daisies, two sorts of *nasturtium* (garden cress), *thymus* (thyme), *lupinus* (lupine), *pisum sativum* (green peas), *canna indica* (Indian-shot) from Clusius. These plants were organised in beds and also along the walls of the garden; for instance, the botanist recommends that Batthyány should plant mallow in a circle, following the line of the walls, while daisies should be planted in small beds just like balsam and paprika, since they all bear nice flowers. Furthermore, he had hyacinth, most probably potato,[101] and also daffodils, violets and perhaps even tulips in his garden, because Homelius asks Batthyány to send him some of those.

Conclusion

Further crumbs of similar information can be recovered from the vast and mostly unpublished and not yet researched letters of the Batthyány family, although the information that they concern would not change our conclusions substantially. The presence in Batthyány's garden for the first time in Hungary of paprika,[102] his enthusiasm for rarities and new and exotic plants, and his collecting spirit reveal a different attitude from that of his ancestors. Through the creation of his garden in Szalónak he took part in an international, and

---

[100] Ibid.
[101] Stirling claims that it was Clusius who distributed the potato in Hungary. Stirling, *Magyar reneszánsz kertművészet*, 27. If so, he probably gave also Batthyány samples of this new plant.
[102] A. Ubrizsy-Savoia, 'Carolus Clusius és a termesztett növények' (Clusius and the cultivated plants), *Botanikai Közlemények*, 62 (1975), 225, and Rapaics, *A magyarság virágai*, 236.

very lively, exchange of plants, and joined the exciting experiments of introducing new species to Hungary.

As was mentioned earlier, Hungarian scholarship has dedicated much attention to the *Stirpium nomenclator Pannonicus* written in co-operation with Clusius and István Beythe, and the circumstances of the making and the illustrations of the *Codex Clusii* are also well known.[103] Furthermore, it can be assumed that apart from the financing of the watercolours in the *Codex Clusii*, Batthyány may well have supported the publication (or at least the illustration) of some other works by Clusius. On the basis of the published letters and the unpublished ones that I have come across in the Hungarian National Archives, I attempted to reconstruct the garden of Boldizsár Batthyány. This reconstruction showed that the focal point of sixteenth-century Hungarian erudite botany is beyond doubt within Batthyány's circle, where the scholarship of Clusius met an apprehensive and supportive public, one that was welcoming to the new discoveries made or transmitted by him. Thus the count's hobby reached beyond his individual scope, something he was well aware of.

However, Batthyány was not alone in this enterprise: not only did he receive suggestions and plants from one of the most prominent botanists of his time, Carolus Clusius, but he was also supported by István and András[104] Beythe, a father and son, who shared his passion and who were educated in this field. István Beythe, before coming to live in the court of Batthyány, spent years at Sárvár,[105] at the residence of the Nádasdys, where Fraxinus worked as the doctor to the Palatine for years, and, as mentioned earlier, supposedly created a great herbal garden there. Both István Beythe and Péter Melius Juhász[106] probably acquired their notions of botany from Fraxinus: while the former's expertise manifested itself in co-operation with Clusius, the latter was the first in Hungary to write a herbarium in Hungarian (Kolozsvár [Cluj]: 1578). If we add to this the fact that another prominent figure in the history of Hungarian

---

[103] Istvánffi, *A Clusius-Codex*; Gombocz, *A magyar botanika története*; *Festschrift anläßlich der 400jährigen Wiederkehr der Wissenschaftlichen Tätigkeit von Carolus Clusius*; Ubrizsy-Savoia, *Die Beziehungen des Lebenswerkes von Carolus Clusius zu Italien und Ungarn*.

[104] A. Beythe, *Fives Könüv* (Güssing, 1595). This work translates parts of Matthiolus, and otherwise relies on Melius Juhász and the Hungarian plant names from the *Stirpium nomenclator Pannonicus*. See Fazekas, 'A magyar nyelvű herbárium-irodalomról', 45, 55.

[105] Ibid., 52; T. Grynaeus, '(Gyógy)növényismeretünk a reneszánsz és a reformáció korában' (Our notions of [herbal] plants from the Renaissance and Reformation), *Orvostörténeti Közlemények*, 109-112 (1985), 108; Szlatky, 'A magyar nyelvű természettudományos és orvosi irodalom', 97; Stirling, *Magyar reneszánsz kertművészet*, 30.

[106] Fazekas, 'A magyar nyelvű herbárium-irodalomról', 52-53; and see the introduction of A. Szabó to the reprint edition of Melius' *Herbárium* (Bucharest, 1978), 46.

medicine and botany, namely Gergely Frankovics[107] (or Frankovith), through his friendship with István Beythe, was also connected to the Németújvár scientific circle[108] and his work entitled *Hasznos es fölötte szikseges könyv* (A useful and particularly necessary book) was printed by Manlius (Monyorókerék [Eberau], 1588), we can see how in this period, which sees the emergence of Hungarian botany, all the most important participants, botanists, patrons, fruit producers, printers, doctors and herbalists were connected, creating a small elite. It seems, furthermore, that while Sárvár was both the main fruit producer and also the alma mater for a generation of scholars with medico-botanical education, it is in the circle of Boldizsár Batthyány that their knowledge and talent found real expression. The reason why Szalónak and Németújvár became the centre of very sophisticated botanical work, unique in its intensity and its highly scholarly nature in the region, can be found in the person of Carolus Clusius and his successful co-operation with the small circle formed around the figure of Count Batthyány. It is due to his indefatigable desire to learn that we have the first work on the 'Pannonian' flora, the first Hungarian nomenclature of plant names, and the absolutely pioneering work on mushrooms, while it is through the support of Batthyány (both financial and scholarly) and the Beythes (scholarly) that his efforts were productive.

What Batthyány gained through his friendship and co-operation with Clusius was not only a series of smaller or bigger triumphs connected to the new plants in his garden, planted there for the first time in the region, but also the achievement of having his name recorded[109] forever. Apart from being remembered as a powerful aristocrat and successful warrior, Batthyány finally made his mark in the history of botany as well, since Clusius did not forget about his friends and patrons. He took advantage of every occasion to mention them in various anecdotes in different works and to accentuate their role in his own discoveries, thus illustrating his unique modesty and friendliness. Besides the dedication[110] of the *Aliquot notae*, and the anecdote that was referred to in the introduction, we come across the name of Batthyány in

---

[107] K. Alföldi-Flatt, 'Frankovith Gergely és orvosbotanikai műve' (Gergely Frankovith and his Medico-Botanical Work), *Természettudományi Közlöny*, 37 (1895), supplement no. 2, 49-59.
[108] Stirling, *Magyar reneszánsz kertművészet*, 22.
[109] See the dedication in Clusius, *Aliquot notae in Garciae Aromatum historia* (Antwerp, 1582), and the innumerable references to things the botanist saw while staying with Batthyány such as local anecdotes, legends concerning plants, and so on.
[110] Ibid., '[…] since I know that Your Magnificence finds pleasure in such things, it seemed like a good idea to call it after you, hoping that in the outstanding library you create, there will be some place also allowed to these among the other books, and you will think of this as a humble expression of my infinite gratitude for all the benefices you have bestowed upon me'.

numerous works by Clusius as well. Yet the simplest and warmest formulation of their multiply decennial relationship may be found in a dedication to another friend, Giovanni Vincenzo Pinelli, in which Clusius calls the late count *amicus* and remembers him as 'The distinguished hero Balthasar Batthyan, hereditary master of His Majesty the King of Hungary's, Lord High Steward, who loved me in a unique manner'.[111]

---

[111] C. Clusius, *Fungorum in Pannoniis observatorum brevis historia*, in appendix to *Rariorum plantarum historia* (Antwerp, 1601), cclxii.

Lilies to Norway and cloudberry jam to the Netherlands: On the correspondence between Carolus Clusius and Henrik Høyer, 1597-1604[1]

Kjell Lundquist

> [...] therefore, but a moment ago, leisurely wandering in the garden, I burst into a song of praise as I beheld your [Clusius'] bulbs and in a sort of a poetic rapture I put the few lines on paper I now send you. Not because I deem them worthy of your discerning ears, but so that you may recognize that I think of you more often than you realize, and how I uphold your name, honour and reputation.[2]
>
> Henrik Høyer, 17 August 1597

Introduction

On 19 October 1593 the famous botanist Carolus Clusius (1526-1609), who was then 67 years old, moved from Frankfurt to Leiden to lay out and manage the newly (1590) established botanical garden, the second modern botanical

---

[1] This article is based on my paper at the conference 'Clusius in a new context' at the Scaliger Institute / Leiden University Library in Leiden, 23-25 September 2004. A different version was published in Swedish, in *Svenska Linnésällskapets Årsskrift 2004-2005* (Yearbook of the Swedish Linnaeus Society 2004-2005). The relationship between Høyer and Clusius, mainly in the context of lilies and the introduction of new garden plants in Scandinavia, is also discussed in my doctoral dissertation: K. Lundquist, *Lilium martagon L. Krolliljans introduktion och tidiga historia i Sverige intill år 1795 – i en europeisk liljekontext* (Alnarp, 2005) [Acta Universitatis Agriculturae Sueciae], 265-286.
I would like to thank John Robert Christianson (Iowa), Per Magnus Jørgensen (Bergen), Johan Lange (Søborg, Denmark), and Venke Åsheim Olsen (Trondheim), Finn-Egil Eckblad († 2000) and Madeleine von Essen (Oslo) for additional information and an exchange of views. Special thanks go to Urban Örneholm of the Department of Linguistics and Philology, Classical Languages, at Uppsala University for transcribing and translating most of the Latin letters of Høyer into Swedish (2003-2005), his corrections of the final version, and to Christian Idström, for the translation of my paper into English, and for checking and correcting the present article.

[2] '[...] atque ideo nuper cum animi causa in horto deambulans in bulbos tuos forte inciderem, desubito raptus in laudes tuas, furore quodam uti fit, poëtico, paucula quaedam fudi, quae nunc mitto: non quod existimem ea politissimis tuis auribus digna: (talia enim vix esse possunt schediastica) sed ut intelligas saltem: me opinione tua saepius de te cogitare: tuique nominis, laudis, honorisque esse studiosissimum.' Leiden University Library, VUL 101.

Ill. 11. Title-page of a late, revised edition of Rembert Dodoens, *Cruydt-boeck* (Antwerp, 1644). The 'White lily' (*Lilium album*) and the 'Red lily' (*Lilium chalcedonicum*) on the sides of the gate into the garden, to paradise, are given symbolic prominence.

garden north of the Alps. In the same year, the German physician Henrik Høyer (ca. 1565-1615/16) moved to Bergen in Norway, which was at that time a province of Denmark. Only three years later, in 1596, Høyer too arrived in Leiden, in order to fully qualify as a doctor of medicine.[3] The meeting of Clusius and Høyer in the university town was to lead to a lively correspondence and scientific exchange between North and South. It also led to a profound friendship based on mutual respect. The relationship between Bergen and Leiden that had been established already in earlier years deepened, and for nearly a decade, natural-history objects from the three natural kingdoms (minerals, bulbs and other parts of plants, and birds and other animals) were exchanged between the Netherlands and Norway. In several cases, the plants which arrived from Leiden were the first of their species to be introduced in Scandinavia (Ill. 12).

This article is based on seven letters in Latin from Henrik Høyer to Clusius, written between the years 1597 and 1604, and partly annotated by Clusius.[4] It concentrates on the botanical exchanges between these two men, although these letters are extremely relevant to historians of zoology and ethnology as well, as will be briefly discussed below. The letters, which concern a unique part of natural history that branches off into several biological disciplines, were first commented on by Finn-Egil Eckblad (1991) in an article about Henrik Høyer and the first tulips in Norway.[5] It primarily concerns the introduction of the tulip in Norway and focuses on the first three letters. The exchange between Clusius and Høyer had, however, already been referred to in earlier publications. Clusius himself mentions Høyer and his informants in his *Exoticorum libri decem* (1605) in connection with the 'sea-shrub' called 'Erica marina' and the geese on the Faeroe and Orkney Islands.[6] The exchange between the two men is also briefly mentioned by Clusius' biographer, Hunger (1927), and by Pavord (1999).[7]

---

[3] The University of Leiden, the first university in the Netherlands, was founded in 1575 by William of Orange.

[4] The original seven letters from Henrik Høyer to Carolus Clusius (1597-1604) are kept in Leiden University Library, VUL 101.

[5] See F.E. Eckblad, 'Henrik Høyer og de første tulipaner i Norge', *Blyttia*, 3 (1991), 145-150. The first tulips in the Leiden botanic garden blossomed in 1594, in Norway in Bergen in 1597. Evidence so far for tulips in Denmark dates from the 1630s and for Sweden from 1629. Eckblad (1923-2000) held a chair at the Biologisk Instituttet at Oslo University. For an overview of his life and work, see G. Gulden and K. Høiland, 'Finn-Egil Eckblad 12.8.1923 – 14.7.2000', *Blyttia*, 3-4 (2000), 147-152.

[6] C. Clusius, *Exoticorum libri decem: quibus animalium, plantarum, aromatum, aliorumque peregrinorum fructuum historiae describuntur: Item Petri Belloni observationes eodem Carolo Clusio interprete* (Leiden, 1605), 122, 368.

[7] See F.W.T. Hunger, *Charles de l'Escluse (Carolus Clusius), Nederlandsch kruidkundige 1526-1609*, vol. I (The Hague, 1927), 268; and A. Pavord, *The Tulip* (London, 1999), 63. Thanks to John Robert Christianson who drew my attention to the latter. Pavord's source is a lecture by Agnes Stork to

Ill. 12. Map of Scandinavia and Northern Europe from Abraham Ortelius, *Theatrum orbis terrarum* (Antwerp, 1570). (See also colour plate 17).

Henrik Høyer – A short biography

Høyer was born in Stralsund around 1565. He began his studies in medicine in Rostock in March 1586. After obtaining his licentiate, he moved to Bergen in 1593 in order to take up a post as physician, thereby becoming the first physician in Norwegian history to be known by name. Høyer stayed with the merchant Nicolaus de Freundt, who obtained an apothecary's licence in 1595

---

commemorate the 400th anniversary in 1994 of the cultivation of tulips in the Netherlands (personal communication from Pavord, 2003).

and who acted as Høyer's pharmacist once Høyer had been appointed as city physician in Bergen in 1599.⁸ Høyer was a gifted scientist with a broad range of interests. Already during his first year in Bergen he sent plants – in particular cloudberries (*Rubus chamaemorus*) – to Clusius, and birds from the Faeroe Islands to the Leiden anatomist Pieter Paaw.⁹ The cloudberries (Ill. 13) play a prominent role in these early exchanges between north and south. Høyer sent berries to Clusius already in 1593, then plants with leaves and flowers during the following year. In 1596 the two men discussed the taxonomy of the species, the ways to prepare cloudberries and their medicinal effects against scurvy. Finally, in August 1597, Høyer sent some cloudberry jam to Clusius.¹⁰

In Bergen Høyer met the learned and renowned historian and collector of books Anders Foss (1543-1607), who was born in Denmark. He was Lutheran bishop in Bergen from 1583 to 1607, wrote a critique of Saxo Grammaticus' history of Denmark, and was a good friend of the famous astronomer Tycho Brahe (1546-1601), whom he often visited. Foss' German wife, Marine Ruppertsdatter, was accused of witchcraft by the head of the county constabulary Peder Thott in 1590, but she was finally found not guilty in 1598. Bishop Foss was, in fact, known as a fighting man, and the couple often had to go to the capital, Copenhagen, to defend themselves against accusations and in connection with other legal disputes.¹¹ Their youngest daughter would become Høyer's first wife. In the spring of 1596 Høyer set off on his journey to Leiden, together with bishop Foss, his wife and daughter. They travelled via the island of Ven, where they visited Tycho Brahe at Uraniborg on 12 May. In connection with this visit Høyer even wrote an ode to Tycho Brahe.¹² Høyer obtained his doctor's degree in Leiden

---

[8] The biographical information concerning Høyer is taken from Eckblad, 'Henrik Høyer og de første tulipaner i Norge', and its references, i.e. A. Holtsmark, 'Henrik Høyer', *Norsk biografisk leksikon*, vol. VI (1934), 444-445.

[9] He is also known as Petrus Pavius (1564-1617).

[10] Lundquist, *Lilium martagon* L., nn. 116, 268. For a summary of Høyer's shipments, see P.M. Jørgensen, 'Byen er Bergen – faget er botanikk', *Bergens Museums Årbok* (2003), 37-39; and P.M. Jørgensen, 'Tulipanen og Bergen i 400 år', *Bergens Tidende*, 28 May, 1997.

[11] Bishop Foss' predecessor Geble Pedersen had been involved in a case of witchcraft too, but as a victim: Anne Pedersdotter was accused of setting illness on him in order to have her own husband made bishop instead. She was burned as a witch at Nornes in Bergen in 1590, the same year that Marine was charged. Anders Foss and Marine were married around 1569 and had at least four children. Concerning Anders Foss, see also A.C. Bang, *Den norske kirkes historie i Reformations-aarhundredet* (Oslo, 1895), 63-85.

[12] Concerning Tycho Brahe, Uraniborg and the contemporary activities on Ven, see J.R. Christianson, *On Tycho's island. Tycho Brahe, science, and culture in the sixteenth century* (Cambridge, 2003). According to Christianson, Brahe may have asked Henrik Høyer to write this poem when the latter was staying on

Ill. 13. Ripe cloudberries (*Rubus chamaemorus*), ready to pick, growing on boggy ground in the mountains of the province of Jämtland, Sweden, close to the Norwegian border. (See also colour plate 5).

in the course of the same year and returned to Bergen, where he married the bishop's daughter.[13]

During this short stay in Leiden Høyer became close friends with Clusius. When departing from Leiden to return to Norway, Høyer received many bulbs and tubers. Apparently, Clusius was interested in whether they could prosper in the harsher climate of the north. Høyer describes his efforts as follows:[14]

---

the island as a guest (personal communication, 2003). The poem was included in the publication of Brahe's astronomical correspondence by Uraniborg's printing office in 1596. The date of the Foss and Høyer visit is mentioned in Brahe's meteorological diary. Cf. T. Brahe, *Opera omnia*, ed. I.L.E. Dreyer, vol. IX (Copenhagen, 1913-29), 139.

[13] His second wife, Thyri Anfinnsdatter Soop (1550/1555-1628/33), was ten years older than Høyer and outlived him by fifteen years. See T. Nygaard's genealogical website http://hem.bredband.net/nygtor/troms/3419.htm (2005) and P. Nermo's 'Nordic genealogy site' (Norge) website http://www.nermo.org (2005).

[14] 'Quandoquidem vero exotica haec tua caelo nostro rigidiore assuefiunt, ad plura tentanda in posterum invitor, ac propterea maiori studio atque animi alacritate exoticarum culturam aggrediar: et si Deus me voluerit diutius salutem, et superstitem, adhibebo curam, ut Belgici atque Italiae delicias, quantum ejus fieri potest, in Norwegiam transferam, ut omnino sciri possit genus huius terrae.' Høyer to Clusius, 9 April 1597.

All the while your exotic plants become accustomed to our colder climate, I feel the urge to try out more of them. I shall therefore with great zeal and attention take to the cultivation of exotic plants and, God sparing my life and health, I shall endeavour to bring loveliness here from Belgium and Italy, so that the characteristics of our soils be known.

Høyer was very grateful for these gifts. 'My Honourable and Most Excellent man, My Revered Friend', his first letter to Clusius begins, 'I owe you undying gratitude for the outstanding benevolence you bestowed on me when I was in Leiden last year; for your services and your merits and, above all, for the precious bulbs'.[15]

These bulbs were probably planted in De Freundt's medicinal garden in Bergen and in the bishop's garden. The latter had been established before 1557 by Foss' predecessor Geble (Gjeble) Pederssøn (1490-1556/57), the first Lutheran bishop of Norway and the man who should perhaps be regarded as the first real botanist of Norway. In his *Oration about master Geble* (1571) Pederssøn's student and adopted son Absalon Pederssøn Beyer (1528-1575) describes a pleasant and extensive garden, a Flemish gardener called Adrian, and the bishop's herbarium.[16] Especially the latter is remarkable, as J. Ingar I. Båtvik has already pointed out in 2000: 'Thus, there are good grounds to believe that the Norwegian Gjeble Pederssøn had a herbarium as old as the person who is credited with founding the art of making herbaria, Luca Ghini.'[17]

Thus, the Bergen to which Henrik Høyer moved from Germany was well prepared for him. In terms of the medicinal use of plants, modern botanical interests, gardening practices and the artistic potential of gardens, it helps to explain Høyer's threefold interest in plants: pharmacy, botany and horticulture.

---

[15] 'Nobilissime et Praestantissime Vir, amice Venerande, immortales tibi debeo gratias pro singulari benevolentia tua, qua me complexus es superior anno Leidae cum essem: tum pro officiis atque bene meritis: inprimis etiam pro Bulbis preciosis'. Høyer to Clusius, 9 April 1597.

[16] See Jørgenson, 'Byen er Bergen', 37. Cf. C.A. Lange, *De norske klostres historie i middelalderen* (Oslo, 1847), 256: 'During the time of Bishop Geble, when the monastery and the church were transformed to a Bishopric and Cathedral, in this garden, apples, pears, cherries and grapes were grown by the gardener Adrian from Flanders, later even sweet chestnuts, coriander, figs and laurel. The garden is now almost gone' (my translation). Cf. K. Fægri, 'Klostervesenets bidrag til Norges flora og vegetasjon', in *Foreningen til norske fortidsminnesmerkers bevaring, Årbok* (1987), 225-238, here 230, disagrees with Lange about the continuity of the monastery gardens in Bergen.

[17] My translation. For quotations from the oration about Master Geble and further references, see J.I.I. Båtvik, 'Gamle bevarte herbarier, og Østfolds eldste herbariebelegg', *Natur i Østfold*, 19 (2000), 17-27, esp. 18.

The Letters[18]

Letter I

The first letter from Høyer to Clusius is dated 9 April 1597 (JC), Bergen.[19] After reverently and profusely expressing his gratitude, Høyer elaborates lyrically and in great detail on the state of affairs in cultivation. He has deferred writing until he really had something substantial to report.

This autumn I planted all [bulbs] but the *Tychonian,* in three places[20], where most of them wintered well, many without any tending at all. The tulips showed around the 14 or 15 of February, if I remember correctly. There was a good average temperature then and in March came the crown imperials and narcissi. At the beginning of April came the crocuses. Presently they are impaired considerably by chill nights and northerly winds: when these come to an end, I am convinced they will bloom as desired. I shall report to you as soon as this comes about.[21]

The passage about the *Tychonian* bulbs is noteworthy in the sense that it can be interpreted as evidence that antedates the introduction of the tulip into Denmark by several decades. Before this evidence was known, the earliest written and generally accepted sources for tulips in Denmark dated from the 1630s.[22] My interpretation is that Høyer was given these 'Tychonian' bulbs in Leiden during the autumn of 1596 as a present for Tycho Brahe. Høyer must have taken them back with him to Bergen with the intention of later sending them on via Bishop Foss to Tycho Brahe on Ven. He probably never got the chance to do this, as Tycho Brahe left Ven for good in March 1597 (Ill. 14).[23]

Høyer not only acquired new plant material from Clusius but also from Paaw, as he writes in the same letter. 'The splendid doctor Pavius has given me

---

[18] For a more extensive presentation, see Lundquist, *Lilium martagon* L.
[19] JC stands for Julian Calendar. The Roman-Catholic part of the world introduced the Gregorian calendar in 1582; in Sweden it was introduced in 1753. Høyer exclusively uses the Julian calendar, Clusius the Gregorian one, but most often Clusius uses both to avoid confusion.
[20] The sentence may be interpreted as meaning in three different places in de Freundt's garden, or in three different gardens. According to the Bergen botanist Per Magnus Jørgensen (personal communication 2004) it refers to three different gardens: De Freundt's, bishop Foss', and another one.
[21] 'Posueram vero superiori autumno omnes praeter *Tychonianos* tribus in locis, ubi et plerique feliciter perennarunt, nonnulli sine omni custodia. Tulipae apparucrunt circa mensis Februarii diem 14 aut 15 ni fallor. Erat autem aura tunc temperata mediocriter. Martio corona Imperialis et Narcissi. Initio Aprilis crocus. Nunc vero nocturno frigore et flatibus borealibus non parum praepediuntur: quibus cessantibus deinceps non dubito habituros flores ex voto. Quod mox ubi fiet tibi significabo.' Høyer to Clusius, 9 April 1597.
[22] See also note 5 above.
[23] For further reading, see Lundquist, *Lilium martagon* L., 272-273.

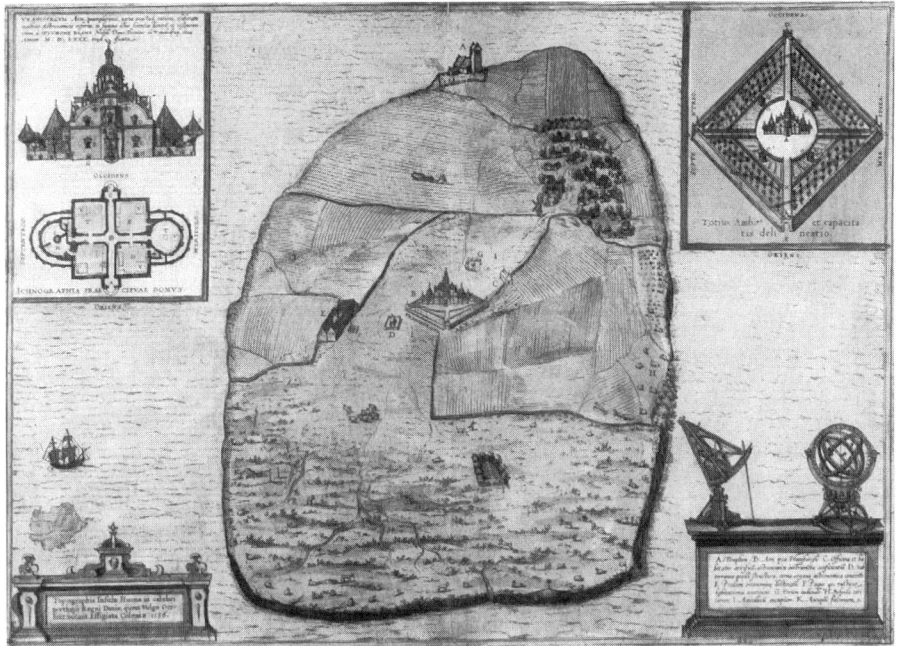

Ill. 14. The island of Ven with Uraniborg castle. Tycho Brahe's second map of the island. From G. Braun and F. Hogenberg, *Civitates orbis terrarum*, part IV (Cologne, 1586).

some garden plants, a couple of which have put forth new shoots'. He also acquired seed from Amsterdam:

Almost all seed died, to my great distress. Petum mas, which was sent from a pharmacy in Amsterdam last autumn, died from the winter's attack. I wish that you [Clusius], if good and fresh seed is available, would send me some, to see if it would sprout with me.[24]

The letter concludes with some remarks about the cultivation of the new plants, the necessary gardening skills, and the gap between what could and what *should* be found in contemporary handbooks.

---

[24] 'Donaverat me optimus vir D. D. Pawius hortensibus quibusdam plantis, quarum pares repullulascit. Semina vero perdita pene omnia. Magno meo cum dolore. Petum mas a pharmacopaeo Amstelodamensi mihi in autumno missum, hyemis iniuria interiit. Velim si bonum ac recens semen haberi apud vos possit, nonnihil mihi mitti, si forte haec possit provenire.' Høyer to Clusius, 9 April 1597.

I wish someone would emerge who could write in a professional way about simple cultivation: I do possess such books, but they are but brief and contain nothing out of the ordinary, nothing which is not known to every kitchen gardener.[25]

Høyer was clearly looking for advice about the cultivation of his exclusive new introductions:

I want something complete, something that would first introduce gardening in general and then deal specifically with the cultivation by itself, of each species or form. I know of no one in our times who could perform such a piece better than you [Clusius], you who have spent not only years, but an age, on these studies, so that among all who are devoted to botany, even with Momus[26] himself as a judge, you are attributed the foremost.[27]

It would take almost a century, until 1694, before Norway had its first gardening book: *Horticultura* by Christian Gartner.[28]

Clusius has listed some names of plants on the envelopes of both the first and second letter from Høyer. He made a note of the plants, mainly bulbs, that he intended to send to Høyer in a new shipment in the autumn (Ill. 15).[29]

Letter II

As early as 12 May 1597 (JC) Høyer wrote again to Clusius to tell him about the first flowers in the garden and, more generally, in Bergen and Norway: tulips, crown imperials and hyacinths. Høyer missed his narcissi, though.

All tulips, which I have planted in different locations, are flowering with the exception of a few […] All flowers but two are in different shades of red; one is yellow, the other white. The Crown Imperial (of which I only have one) shines amidst the tulips. Both the oriental hyacinths have almost finished blooming. I have no narcissi but the

---

[25] 'De simpliciori cultura velim aliquem in palaestram descendere, qui ex professo scribat: Extant enim mihi tales libelli, sed schediastici, praeter vulgaria, et nullis non olitoribus cognita, nihil continentes.' Høyer to Clusius, 9 April 1597.

[26] Momos (or Roman Momus): in Greek mythology the god of mockery, faultfinding, scorn and (un)fair criticism, son of the Night. He is also the patron of writers and poets. The Hellenistic poet Kallimachos used Momos as a name for his critics.

[27] 'Ego vero absolutum aliquid desidero: quod hortorum curam omnem in genere primum tradat: deinde in specie omnium simplicium cultum per singulas species pro cuiusque genio deductum. Sed id nemo quod sciam hoc seculo rectius posset, quam tu, qui non annos, sed aevum in isto studio trivisti, ut propterea palmam, vel ipso Momo judice, tibi concedant [?] necesse sit, quicunque Botanicam sectantur.' Høyer to Clusius, 9 April 1597.

[28] See G. Balvoll and G. Weisaeth, *Horticultura. Norsk hagebok frå 1694 av Christian Gartner* (Landbruksforlaget, 1994).

[29] The discrepancies between Clusius' lists and Høyer's documentation of the plants he cultivated are beyond the scope of this essay. See Lundquist, *Lilium martagon* L., 276-277.

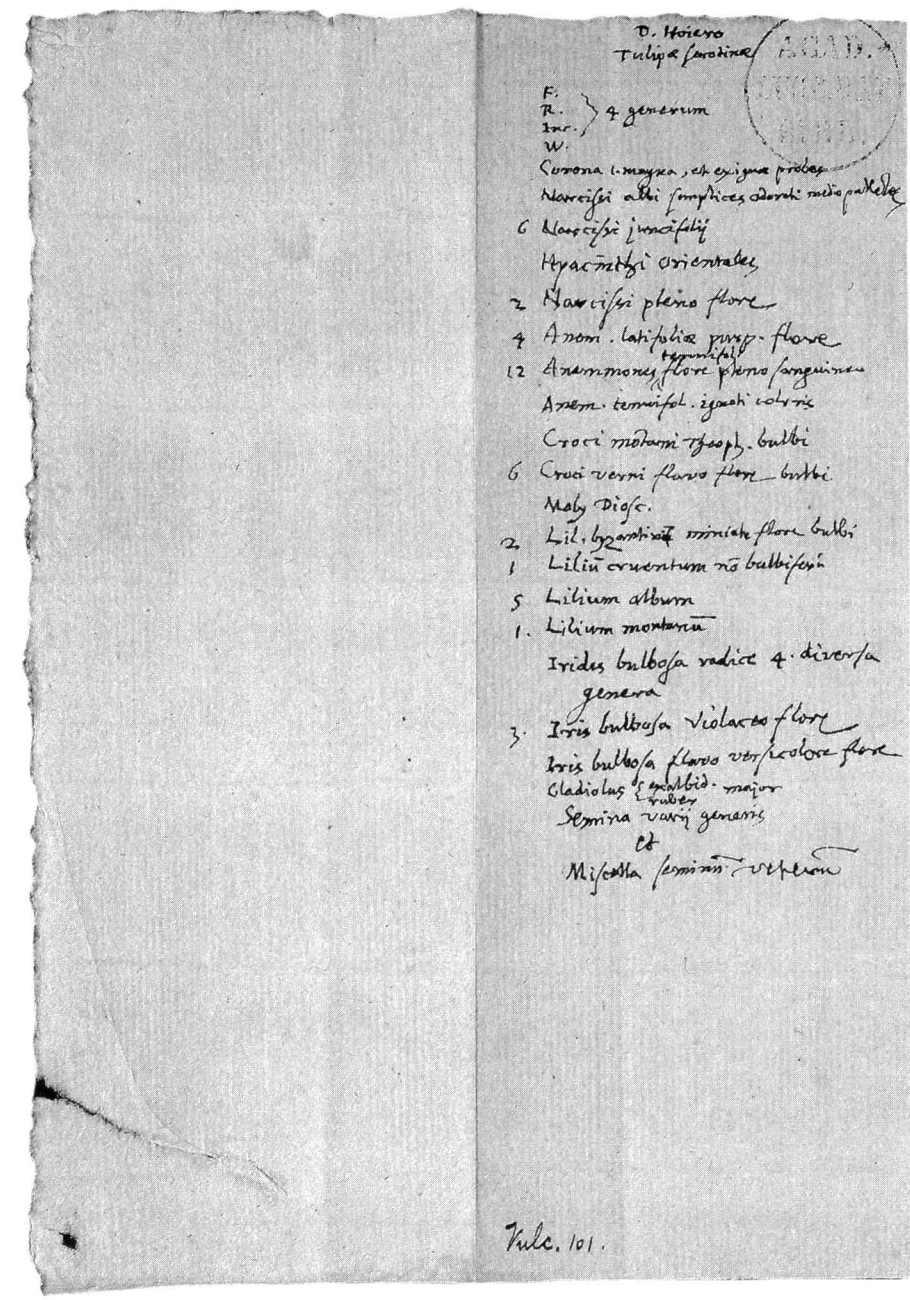

Ill. 15. List of plants – bulbs, tubers and seed – which Clusius already had sent or intended to send to Henrik Høyer, written by Clusius on the envelope to the first letter from Høyer of 9 April 1597, which Clusius received in Leiden on 10/18 May 1597 and answered on 22 May.

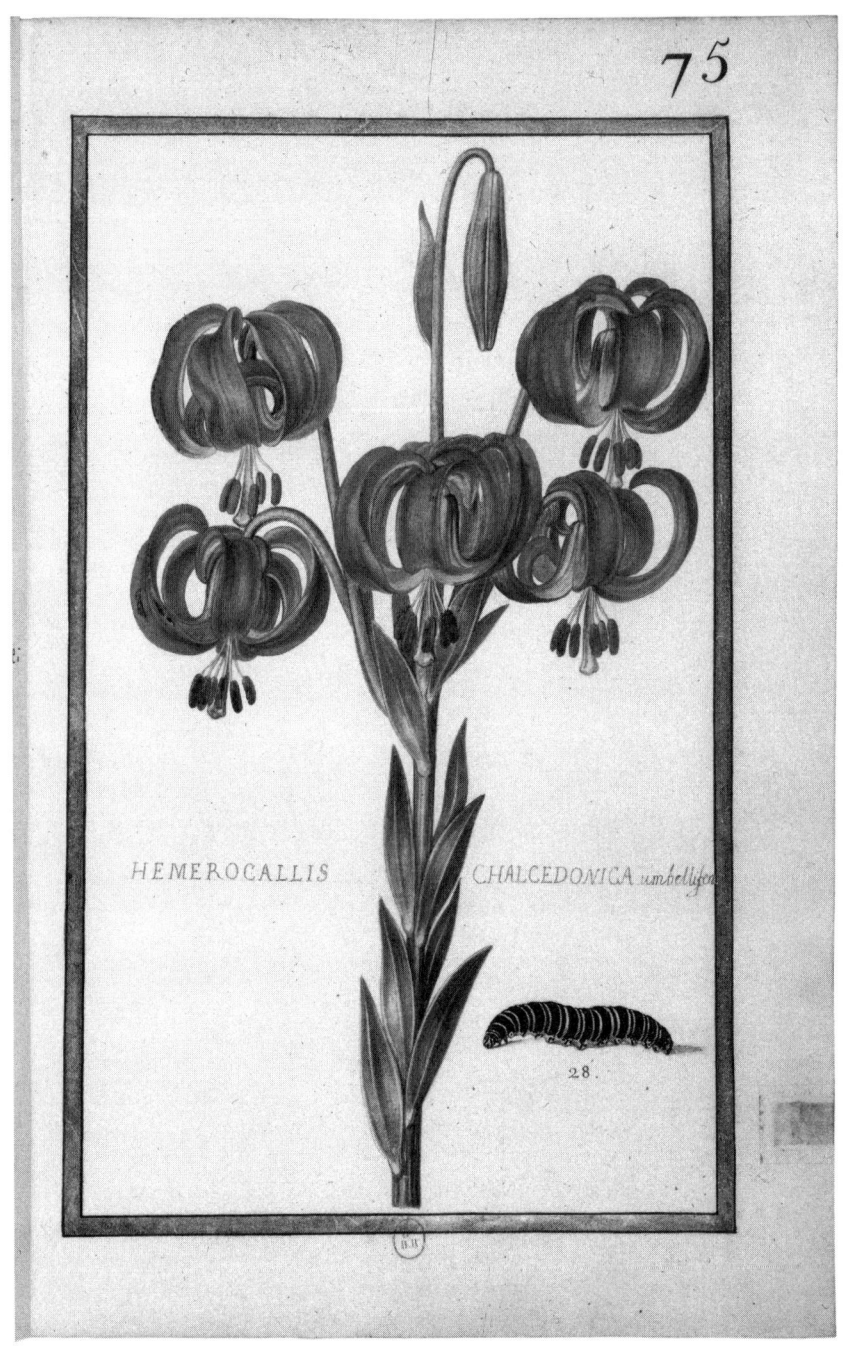

Ill. 16. 'Gufiniare/Zufiniare/Zufiniaris' = Scarlet Martagon (*Lilium chalcedonicum*), here called 'HEMEROCALLIS CHALCEDONICA umbellifera', on a contemporary painting by Daniel Rabel, the original of the engraved florilegium *Theatrum florae* (1622). (See also colour plate 4).

only crocus shows at least some leaves. The anemone showed up two days ago. My only Gufiniare³⁰ showed good form.³¹ (Ill. 16).

Høyer expressed both gratitude and a willingness to reciprocate:

I wish there were something in our country by which we could convey our appreciation: I doubt that you can be as eager to receive what we would be eager to send; but if such a thing exists: You need but to ask, nay command!³²

The letter also contains a suggestion that plant exchange with the Netherlands may also have occurred in Sweden, Denmark and Germany. The sentence is not easy to interpret, however:

I grant you – all that is mine is yours by which we Norwegians not only have elevated ourselves, by no means insignificantly, but also have triumphed over Sweden, Denmark and even some coastal areas of Germany. It is thy good deeds, my Clusius, that have made fine folk out of peasants, people out of barbarians, in a manner of speaking.³³

Letter III

[…] in particular I cherish your [Clusius'] bulbs. In the same way as Bosphorus outshines other stars, these bulbs surpass other bulbs by the variations, colours, luminance and elegance of their flowers.³⁴

---

³⁰ 'Gufiniare' refers to a Scarlet Martagon Lily (*Lilium chalcedonicum*). It must be Høyer's attempt to write 'Zufiniare'/ 'Zuphiniare', one of Clusius' names for *Lilium chalcedonicum*, which he uses in his Hungarian flora of 1583 according to Caspar Bauhin in *Pinax theatri botanici* (Basle, 1623). The Scarlet Martagon Lily seems to have been introduced in the Netherlands in the middle of the 1570s. See further Lundqvist, *Lilium martagon* L., n. 132, pp. 275-277 and 180-196.
³¹ 'Tulipae omnes, quas diversis in locis posueram omnes florent pauculis exceptis serotinis. Flores omnes varia rubedine ludunt. Duobus saltem individuis exceptis, quorum illud luteum, hoc vero candidum habet florem. Corona item imperialis (quae mihi unica) in medio Tuliparum superbit. Hyacinthi orientales bini pene defloruerunt. Narcissus verus mihi nullus. Ex radicibus croci unica superstes folia saltem ostentat. Anemone item ante biduum prodiit. Gufiniare unicum etiam saluum.' Høyer to Clusius, 12 May 1597.
³² 'Utinam vero apud nos vicissim aliquid sit, quo possimus gratitudinem tibi nostram declarare: profecto non tam cupide expetiturus esses, quam nos facile missuri. Itaque, si quid est mone, aut potius jube.' Høyer to Clusius, 12 May 1597.
³³ 'Habes omnia mea – tua quibus non mediocriter nos non tantum efferimus Norwegiani, sed sed et Sueciae, Daniae, atque Saxoniae littoraliori quodammodo insultamus: tuo mi Clusi beneficio, qui ex agrestibus iam cultos ex barbaris humanos (ut ita loquar) reddidisti.' Høyer to Clusius, 12 May 1597.
³⁴ 'Imprimis vero magno in precio apud me sunt eruntque bulbi isti tui; Nam non secus ut bosphorus reliqua astra luminis sui splendore vincit: ita hi bulbi floris varietate, colore, nitore, elegantia, […].' Høyer to Clusius, 17 August 1597.

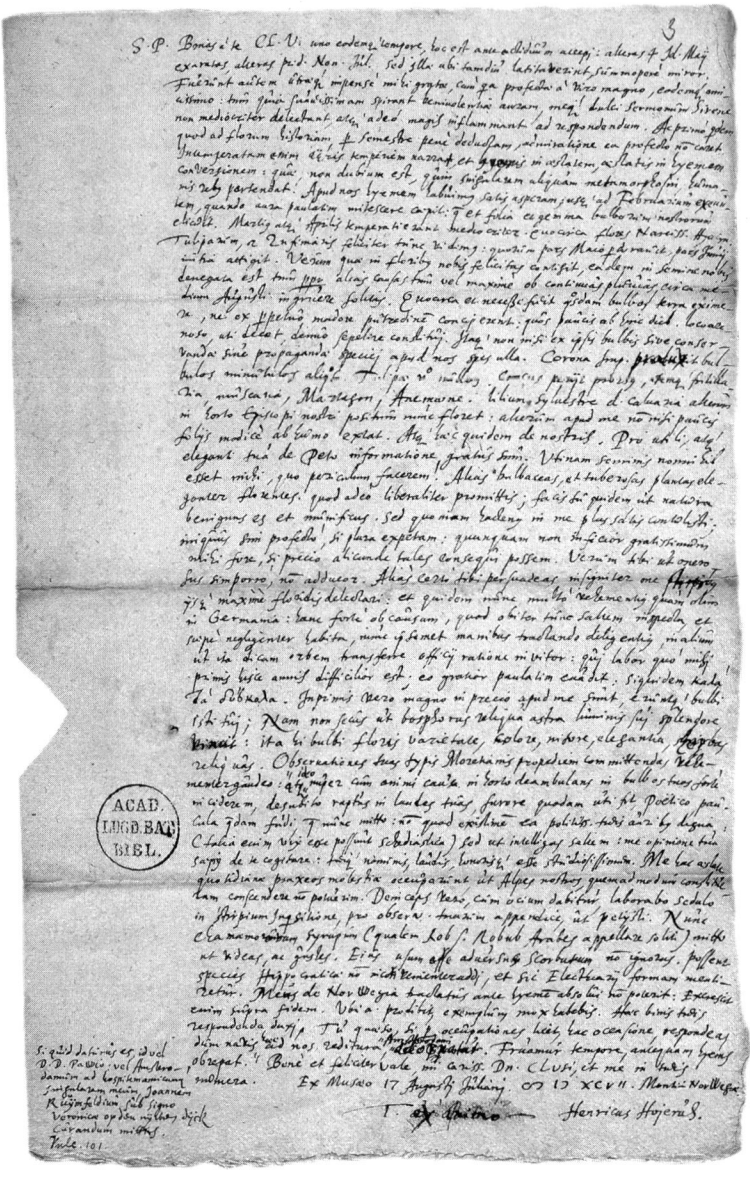

Ill. 17. 'Only the crown imperial had produced new bulbs, none of the tulips. The losses of bulbs and tubers were also great. The crocuses perished completely, as did the Fritillaries [*Fritillaria meleagris*], grape hyacinths, 'Martagon' [*Lilium bulbiferum/ L. bulbiferum* var. *croceum*] and the anemones. One Lilium sylvestre, so-called Calvaria [*Lilium martagon*], planted in our bishop's garden is blooming now, another: [Turk's-cap lily], with me, sprouts only a few leaves, just above ground.' Extract from the third letter from Høyer to Clusius, 17 August 1597.

In his third letter, dated 17 August 1597 (JC), Høyer thanks Clusius for his two replies of August 9 and summarizes the past half-year's flower account ('florum historiam'). The winter had been severe. March and April offered only average temperatures. Narcissi, hyacinths, tulips and the Scarlet martagon lily thus bloomed during the whole of May and well into June.

In terms of contents and subjects this letter is relatively disorganized. The results of bulb cultivation are mixed up with the outcome of sowing; instructions for cultivation mingle with memories of Germany, while abundant expressions of gratitude are interspersed with new wishes and requests. Persistent rains in the middle of August had made Høyer dig up some bulbs and replant them in sandy soil to prevent rotting. Only the Crown Imperial had produced new bulbs, but none of the tulips had done so. The losses of bulbs and tubers were great as well. The crocuses, fritillaries (*Fritillaria meleagris*?), grape hyacinths, 'Martagon' (*Lilium bulbiferum / L. bulbiferum* var. *croceum*) (Ill. 18) and anemones had perished completely.[35]

The interpretation of one particular sentence in this third letter – perhaps the most significant one of the whole Høyer-Clusiuscorrespondence with respect to the introduction of the Turk's-cap lily (*Lilium martagon* L.) into Sweden and Scandinavia – has raised some discussion:

Lilium sylvestre d.[ictum] calvaria alterum in horto Episcopi nostri positum nunc floret: alterum apud me non nisi paucis folijs modice ab humo extat.

'Lilium sylvestre' is the well-known and widespread name for the Turk's-cap lily, *Lilium martagon*. The name was used at the time by Rembert Dodoens and several German botanists, among others. Clusius himself used it too in his Hungarian flora. Somewhat inconsistently, Clusius uses a different pre-linnéan name for the species in the list on the envelope of the first letter, i.e. 'Lilium montanum'. To complicate matters, 'Lilium sylvestre alterum' is one of Dodoens' names for the Scarlet martagon lily (*Lilium chalcedonicum*), the Gufiniare mentioned above. My interpretation of this phrase reads:

One Lilium sylvestre, so-called Calvaria [Turk's-cap lily (*Lilium martagon*)], planted in our bishop's garden is blooming now, another [Turk's-cap lily], with me, sprouts only a few leaves, just above ground. (Ill. 17)

Høyer probably brought the Turk's-cap lily bulbs directly from Leiden in 1596 and actually tells Clusius about them. However, there is nothing to contradict that the species could have been brought to Norway even earlier. The

---

[35] 'Crocus perijt prorsus, itemque fritillaria, muscaria, Martagon, Anemone.' Høyer to Clusius, 17 August 1597. Concerning the interpretation of 'Martagon', see Lundquist, *Lilium martagon* L., 277, n. 140.

Turk's-cap lily is found today in great profusion in Gamlehagen outside Bergen, on the former church estate.[36] Here it has grown wild outside the former formal parterre, together with spring snowflakes and crocuses. These lilies may well be direct descendants of the wards of the bishop, via Høyer and

Ill. 18. Orange lily or Fire lily (*Lilium bulbiferum*) was one of the most commonly grown species in the Lily-genus in Norway late into the nineteenth century.
The species is hardy all over the country and can be grown in the mountain fringes. The variety *Lilium bulbiferum* var. *croceum* was listed by Clusius on the cover of the first letter ('Lilium cruentum non bulbiferum') to Høyer. The species also occurred under the confusing name 'Martagon' in the first letter by Høyer. (See also colour plate 6).

---

[36] Near the Fana School at Store Milde.

Clusius.[37] As Per Magnus Jørgensen argues, there are good grounds to assume that they stem from Clusius' deliveries, since Foss was Høyer's father-in-law and Høyer had no garden of his own. If so, they are a living cultural heritage as well as visible proof of the type of the late sixteenth-century Turk's-cap lily in Northern and Western Europe.[38] Høyer writes:

> Other bulbs and tubers flower beautifully, such you send me freely, because you are by nature benevolent and generous. As you have so far given me more than enough, it would be greedy of me to ask for more even if I would be happy to get some. I do not wish to trouble you, but you know I do like flowering plants best.[39]

Høyer's interest in gardening is obvious; it was not a doctor's interest, but much more a gardener's one. Bulbs had captured Høyer's heart.

> In Germany, I saw them in passing, often neglected, but now I am tempted to carefully and with my own hands, bring them to a new world; a task which, over time, has become the dearer to me […] in particular I cherish your bulbs.[40]

Høyer further explains that he is working on the 'plant survey' for Clusius' appendix which he had been asked to make. His burdensome practice in Bergen had prevented his going to the mountains and he explains that his treatise on Norway will not be ready before the winter. During this same August he also sent cloudberry jam ('chamaemorum syrupum') to the Netherlands: '(what the Arabs call *rob* or *robub*) so that you may see and taste it. You know of its use

---

[37] A. Dietze, '1600-talls kjøkkenhagetradisjon på Baroniet Rosendal, Kvinnherad, Norge', in D. Moe, P.H. Salvesen and D.O. Øvstedal (eds.), *Historiske hager. En nordisk hagehistorisk artikkelsamling ved 100-årsfeiringen av Muséhagen i Bergen maj 1999* (Bergen, 2000), 40-45. She reports several kinds of lilies from the Rosendal Barony, 100 km south of Bergen, in plant lists from 1666 and 1667. The possible connection between Høyer's Turk's cap lilies in Bergen and the bulbs at Rosendal remains to be investigated.

[38] Personal communication from Jørgensen (2004). See also P.H. Salvesen, 'Levende kulturminner i Gamlehagen på Store Milde. Rosene', *Årringen* (2002), 4-12.

[39] 'Alias bulbaceas, et tuberosas plantas eleganter florentes, quod adeo liberaliter promittis; facis tu quidem ut natura benignus es et munificus. Sed quoniam hactenus in me plus satis contulisti iniquus sim profecto, si plura expetam: quunquam non inficior gratissimum mihi fore, si precio alicunde tales consequi possem. Verum tibi ut onerosus sim porro, non adducor. Alias certo tibi persuadeas insigniter me stirpiisque maxime floridis delectari'. Høyer to Clusius, 17 August 1597.

[40] '[…] et quidem nunc multo vehementius quam olim in Germania: hanc forte ob causam, quod obiter tunc saltem inspecta, et saepe negligenter habita, nunc ipsemet manibus tractando diligentius in alium ut ita dicam orbem transferre officii ratione invitor: qui labor quo mihi primis hisce annis difficilior est […] Inprimis vero magno in precio apud me sunt eruntque bulbi isti tui.' Høyer to Clusius, 17 August 1597.

against scurvy.'⁴¹ He added jokingly or not, 'I could add a suitable Hippocratic example and make it an electuary'⁴², concluding with the following:

This is my answer to both your letters. I beg you to reply, if you can spare the time, when this ship calls at Amsterdam on its way back. Let us make use of the time before winter closes on us.⁴³

Letter IV

Høyer's plea was answered. In his fourth letter from Bergen, dated 16/26 October 1597, Høyer confirms the receipt on 4/14 August and 10 September of two shipments of bulbs and seeds from Leiden.

Many greetings to you, my most illustrious and honourable friend. The [letters] you sent me, in August and September, have been received in good order together with the bulbs and seeds of different kinds; for these I declare myself to be greatly indebted to you. Now, I only wish I could give you something in return, appreciating, of course, which I am sure you understand, your good deed. My resources are currently limited, however. In the future, God willing, my situation may improve and my chances of reciprocating be greater. All bulbs have been planted and some of the seeds as you ordered. I will not hide from you that I have not found the 'xyphium' [Spanish iris/Dutch iris) (*Iris xiphium*)] mentioned on the list. May God grant us our delight, next Spring, in all the bulbs, well kept and blossoming; a greater joy cannot be had in this pursuit.⁴⁴

---

⁴¹ 'Nunc chamaemororum syrupum (qualem Rob sive Robub Arabes apellare soliti) mitto ut videas, ac gustes. Eius usum esse adversus scorbutum non ignoras.' Høyer to Clusius, 17 August 1597.

⁴² 'Possem species Hippocratica non inconvenienter addi, et sic Electuarii formam mentiretur'. A medicinal substance mixed with honey or syrup.

⁴³ 'Haec binis tuis respondenda duxi. Tu quaeso, si per occupationes licet, hac occasione, respondeas dum navis haec ad nos reditura. Amsterodami adoriatur [?]. Fruamur tempore, antequam hyems obrepat.' Høyer to Clusius, 17 August 1597. Cf. Lundquist, *Lilium martagon* L., nn. 144, 280. See also F.E. Eckblad, 'Molter som skjørbuksmiddel i skriftlige kilder', *Blyttia*, 4 (1988), 177-178, about the 'cloudberry-exchange' between Høyer and Clusius, and the preparation of cloudberry jam. Eckblad does not, however, mention any shipping of actual jam from Norway.

⁴⁴ 'Salutem Plurimam Clarissime Vir, amice honorande. Quas tu mense Augusto primum, deinde Septembri ad me dedisti: illas ego heic rectissime accepi una cum adiunctis bulbis, et seminibus variorum generum: pro quibus maximas me tibi gratias debere lubens profiteor. Utinam facultas mihi nunc esset aliquid rependendi reipsa profecto experiaris, accipere me beneficia nosse, et aestimare. Nunc autem curta mihi supellex. Sed in posterum Deo volente maior commoditas erit, forte et facultas. Bulbos omnes terrae commisi: quaedam et semina quemadmodum jussisti. Hoc te celare nolo, xyphium cuius index meminit, mihi non fuisse visum. Faxit Deus, ut bulbos omnes, nostras delicias futuro vere, salvos et florentes videamus: ita maior nobis felicitas heic hocque [?], in isto quidem studio contingere non poterit.' Høyer to Clusius, 16/26 October 1597.

According to Eckblad 'we may safely assume that Høyer was able to plant these new bulbs and tubers and that most of them bloomed the next year, i.e. in 1598'.[45] Unfortunately there are no more letters about the handling of the plants to confirm this.

Høyer's October letter reached Clusius on 18/28 November. It was delivered by a Norwegian student who travelled to the Netherlands to continue his studies at the academy in Leiden.

> The person carrying this letter is a student, born here, in a respectable house: he is to spend some time at your academy for his studies. He has kinsmen in royal service governing the provinces Upper Nordland and Finnmark. From these kinsmen, he will without problems be able to acquire whatever you may wish to have from the people of the North. Ask freely for whatever you see fit to ask for when you see him: I have instructed him that if given to understand that he could do you a service, he should make sure to diligently do it.[46]

A certain sense of rank in the relationship between Høyer and the student is noticeable: 'I do not doubt he will follow my admonitions', Høyer concludes. Høyer positions himself in a similar way in relation to Clusius: 'I too, will endeavour to fulfil my duties.'[47]

Clusius may have been surprised by the presents accompanying the letter: two reindeer skins, a pair of Finnish fur boots and some 'false rubies'. The student carried these together with some pressed plants or herbarium sheets.

> This student brings you a reindeer skin (some call the animal 'tarandum'), and some 'false rubies' with a list [?], in some strange way formed by nature to cubes; they have, as you can see, emerged in the rock-strewn, stony grounds around Nidaros/Trondheim. He also brings a pair of Finnish fur boots and other things from me, which he will show you […] I ask you, please, to accept these things until I can find something better.[48]

---

[45] Eckblad, 'Henrik Høyer og de første tulipaner i Norge', 149-150.

[46] 'Qui hasce exhibet, studiosus est honesto apud nos loco natus: studiorum gratia aliquamdiu in vestra Academia commoraturus. Affines habet, qui in ulteriori Norlandia, et Finmarckia multos iam annos Praefectos agunt Regios. Ex illis curare hic tibi poterit, sine negocio, quicquid rerum septentrionalium volueris. Itaque coram cum ipso conferens quicquid visum fuerit, audacter petere debes: ita enim instruxi, ut si qua in re gratificari se tibi posse intelligat, in eo obnixe studeat.' Høyer to Clusius, 16/26 October 1597.

[47] 'Nec dubito quin meis praeceptis sit pariturus. Ego quoque si quid potero, meo non defuturus sum officio.' Høyer to Clusius, 16/26 October 1597.

[48] 'Habet idem hic studiosus pro te pellem Rangiferae (alii Tarandum vocant, nostri Reensdiur) et nonnullos Rubinos spurios cum matricula in cubos mirabiliter a natura elaboratos: qui circa Nidrosiam in scopulo sabuloso uti vides, nati. Habet item calceos Finnicos hirsutos et alia quae tibi meo nomine exhibiturus est. Te rogo, ut hilari vultu accipias, usque dum meliora persolvam.' Høyer to Clusius, 16/26 October 1597. Clusius made a note of the reception of the hides on the envelope: 'Leyden 18/28 Novembris cum 2 pellibus Rangiferae […] fin.'

Plans for collaboration, specifically concerning the exploration of the Norwegian plant kingdom, and Clusius' aspirations for Høyer's excursions, are expressed in a few sentences in this fourth letter, which also reveals the existence of an early draft of a Norwegian flora.[49]

Your suggestion on plant observations is wise, including the specifics of each country. But our Norway, which is relatively barren and sometimes daunting in terms of accessibility, is not easy to explore. One who has only partly and superficially investigated the countryside, would still be considered to have performed a Herculean quest [...].[50]

Then Høyer apologises.

For myself, I have let myself be prohibited by my office from venturing forth. Later, however, God granting me continued right to the use of life, I will have more opportunities to go to the northern and upland regions on private business [...] Then, at the same time, I will be able to investigate the plants in these areas.
If I then find something special, I will send you either the plants themselves in paper, seed, or even bulbs if I can find any. Now [...] I have completed a brief account of nearly all plants common in our country – I have also entrusted the student with some [pressed] specimens folded in paper. You will have to put up with these little things as a token of my wish to prove to you my concern for botany [?].[51]

No herbarium, 'catalogue' or other draft of a Norwegian flora by Høyer has, however, been found.

The greater part of this long, fourth letter deals with the sea and its plants, primarily seaweeds. Høyer's speculations on 'Erica marina'[52] are fantastic and

---

[49] Concerning the issue of the oldest herbarium and herbarium plants in Norway, see Båtvik, 'Gamle bevarte herbarier', and the publications of Eckblad cited in this article. In my view, however, Høyer's efforts and work have not been thoroughly examined in either the Norwegian herbarium tradition or the exploration of the Norwegian flora.

[50] 'De observationibus plantarum recte suades, cum unaquaeque terra peculiare quid obtineat. Sed Norwegia nostra, quae vastuosa admodum est, et inviis accessibus passim formidabilis: non ita facile perlustrari potest. Imo Herculeum laborem praestiterit, qui vel aliquam eius portionem mediocriter perspexerit.' Høyer to Clusius, 16/26 October 1597.

[51] 'Mihi quidem hactenus ad interiora penetrari, officii ratione non licuit. Deinceps vero, si vitae usuram Dominus Deus concesserit, saepius mihi propter negocia privata, in partes septentrionaliores et montana proficiscendum erit. Atque eodem saltu in plantas simul inquirere potero. Tum si quas peculiares invenero, vel ipsas inter chartas repositas, vel ipsarum semina, vel etiam bulbos si forte aliquos offerentur, ad te mittam. Nunc [...] brevi catalogo complexus sum stirpes pene omnes, quae apud nos vulgo nascuntur: plantas etiam quasdam chartis insertas, huic studioso commisi. Tuum erit singula boni consulere, atque certo sic statuere nihil mihi magis volupe futurum unquam, quam meam in Botanico studio diligentiam tibi probare.' Høyer to Clusius, 16/26 October 1597.

[52] It is possible that one of the herbarium sheets consisted of the 'Frutex marinus Ericae facie' which Clusius published in *Exoticorum libri decem* (1605), 122. This is suggested by the text: 'The 'sea

would merit a separate treatise. Høyer writes lyrically about the fruits of the sea and its creative force:

This our amazing sea produces so many wonders, who would without marvelling behold so many, and such different things, created as if in play?[53]

In a *post scriptum* a request for yet another plant recurs:

If you could spare me one or two cyclamen tubers and send them together with what else you present my garden with, you will be doing me a fervently desired service.[54]

Clusius replied the same day and also sent a little box with four of the requested cyclamen tubers (*Cyclamen purpurascens* or possibly *Cyclamen hederifolium*) to Høyer.[55]

LETTER V

The next (preserved) letter by Høyer dates from almost six years later (28 June 1603). His handwriting has degenerated. Half of the first page (of two) is taken up by apologies for not having written for so long and promises to do better in future. The letter then goes on to deal mainly with birds. Høyer undertakes to give Clusius more information about the birds believed to be hatched in the trees on the Shetland and Orkney Islands. Notably, Høyer had asked a native Shetlander to spend the winter with him in Bergen.

The end of the letter is sad reading. Like Clusius in Leiden, Høyer suffered thefts of bulbs, particularly of bulbs sent to him by Clusius. Høyer does not mince words in his anger:

---

shrub' I believe should be called 'Erica marina' sent to me AD 1597 by the very learned Henrik Høyer, physician in Bergen, Norway; a representation of this Erica is found in the following illustration.' ('Marinum hunc fruticem, quem non incommode Ericam marinam appellari posse arbitror, mittebat ad me anno à Christi nativitate millesimo quingentefimo nonagesimo septimo Doctissimus vir Henricus Hoierus, Bergensis in Nowagica Medicus; cujus Ericae effigies in sequente tabella expressa.')

[53] 'Hoc certe admirandum mare nostrum tam varia proferre miracula [...] quis illas tot tamque varias, veluti per ludum fabrefactas, sine admiratione contempletur?' Høyer to Clusius, 16/26 October 1597.

[54] 'Radix una atque altera cyclaminis si carere posses, atque una cum continenti mittere pro horto posses; rem mihi impense gratam praestares.' Høyer to Clusius, 16/26 October 1597.

[55] Clusius himself listed the plants on the envelope of the letter: 'Respondi eodem die et misi Pyxidulas [?] et misi 4 Cyclaminis radices.' *Cyclamen purpurascens* is known from manuscripts of the fifteenth, sixteenth and seventeenth centuries in Denmark. *C. hederifolium* was in Denmark first depicted in the seventeenth century, but it was already known and appreciated on the Continent at the time.

For the rest, I throw myself at your feet, you father of herbs and Flora's illustrious Captain and bemoan weeping the misfortune of my garden, the loss of tulips, crown imperials, fritillaries, anemones and other fine bulbs, which a dastardly villain recently by stealth [robbed me of], even all those that I owned by your benevolence and [had been] successfully [growing] for seven whole years.[56]

The final lines emphasize what bulbs really meant to Høyer.

I demand punishment and pray and declare that this scoundrel who would deprive me of these my darlings, more precious to me than gold, shall be struck by eternal suffering.[57]

Letter VI

The long sixth letter from Bergen, dated 22 March 1604, deals above all with Atlantic birds and bird life on the Faeroe Islands. It appears to have been used by Clusius as a rough draft for his report on these topics in *Exoticorum libri decem* (1605).[58] The letter contains no information about plants.

Letter VII

In the seventh and last letter, dated 9 August 1604, Høyer informs Clusius that he is sending some different naturalia to Paaw, among them reindeer cheese and a booby[59] from the Faeroe Islands. He also tells Clusius that he has agreed with the bishop of Nidaros (Trondheim) – whose bishopric borders on Sweden, Russia and Novgorod – that the bishop would inform him (Høyer) about any new and interesting discoveries in the field of natural sciences. Høyer further discusses the seaweed 'Phaseolum marinum' and Clusius' views on the cloudberries in Norway. Clusius maintained that the Norwegian cloudberries were closely related to the English variety, 'Chaemaemorus angliana' which he

---

[56] 'Ceterum ad genua tua Herbarum pater et Florae Praeses Clarissime procumbo ac querelam meam de calamitate Horti mej ob amissos Tulipas, Coronas, Fritillariam, Anemones nec non alios elegantissimos bulbos cum lacrimis fundo: quos perdidissimus nebulo furto mihi nuper omnes quoque tuo beneficio possedi, totoque sep[t]ennio non infeliciter magna [?].' Høyer to Clusius, 28 June 1603.

[57] 'Prout [?] vindictam exposco, et ut male, imo pessime, hominis scelesto sit perpetuum qui frustratum me voluit his deleciis meis, auro longe mihi carioribus, precor atque voveo.' Høyer to Clusius, 28 June 1603.

[58] Clusius, *Exoticorum libri decem*, 368. Høyer's letter reached Clusius on 24 April 1604, and Clusius answered already on Walpurgis night.

[59] A sea bird related to the gannet.

had received from the British botanist and physician Thomas Penny, whereas Høyer regarded them as identical with the Swedish ones and as belonging to one and the same species.[60]

Epilogue

Bishop Foss died in 1607. The company of his son-in-law, Høyer, had probably formed one of the brighter spots in his life.[61] Høyer inherited the bishop's large collection of books. He was also appointed to manage the estate left by

Ill. 19. Cloudberry (*Rubus chamaemorus*) in flower and fruit. Woodcut from Johannes Palmberg, *Serta florea Svecana, eller Swenske örtekrantz* (Strängnäs, 1684).

---

[60] Cf. Eckblad, 'Molter som skjørbuksmiddel i skriftlige kilder', 177-178, and Clusius' description of the cloudberry (based on Høyer's earlier shipments and discussions) in *Rariorum plantarum historia* (Antwerp, 1601), where Clusius regards the two varieties as two different species. Eckblad's essay includes a translation (into Norwegian) of Høyer's clear claim that the two 'species' are one and the same.
[61] Bang, *Den norske kirkes historie*, 684-685.

another collector, the Norwegian county governor Erik Rosenkrantz. In the latter's collection he found a series of old Norwegian letters and documents concerning farms and estates, which Høyer included in his own collection of manuscripts. Høyer, who also did historical research himself, thus gradually acquired a considerable collection of both manuscripts and books. In fact, according to his biographer Anne Holtsmark he primarily deserves to be remembered not as a physician, but as a historian and collector of manuscripts.[62] Høyer also entertained the idea of compiling an 'Atlas of Norway' and mentioned it in several of his letters to Clusius, but the project was constantly delayed because of the increasing amount of material. Høyer's Norwegian history was said to be nearly complete, but it was never published.

When Høyer himself died in 1615/16, the King ordered his official in Bergen, Urne, to have Høyer's collection catalogued and copied and all originals and handwritten copies sent to Copenhagen. There the entire collection was destroyed in the big fire in 1728. Thus, ironically, it is due to Høyer's great historical ambitions and diligence that all return letters from Clusius to Høyer have gone up in smoke: letters which would have confirmed the first and early blooming of the Turk's-cap lily in Norway, and must have contained much additional documentation concerning the relationship between the two natural scientists.[63]

To publish a facsimile edition of all seven letters from Henrik Høyer to Carolus Clusius, transcribed in Latin, with translations in English and Swedish, and annotation concerning both plant introductions and other aspects of natural history, would mean fulfilling a wish of Høyer for Clusius:

I beg you, my lord Clusius, by Apollo himself, to bring what you possess to maturity, so that all monuments of your intellect may come to light before age crushes them. You may rest assured of the gratitude of all good men and your name made immortal for posterity.[64]

Henrik Høyer 9 April 1597

---

[62] See Holtsmark, 'Henrik Høyer'. This is emphasized by Bang, *Den norske kirkes historie*. Høyer was 'a man of great talents and all-round interests, known as a professional historian' (my translation).

[63] R. Stien, 'Legitimis libri possessor – norske leger som boksamlere i eldre tid', *Axonet, Medlemsblad for norsk nevrologisk forening*, 1 (2002), 12-13.

[64] 'Quaeso te, mi domine Clusi per ipsum Apollinem, matura quantum in te est, ut omnia quae in promtu habes ingenii monumenta, prodeant in lucem, antequam senectus obrepat. Gratiam certe referes ab omnibus bonis, et a posteritate nomen immortale.'

Appendix: Høyer's plant introductions to Norway – A summary

Starting from Høyer's letters and Clusius' plant lists, Finn-Egil Eckblad has established the following datings for the introduction of the following garden plants in Norway.[1] The most accurate contemporary representations of the species can be found in Carolus Clusius, *Rariorum plantarum historia* (Antwerp, 1601), which has been used by Eckblad to illustrate the early tulips and narcissi in Bergen.

In 1597, the following plants bloomed in Bergen:
- hyacinths (*Hyacinthus orientalis*)
- anemones (*Anemone* spp.)
- Crown Imperial (*Fritillaria imperialis*)
- different kinds of tulips (*Tulipa gesneriana* / *Tulipa* spp.)

In 1598 bloomed, in addition:
- Pheasant's Eye Daffodil (*Narcissus poëticus*)
- Daffodil (*Narcissus pseudonarcissus*)
- Jonquil (*Narcissus jonquilla*)
- Crocus (*Crocus vernus*)
- Yellow Crocus (*Crocus x stellaris / aureus*)
- irises (*Iris bulbosus*) [interpreted as Spanish Iris / Dutch Iris (*Iris xiphium*)]
- anemones (*Anemone coronaria*, *Anemone* spp.)
- gladioli (*Gladiolus* sp.)
- Madonna Lily (*Lilium candidum*)
- other lilies (*Lilium* spp.).

On the basis of a further study of the Høyer-Clusius letters, we may now add the following plants to the 1597 list:[2]
- another species of narcissus (*Narcissus* sp. / spp.)
- another species of crocus (*Crocus* sp. / spp.)
- Scarlet Martagon Lily (*Lilium chalcedonicum*)
- Turk's-cap Lily (*Lilium martagon*)
- other bulbs and tubers, which 'bloom beautifully'.[3]

For 1598 we may add, with a slight reservation:
- *Allium subhirsutum* ('Moly Diosc.[orides]')
- Orange Lily or Fire Lily (*Lilium bulbiferum* / *L. bulbiferum* var. *croceum*), if it/they survived better in its/their second try
- cyclamen (*Cyclamen purpurascens* / *C. hederifolium*)

---

[1] Eckblad, 'Henrik Høyer og de første tulipaner i Norge', 150.
[2] The list is mainly compiled from the information in Høyer's letter, not from Clusius' 'delivery lists' on the envelopes.
[3] The bulbs and tubers which bloomed already in 1597 must have been brought directly from Leiden by Høyer himself, or have arrived by shipment later during the autumn of 1596. They must have been planted that same year.

Colour plates

Plate 1 (illustration 1 on p. 2). Portrait of Carolus Clusius, at the age of 59. Canvas, painted in Vienna in 1585 by an unknown artist, possibly Jacob de Monte. Universiteitsbibliotheek Leiden.

Plate 2 (cover). Drawing of a daffodil, accompanying a letter from Carolus Clusius in Leiden to Matteo Caccini in Florence, dated 10 October 1608. Universiteitsbibliotheek Leiden.

Plate 3 (illustration 3 on p. 21). A watercolour of a clematis in *Libri Picturati*, vol. A 23, f. 18v. Jagiellon Library, Kraków.

Plate 4 (illustration 16 on p. 156). 'Gufiniare/Zufiniare/Zufiniaris' = Scarlet Martagon (*Lilium chalcedonicum*), here called 'HEMEROCALLIS CHALCEDONICA umbellifera', on a painting by Daniel Rabel, the original of the engraved florilegium *Theatrum florae* (1622). Bibliothèque Nationale de France, Paris.

Plate 5 (illustration 13 on p. 150). Ripe cloudberries (*Rubus chamaemorus*), ready to pick, growing on boggy ground in the mountains of the province of Jämtland, Sweden, close to the Norwegian border.

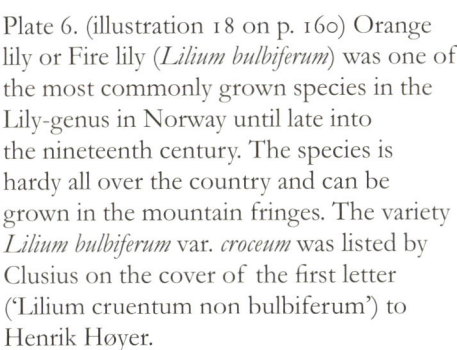

Plate 6. (illustration 18 on p. 160) Orange lily or Fire lily (*Lilium bulbiferum*) was one of the most commonly grown species in the Lily-genus in Norway until late into the nineteenth century. The species is hardy all over the country and can be grown in the mountain fringes. The variety *Lilium bulbiferum* var. *croceum* was listed by Clusius on the cover of the first letter ('Lilium cruentum non bulbiferum') to Henrik Høyer.

Plate 7 (illustration 54 on p. 309). *Spartium junceum* L.

Plate 8 (illustration 56 on p. 310). *Muscari botryoides* L.

Plate 9 (illustration 55 on p. 310). *Hibiscus rosa-sinensis* L.

Plate 10 (illustration 50a on p. 280).
Table 6 in the *Clusius Codex*, Universiteitsbibliotheek Leiden.

Plate 11 (illustration 50b on p. 280).
A page of the Oxford album, copied after
table 6 in the Leiden *Clusius Codex*.
University of Oxford, Department of
Plant Sciences, Sherard Collection, Ms. 43.

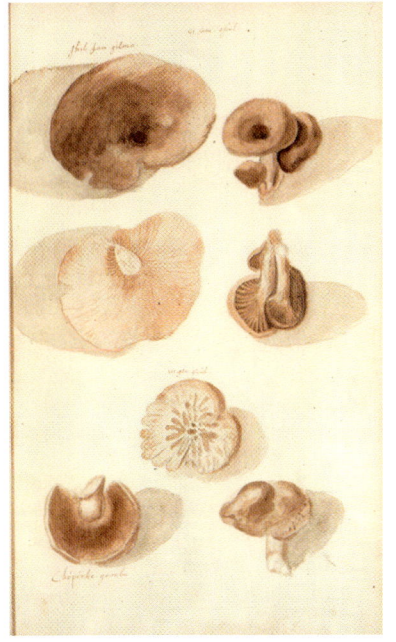

Plate 12 (illustration 50c on p. 280). Page n. 7 in
the Brussels album, which combines table 6 of the
Clusius Codex in Leiden with the figures on table 10 of
that same *Codex*. Royal Library Brussels.

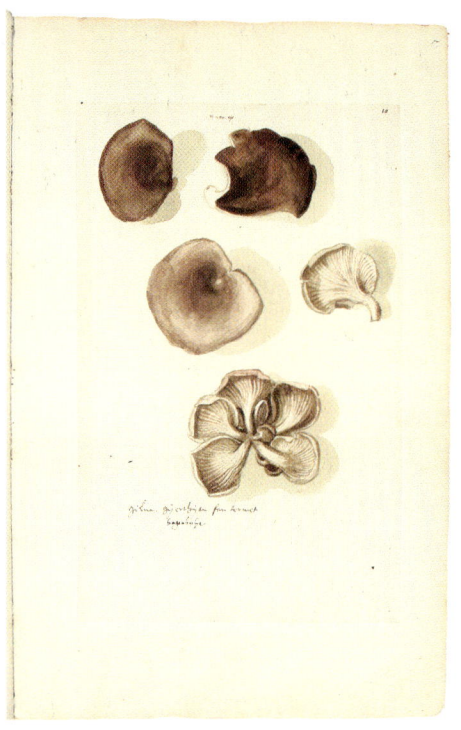

Plate 13 (illustration 51a on p. 281). Table 5 in the *Clusius Codex*, Universiteitsbibliotheek Leiden.

Plate 14 (illustration 51b on p. 282) A page of the Oxford album, copied after table 5 in the Leiden *Clusius Codex*. University of Oxford, Library of the Department of Plant Sciences, Sherard Collection.

Plate 15 (illustration 51c on p. 282). Page n. 6 in the Brussels album, which combines table 5 in the *Clusius Codex* in Leiden with figures on table 9 of that same codex. Royal Library Brussels.

Plate 16 (illustration 8 on p. 121). Map of Hungary from Abraham Ortelius, *Theatrum orbis terrarum* (Antwerp, 1595). Library of the Hungarian Academy of Sciences, Budapest.

COLOUR PLATES 11

Plate 17 (illustration 12 on p. 148). Map of Scandinavia and Northern Europe from Abraham Ortelius, *Theatrum orbis terrarum* (Antwerp, 1570). Lund University Library.

PART III

Clusius' translations and illustrations: Processing information

Two glimpses of America from a distance: Carolus Clusius and Nicolás Monardes

José Pardo Tomás

Introduction

In 1611, Sebastián de Covarrubias included an entry on tobacco in what is considered to be the first Spanish dictionary, his *Tesoro de la lengua castellana*. Covarrubias defined tobacco as a 'well-known plant', and mentioned both its medicinal use and its popular abuse, which he categorised as 'pure vice', though discounting the opinion of those who believed that there was diabolical intervention (*hechizo*) in the fervour with which the Christians avidly sought tobacco. He does not mention the American origin of the plant; in fact, he even declared that tobacco 'was used in Pliny's day', citing Book 25 of the *Historia naturalis* as evidence. But that Covarrubias knew and had read somewhat more than Pliny is evident when he continues:

The first to discover it was the devil, making his priests and ministers take it when they had to make prophecies to those who consulted them, and the demon revealed to them what they understood by conjecture through that stupefied state.[1]

This is clearly an allusion to the significance given to the consumption of tobacco and related practices among certain Amerindian cultures, although here the Indian has been transformed directly into the devil. There can be no doubt that Covarrubias' description is based on a reading of one of the texts that offered a similar account of the use of tobacco by the Amerindians. By 1611, some twenty to thirty texts that contained a description of this practice were circulating in Europe. One of the most accessible of them (because there were various editions in Spanish, Latin, Italian, English, and French) was the *Historia medicinal* by Nicolás Monardes, which contains the following passage:

[…] when there was a matter of great importance among the *indios*, for which the *caciques* or leaders of the people needed to consult […] the priest took some leaves of tobacco in their presence […] and when the plant had taken effect he remembered

---

[1] S. de Covarrubias, *Tesoro de la lengua castellana* (Madrid, 1611).

and replied to them in accordance with the fantasies and illusions that he saw while in that state; and he interpreted them as he chose, or as the Devil advised him.[2]

Irrespective of whether Covarrubias used this or a similar source, the important point here is the peculiar way in which the author made use of the text, attributing the role of the discoverer of tobacco to the devil. Although the devil was already mentioned in Monardes' text – and in Clusius' Latin version of it[3] –, Covarrubias actually turned him into the sole protagonist of the discovery of tobacco and suppressed any explicit reference to the Amerindians, or even to the American origin of the plant.

Nora Catelli and Marietta Gargatagli have drawn attention to the scholarly neglect of what they called 'scenes of translation' between Spain – and by extension, the whole of Europe – and America. The anthology of texts with which they began to fill this lacuna – whose very title is taken from Covarrubias' text on tobacco[4] – certainly contains much that is noteworthy, but if there is any common denominator among those texts, it is that (to quote the authors) 'the scenes of translation, near or far, Spanish or American, display a repetitive and shared mechanism. They are articulated as a continuous series of strategies of omission of the other, who is always an enemy who has already been characterised as satanic and anthropophagous, depraved, murderous, selfish or cruel'.[5] To a greater or lesser extent, the discursive and mental strategies that favour that 'omission of the other' can be found in practically the entire European literature of the sixteenth century on the New World,[6] even though, as James Amelang recently pointed out, that immense textual corpus constitutes 'the most extensive as well as innovative ethnographic project of the sixteenth century'.[7] It is an intellectual project that the Europeans developed above all to present an account of themselves rather than to

---

[2] N. Monardes, *Primera y segunda y tercera partes de la historia medicinal de las cosas que se traen de nuestras Indias Occidentales, que sirven en medicina* (Seville, 1580), 36v-37r. The first edition of Part II, containing the passage cited, was published in Seville in 1571; all of the present citations are taken from the 1580 edition because it is the most complete and accessible now that there is a facsimile with an introductory essay by J.M. López Piñero (Madrid, 1989).

[3] His Latin version runs: 'Etenim Indos moris erat Sacerdotes de bellorum evetu, aliisque magni momenti negociis consulere. Consultus sacerdos, istius plantae folia sicca urebat [...] Discussa eius fumi facultate, ad se redibat, referebatque negocium cum daemone contulisse'; N. Monardes, *De simplicibus medicamentis ex Occidentali India delatis, quorum in medicina usus est [...] interprete Carolo Clusio* (Antwerp, 1574), 25.

[4] N. Catelli and M. Gargatagli, *El tabaco que fumaba Plinio. Escenas de la traducción en España y América: relatos, leyes y reflexiones sobre los otros* (Barcelona, 1998). The texts by Covarrubias: 245-248.

[5] Ibid., 18.

[6] T. Todorov, *La conquête de l'Amérique. La question de l'autre* (Paris, 1982).

[7] J. Amelang, 'Mourning laments becomes eclectic: ritual, lament and the problem of continuity', *Past and present*, 187 (2005), 3-31, here 23.

understand that recently discovered New World. In other words, the Europeans wrote about America not to translate the world of others, but to translate their own world.[8] This translation was intended primarily for themselves and secondarily for the colonised, whom it was essential to dominate culturally and politically,[9] converting them into subjects, Christians, and labour power to exploit the natural resources of the newly found lands.[10]

In view of this, as Peter Mason showed more than a decade ago, we believe that, in spite of all that has been written on the subject, it is still useful to consider some of the various cultural mechanisms deployed by the Europeans in the complex and exciting process of the construction or invention of America. While Mason's strategy was to analyse the texts and images of 'so-called monstrous human races' produced by the Europeans to reveal 'their place within the European imaginary and their role as translators of the New World',[11] we consider that the same texts and images should also be analysed in relation to plants and other natural products.[12] In our view, the American

---

[8] Although their object of study is the native languages and Spanish, Catelli and Gargatagli seem to defend the same idea when they write: 'The Spaniards did not translate the native languages into Spanish, but they translated Spanish into the native languages', *El tabaco que fumaba Plinio*, 17.

[9] 'Translation is above all political literature, literature of the *polis*: it seeks to intervene in the exising tradition, to modify it, negate it, recreate it, change it. And it simultaneously imagines and negates the *others* as it does so, with the excessive passion discussed by Bloom (*Agon. Towards a theory of revisionism*, New York, 1982), that 'initial and asymmetrical' passion (love and hatred) which marks and embraces its actors in the violence of history'; Catelli and Gargatagli, *El tabaco que fumaba Plinio*, 19.

[10] This is not the place to review the long historiographical debate triggered by the publication of J.H.Elliott, *The Old World and the New, 1492-1650* (Cambridge, 1970); although it is still very stimulating to read works like S.J. Greenblatt, *Marvelous possessions: The wonder of the New World* (Chicago, 1991); A. Grafton, *New worlds, ancient texts: The power of tradition and the shock of discovery* (Cambridge, MA, 1992), and A. Pagden, *European encounters with the New World: From Renaissance to Romanticism* (New Haven, 1993). These pages have also benefited from some of the ideas put forward in the more recent literature: B. Schmidt, 'Exotic allies: The Dutch-Chilean encounter and the (failed) conquest of America', *Renaissance quarterly*, 52 (1999), 440-473; J. Cañizares-Esguerra, 'New World, new stars: Patriotic astrology and the invention of Indian and Creole bodies in colonial Spanish America, 1600-1650', *American historical review*, 104 (1999), 33-68; J. Carrillo, 'Taming the visible: Word and image in Oviedo's *Historia general y natural de las Indias*', *Viator*, 31 (2000), 399-431; and F. Egmond and P. Mason, '"These are people who eat raw fish": Contours of the ethnographic imagination in the sixteenth century', *Viator*, 31 (2000), 311-339.

[11] P. Mason, *Deconstructing America. Representations of the other* (London, 1990), 7.

[12] For some forays into this field, see J. Pardo-Tomás, 'Le immagini delle piante americane nell'opera di Gonzalo Fernandez de Oviedo (1478-1557)', in G. Olmi, L. Tongiorgi and A. Zanca (eds), *Natura-cultura. L'interpretazione del mondo fisico nei testi e nelle immagini* (Florence, 2000), 163-188; '*Imago naturae*: historia natural, materia médica y nuevos mundos', *Historia*, 16/24 (2000), 25-40; *El tesoro natural de América. Colonialismo y ciencia en el siglo XVI* (Madrid, 2002); 'Tra 'oppinioni' e 'dispareri' rappresentazioni della flora americana nell'erbario di Pier'Antonio Michiel (1510-1576)', in G. Olmi and L. Tongiorgi (eds.), *La natura e il corpo. Nature and body. Studi in memoria di Attilio Zanca* (Florence, 2006), 73-100.

plants that feature in the European texts and images can be analysed as another of those efficacious 'translators of the New World', as well as one of the most clear-cut mechanisms, not only of the omission of the other, of his or her culture, and of its role in conferring meaning on those plants, but also of pure and simple cultural expropriation.

Thus, although the main purpose of this contribution is to analyse how Clusius translated Monardes, we should not lose sight of the fact that there is more at stake. For Monardes translated the Amerindians – on a much larger scale than he was prepared to admit – even though he always did so via other translations: those which his many eye-witnesses brought him from there concerning plants, their names (sometimes in the language of the other),[13] their shapes, colours, scents and properties, in the Aristotelian sense of the term, but also concerning the remedies that the other prepared from them and the effects that those remedies produced on the bodies of different individuals. Monardes translated it all into Spanish. That was the text of texts that Clusius took to translate into Latin, adding a series of notes with his own comments. This process of translation was extremely complex and highly mediated by agents and voices that had little connection with Clusius' own work, and which were not even required to give meaning to the erudite and bookish learning of Carolus Clusius, as we shall see in the course of analysing his translation of Monardes into Latin and his commentaries.

Although this cultural operation was not unique at the time, it proceeded in the opposite direction to the others. Monardes' work was edited and translated during the same period into Italian, English and French,[14] but it is important to emphasise that Clusius' translation was from Spanish into Latin. This was the exact opposite of the Renaissance practice of translation, which passed from the prestigious Greek or Latin of the original to the vernacular.[15] In fact,

---

[13] On the practice of recording Amerindian names as a mechanism of appropriation on the part of the coloniser, see the interesting analysis of the work of Gonzalo Fernández de Oviedo by J. Carrillo, 'Naming difference: The politics of naming in Fernandez de Oviedo's *Historia general y natural de las Indias*', *Science in context*, 16 (2003), 489-504.

[14] After some partial translations into Italian (1570, 1574) and French (1572, 1588), Parts I and II of Monardes' work were translated into Italian: *Due altri libri parimente di quelle cose che si portano dall'Indie Occidentali* (Venice, 1575); Parts I-III were translated into English: *Ioyfull newes out of the neve founde worlde, wherein in declared the rare singular vertues [...] with their applications, aswell of phisicke as chirurgerie* (London, 1577) and French: *Histoire des simples medicaments apportés des Terres Neuves, desquels on se sert en la medecine* (Lyon, 1602). On the Italian translations, see J. Pardo Tomás, 'Obras españolas sobre historia natural y materia médica americanas en la Italia del siglo XVI', *Asclepio*, 43 (1991), 51-94, here 55-57 and 71-79. For the others see: J.M. López-Piñero, 'Introducción', in N. Monardes, *La historia medicinal de las cosas que se traen de nuestras Indias Occidentales (1565-1574)* (Madrid, 1989), 9-74, here 23-27.

[15] S.S. Gravelle, 'The Latin-vernacular question and humanist theory of language and culture', *Journal of the history of ideas*, 49/3 (1988), 367-386; F. Waquet, *Latin, or the empire of a sign* (New York, 2001).

Clusius' movement counter to the prevailing tendency entails a further paradox, since one of his purposes was to extend the knowledge of the text of Monardes among those 'who are not familiar with the Spanish tongue',[16] – an audience that was taken to be larger than that of Spanish-speaking readers. Perhaps, however, we should speak not of a simple operation to reach a wider audience, but of an attempt to encode the information contained in the work of Monardes in a language that was the exclusive preserve of the cultural élite of prelates, nobles and scholars, and who are amply reflected in the names, citations and references contained in Clusius' annotations to the text of the physician of Seville.

Nicolás Monardes and the Historia medicinal: *the reduction of Amerindian knowledge to the medicinal and commercial utility of the Christians*

A native of Seville,[17] the son of a Genoese bookseller, Nicolás Bautista Monardes began to practise medicine in his home town around 1533 after returning from the university of Alcalá de Henares, where he had graduated in philosophy and the humanities in 1530, and in medicine in 1533. The grandson on his mother's side of the physician and surgeon Martin de Alfaro, he married Catalina, the daughter of the physician García Pérez de Morales, in 1537. This connection with physicians on both sides of the family enabled him to secure a good position in the world of medical practice in the city, and for the next fifty years his professional, intellectual and personal life was to be bound to Seville and the practice of medicine. It was thus medical practice that enabled Monardes to acquire social prestige and economic prosperity and conditioned his other two main activities: the publication of medical works centred on healing and medicine, and business activities connected with trade with the Castilian colonies on the other side of the Atlantic. The monopoly of commerce with the Americas conferred on the port of Seville by the Spanish Crown made it a privileged position to obtain an advantagous position in this respect.

Monardes' first four publications appeared between 1536 and 1545: a treatise on *pharmacodilosis* (1536) which defended the classical against the Arab medical tradition; a treatise, written from the same perspective, on the debate

---

[16] Dedication 'Generoso virtute et eruditione praestanti viro, dn. Thomae Redigero, dn. suo plurimum observando, Carolus Clusius', in N. Monardes, *De simplicibus medicamentis*, 4.

[17] His precise date of birth is still not known; dates have been proposed between 1493 and 1508. On Monardes: F. Rodríguez-Marín, *La verdadera biografía del Doctor Nicolás Monardes* (Madrid, 1925); F. Guerra, *Nicolás Bautista Monardes. Su vida y su obra* (México, 1961); López-Piñero, 'Introducción'; and Pardo Tomás, *El tesoro natural*, 77-126.

concerning blood-letting in cases of pleuritis (1539); a third on roses and their medical applications; and an edition in Spanish of a medical treatise written by Juan de Aviñón at the end of the fourteenth century (1545). These works were evidently in accordance with the academic medicinal doctrine of the period. Monardes did not publish a single new work in the next twenty years, although there are signs that, from the 1550s on, his intellectual and commercial interests shifted towards the medicinal products of the Americas, to which he was to dedicate his most important work, the *Historia medicinal de las cosas que se traen de nuestras Indias occidentales que sirven al uso de la medicina*, published in Seville in three parts: Part I in 1565, Part II in 1571, and Part III, published together with Parts I and II, in 1574. The complete work was reissued in Seville in 1580. This was the last edition to be published during its author's lifetime.

Monardes' commercial activities connected with the Americas began as early as 1533, when he set up a trading company with Juan Núñez de Herrera, an agent in the settlement of Nombre de Dios, on the continental isthmus of America. The idea of the company was to put African slaves on board for the outbound journey, and to load the ships bound for home with cochineal, a dye that was in great demand by the European textile manufacturers, as well as certain American medicinal products that were also highly profitable, particularly guajacum (used in the treatment of *morbus gallicus*), aromatic resins, balsams, cassia, and the purgative Mechoacán-root (similar to rhubarb). The economic success of Monardes' trading activities, however, appears to have declined around 1563, when his partner died. Four years later, when faced with the claims of his creditors, Monardes sought refuge in the monastery of Regina Coeli in Seville to escape prison. From there he negotiated terms with the authorities to get out of bankrupty, pledging to pay back in instalments the sum of almost twenty-five million *maravedíes* that he owed; he was freed on those conditions in late 1568 or early 1569. Everything seems to indicate that Monardes converted the publication of the *Historia medicinal*, which was initiated in those very years, into a means of recovering his lost fortune, offering a source of sharing in the profits that was more secure than commercial trade in those products. In fact, he appears to have substantially recovered his economic position by the time of his death in 1588.

There can be no doubt of the role that Monardes' experience as a medical practitioner played in the composition of the *Historia medicinal*. Besides waiting upon a group of patients from among the aristocratic and ecclesiastical élites of the city (the Archbishop of Seville and future Inquisitor General, the Duchess of Béjar, the Duke of Alcalá), Monardes also practised among other layers of local society: merchants and businessmen, pilots and mariners, friars and soldiers, and visitors passing through the city, which was the only port for those setting out for or returning from the Americas. Moreover, this all provided him with different occasions for carrying out a type of medical practice that included testing the effects of various medicaments, both of local and of American origin. This wealth of therapeutic experiences had another important consequence, since his patients became his principal informants when, after almost forty years of medical practice, he began to compile the work.

The first part of the *Historia medicinal*, was published in 1565, is dedicated to the Archbishop of Seville. Carolus Clusius had visited Seville only a few months before, although he does not appear to have come across Monardes or his work on that occasion.[18] The first part of the *Historia medicinal* is divided into four main sections, devoted to resins (*caraña*, copal, *tacamahaca*, and the American succedaneum for the classical gum animé), purgatives (especially the Mechoacán root that replaced other common purgatives in the Galenic therapeutic arsenal), 'three medicines celebrated throughout the world' (guajacum, China-root – which is also found in the Americas – and the American varieties of sarsaparilla), which are all basically remedies for *morbus gallicus*, and Peruvian balsam (a succedaneum for classical balsam). It would be no exaggeration to affirm that, from the perspective of a physician based in the metropolis on the banks of the River Guadalquivir, the other side of the Atlantic was fundamentally seen as a sort of 'empire of succedanea', a place from where it was possible to extract medicinal products similar to those included in the classical therapeutic arsenal of Galenic medicine, but less expensive, more abundant, and even – as could be demonstrated, if necessary, on the basis of forty years of experience in treatment – more effective. The success of the publication produced an influx of spontaneous informants who brought 'Doctor Monardes' a piece of root, the seeds of plants with marvellous effects, or, simply, the tale of a cure thanks to some local remedy.

Those spontaneous testimonies are the basis of Part II, published in 1571, and dedicated to Philip II. The twenty medicinal products of vegetable origin and three of mineral origin contained in Part I were supplemented by another dozen, plus those that could be extracted from the armadillo and other animals like sharks and caymans. Part II opened with a comprehensive study of tobacco (the frontispiece was embellished with a woodcut of the plant), and continued with three long chapters on sassafras, *carlo santo*, and *cebadilla* (a type of wild barley), each with its own illustration. The other chapters are devoted

---

[18] F.W.T. Hunger, *Charles de l'Ecluse (Carolus Clusius), Nederlandsch kruidkundige (1526-1609)*, 2 vols. (The Hague, 1927), vol. I, 81; Guerra, *Nicolás Bautista Monardes*, 92-93, who does not reject the possibility, although without any documentary evidence in its favour. See the comments based on the information contained in the correspondence of Clusius by J.M. López-Piñero and M.L. López-Terrada, *La influencia española en la introducción en Europa de las plantas americanas (1493-1623)* (Valencia, 1997), 67 and 89-90.

to certain products of lesser importance and complementary notes on remedies that had already been discussed in Part I. Monardes also decided to include the transcription of a document of one of his informants. This was a long letter that Pedro de Osma, a soldier who had settled in Lima, sent to Monardes, whom he knew only through his work, in 1568. The letter contained a description of the properties of several medicinal products and was accompanied by samples of them.[19] Among Monardes' informants in Part II we also find various soldiers from Florida, who had arrived in Seville between 1567 and 1568, and the Bishop of Cartagena, who arrived with the fleet of 1569 with valuable information about the tree that produced dragon's blood and the properties of the tail of the armadillo; the bishop personally sought out the physician 'because he was fond of the book that we produced on this herbal material' and gave him considerable information as well as some samples for his 'museum'.[20]

Part III, dedicated to Pope Gregory XII and printed in 1574 together with the re-issue of Parts I and II, consisted of thirty-five shorter chapters containing complementary information on products that had already been discussed in the first two parts, as well as adding several new ones. As in the first two parts, most of the products were of vegetable origin, although a few were of animal or mineral origin. Among the latter, the account of bezoar stones is particularly noteworthy,[21] not only because Monardes was a genuine expert on the subject,[22] but also because bezoars form one of the categories of products on which Clusius decided to intervene at greater length in his Latin version, out of both personal interest and that of his friends.[23]

---

[19] Monardes, *Historia medicinal*, 57r-62v.
[20] Ibid., 64v, 65v, 66v.
[21] A concretion made of hair or other material found in the stomach or intestines of animals, especially ruminants. The name derives from the Arabic for 'antidote to poison', the main quality attributed to these 'stones'.
[22] In fact, the first edition of Part I of the *Historia Medicinal* was published together with a treatise on two antidotes, one of which is the bezoar stone: *Dos libros. El uno trata de todas las cosas que traen de nuestras Indias occidentales que sirven al uso de Medicina [...] El otro libro, trata de las dos medicinas maravillosas que son contra todo Veneno, la piedra Bezaar y la yerva Escuerçonera* (Seville, 1565). In Part II of the *Historia medicinal* (1571), bezoar stones are once again the object of attention, this time in the letter of Pedro de Osma (Monardes, *Historia medicinal*, 57v-59v). This Lima-based soldier sent Monardes a dozen of these stones through the intermediary of the 'rich merchant' Juan Antonio Corzo, giving rise to an extensive 'expert' intervention on the part of Monardes to establish the difference between these and 'the ones they bring from the East Indies'. In the process he pointed out something more interesting: the number of people returning in the Indies fleets in those years who brought back bezoar stones taken from American animals with them (Monardes, *Historia medicinal*, 63r-63v). Finally, in Part III (1574) Monardes devoted a whole chapter to the same Peruvian bezoar stones (Monardes, *Historia medicinal*, 90r-92v).
[23] C. Clusius, *Exoticorum libri decem: quibus animalium, plantarum, aromatum, aliorumque peregrin, frustuum historiae describuntur* (Leiden, 1605), 326-330. On the interest of Clusius' friends see, for example, the

The way in which the material contained in the *Historia medicinal* is collected, organised and presented, which was to raise such peculiar problems for Clusius, represents a work in progress that goes into print as it advances. The author felt no need to spend too much time organising the material, nor on cataloguging products, information, informants, etc. On the other hand, this peculiar work in progress reveals in a fairly clear way the ideas and concepts deployed by the physician from Seville, his working method, and the role assigned to the various sources of information from which he constructed his work: the knowledge of the *indios*, the experience of the Christian colonisers, and his own practical experience in Seville with his patients, their illnesses, and the recently imported remedies. This is all far from Clusius' conception of the study of plants and animals, based on scientific practices that were much more closely linked with the tradition of natural history than with medical practice.

The *Historia medicinal* was a rapid and long-lasting success,[24] due fundamentally to three factors: its timeliness; its capacity to transmit credibility in the new medicines, based on the narrative of the author's practical experience; and the coherence and skill of exposition used to found this practice on the rational basis of the Galenic medical system. The latter had to show itself capable of integrating a knowledge considered exclusively empirical, in as much as it was derived from the Amerindian cultures and was thus by definition foreign to the sole 'philosophical and rational basis' that conferred the status of truth on the knowledge of therapeutic effects.

For, as we have emphasised, the *Historia medicinal* is implicitly based on native knowledge, although when this appears explicitly, it is justified from the rhetorical accusation of malevolence towards the *conquistadores* and of secrecy regarding the properties of the plants,[25] by the skill of the Christians

---

case of Arias Montano in J. Gil, *Arias Montano en su entorno [bienes y herederos]* (Mérida, 1998), 85-89; and in the letter from Plantin to Arias Montano, dated 18 September, 1581, published by A. Dávila-Pérez, *La correspondencia de Arias Montano conservada en el Museo Plantin-Moretus, de Amberes* (Madrid/Alcañiz, 2002), 441-446.

[24] At the death of its author, the work had already been translated into four languages and published, in full or in part, in seventeen other editions: six in Italian, five in Latin, three in French, and three in English. The work went through another fourteen editions in the following century: seven in Italian, three in French, two in Latin, one in Englsh, and one in German. See J.M. López-Pinero *et al.*, *Bibliographia medica Hispanica, 1475-1950* (Valencia, 1987-89), vol. I, 150-160; vol. II, 176-180.

[25] Very clearly expressed by Pedro de Osma in his letter from Lima: 'We asked certain *indios* who were travelling with us in our service where those animals got those [bezoar] stones from, but as they were our enemies and did not want us to discover their secrets, they said that they knew nothing about those stones [...]'; Monardes, *Historia medicinal*, 57v-58r.

in extracting information,[26] or by pure chance.[27] In fact, Monardes went to great pains to disqualify the natives' use of the remedies. They were considered merely empirical and ignorant of the 'rational method' proper to the European Galenic physician, the only one authorised to 'experiment' and to pronounce judgement on the remedy.[28]

In spite of all this, practically all of the medicinal remedies discussed in the *Historia medicinal* are of native origin, and this is reflected – directly or indirectly – in Monardes' account. In fact, we have found fifty-three explicit references to the names and uses of the plants by the Amerindians among the sixty-nine chapters that comprise all three parts of Monardes' complete work.

One of the most representative cases is that of the Mechoacán root, the main focus of attention in the first part, as the title of the first edition emphasised.[29] Knowledge of its purgative action is drawn from the illness of a Franciscan friar from the convent in the region of Mechoacán, New Spain, and his contact and 'very close friendship' with 'Cazoncin, *cacique* and lord of all that land'. The Indian chief sent 'one of his *indios* who was a physician', who administered 'some grains of a root'. After the friar had recovered, the Franciscan order distributed the remedy throughout New Spain. It soon reached Seville in the hands of a Genoan who sailed there from Mexico. He

---

[26] For example, the anonymous 'father Francisco', 'taught by an *indio* of that country, who was very knowledgable about these things and was a great expert on the virtues' of plants, on the root of the *Carlo santo*; Monardes, *Historia medicinal*, 50v. But once again it is Pedro de Osma who expresses most clearly the usual ways for the Spaniards to gain access to native knowledge: their relations with the Amerindian women: 'we did not manage to find out [the properties of the plants] because the *indios*, being bad people and our enemy, would not divulge a secret nor a property of a herb, not even when a saw was applied [to their limbs], even though they witnessed us dying; but what we know about those I have described and about others, we know from the Amerindian women; when they get involved with Spaniards, they reveal [their secrets] and tell them all they know'; Monardes, *Historia medicinal*, 62v.

[27] Such as the use of tobacco against wounds from poisoned arrrows: 'As the Carib Indians, who eat human flesh, shoot their arrows with a herb or composition made of many poisons [...] A short while ago, when some Caribs went in their canoes to San Juan de Puerto Rico to fire arrows at *Indios* or Spaniards [...] they killed some [...] and as the farmer did not have *solimán* to heal them, he agreed to apply tobacco juice [...]'; Monardes, *Historia medicinal*, 34v. Solimán was a corrosive powder compound of various substances, including mercury, that was generally used to close open wounds, to cauterise, or to staunch haemorrhages.

[28] Sometimes the disqualification focuses on the criticism that they do not have a precise method (precise weights or measures) for preparing a herbal remedy; for example, in his discussion of sassafras: 'as the *indios* have neither weight nor measure, they have not kept any order in those parts in preparing the water of this tree'; Monardes, *Historia medicinal*, 42r.

[29] Nicolás Monardes. *Dos libros. El uno trata de todas las cosas que traen de nuestras Indias occidentales, que sirven al uso de Medicina y como se ha de usar la rayz del Mechoacan, purga excelentíssima [...]* (Seville, 1565).

found Monardes ill, and upon being requested by the physician to administer a purgative, he replied that 'if there is need of a purgative', it should be the one that he had brought with him from Mechoacán.[30] This narrative pattern recurs time and again: supposedly secret information possessed by the Amerindians is revealed to the colonisers in one of the aforementioned ways, who ensure that it reaches the physician in the metropolis.[31]

There is thus a double transfer – a double translation – from the Amerindians (empirical users lacking the authority to have knowledge) to the colonisers (who are, generally speaking, ignorant of medicinal matters), and from the colonisers to Monardes, an academic physician, the only one qualified to experiment on his patients, to decide on the efficacy of the remedy, its method of preparation, and its comparison with others belonging to the corpus of medicines known at that time. For that is the objective that Monardes has in mind when he embarks on his *Historia medicinal*: to show his conviction that the remedies deriving from the Americas can be used as succedanea for those known from the classical pharmaceutical arsenal. In the last instance, the Americas are a reservoir of substitutes for known substances, though less expensive, more abundant and easier to obtain: copal 'is used instead of incense';[32] the *indios* make oil from the *higuera del infierno* (*Jatropha multifida L.*) 'as Dioscorides teaches to make it from the castor-oil plant';[33] American *ocozotl* is used 'instead of styrax';[34] balsam of Peru is used 'in imitation of real balsam';[35] the chile of the Indies is used 'for everything for which the aromatic spices that they bring from Maluco and Calicut are used';[36] American cassia is 'incomparably better than that which they bring from India to Venice, and which the galleys bring from there to Genoa and from Genoa to Spain';[37] the Mechoacán root ends up 'replacing the use of rhubarb of Barbary', and is even given

---

[30] Monardes, *Historia medicinal*, 22v-23v.
[31] See, for instance, the account of the 'discovery' of guajacum: 'An *Indio* gave notice of it to his master in the following manner: When a Spaniard who had been infected by an *India* was suffering great pains from swollen lymphs [*Bubas*], the *Indio* who was one of the physicians in that country gave him water of guajacum, which not only took away the pains he was suffering, but also cured the affliction: many Spaniards who were infected with the same complaint were cured by the same remedy, which was then brought from there by those who came here to Seville, and from here it spread throughout Spain, and from there throughout the whole world'; Monardes, *Historia medicinal*, 10r.
[32] Ibid., 3r.
[33] Ibid., 5r.
[34] Ibid., 6v.
[35] Ibid., 7v.
[36] Ibid., 19v.
[37] Ibid., 20r.

the name 'Rhubarb of the Indies'.[38] These examples are taken from Part I alone,[39] which ends with the following significant conclusion:

> I consider how many trees and plants there are in our Indies which have great medicinal value [...] without seeking the spices of Maluco and the medicines of Arabia and Persia. For our Indies spontaneously provide them in the uncultivated fields and in the mountains.[40]

Monardes' translation of the American materials into the language of European medicine is even more evident if we consider the authentic editorial context to which the publication of each of the three parts of the *Historia medicinal* belongs. Students of this work have often neglected this fact, hopelessly weakening Monardes' publishing project. For in our view, it is important to bear in mind that each of the three parts of the *Historia medicinal* was first published as part of editions which included other treatises by Monardes which were concerned not with the medicinal properties of products from the Americas, but with the traditional *materia medica* in Galenic medicine. Thus, in 1565 Part I was published together with a treatise on two antidotes;[41] in 1571 Part II was published together with a treatise on the medicinal use of snow;[42] and in 1574 Part III was published together with Parts I and II, the treatise on antidotes, the one on snow, and a new treatise on the use of iron.[43]

Therefore, when Clusius, after more than thirty years' experience of Monardes' work, decided to incorporate the treatises on antidotes, snow and iron in his definitive production,[44] he was simply restoring the editorial plan

---

[38] Ibid., 23r.
[39] There are less examples in Parts II and III, although it can be found in the comparison of tobacco with *solimán* (see annotation in note 27) and oriental *bague* (Monardes, *Historia medicinal*, 34v and 37v, respectively); of *guacatane* with European 'mountain mint' (*Teucrium polium*) (Monardes, *Historia medicinal*, 54r); or of *cebadilla*, once again with *solimán* (Monardes, *Historia medicinal*, 55r).
[40] Monardes, *Historia medicinal*, 30r.
[41] Monardes, *Dos libros*.
[42] N. Monardes, *Segunda parte del libro de las cosas que se traen de nuestras Indias Occidentales que sirven al uso de la medicina [...] Va añadido un libro de la nieve, do verán los que beven frío con ella, cosas dignas de saber y de grande admiración acerca del uso del enfriar con ella* (Seville, 1571).
[43] Nicolás Monardes, *Primera y segunda y tercera partes de la Historia medicinal [...] Tratado de la piedra bezaar y de la yerva escurçonera. Diálogo de las grandezas del Hierro y de sus virtudes medicinales. Tratado de la nieve y del bever frío [...] Van en esta impressión la tercera parte y el diálogo del hierro nuevamente hechos [...]* (Sevilla, 1574).
[44] Placed at the end of Book X of the *Exoticorum libri decem*, with frontispiece and separate pagination (Clusius, *Exoticorum libri decem*, 1-52), entitled: *Nicolai Monardi Hispalensis medici praestantissimi libri tres, magna medicinae secreta et varia experimenta continentes. Et illi quidem Hispanico sermone conscripti; nunc verò recens Latino donati à Carolo Clusio Atrebate. Horum seriem proxima pagina indicabit [...] Primus agit de lapide bezaar & herba scorzonera, duobus praestantissimis adversus venena medicamentis. Alter, de ferro, & eius insignibus facultatibus. Tertius, de nive, eius commodis.*

of Monardes in its entirety, although in a context, the *Exoticorum*, which qualified Monardes' entire oeuvre, covering both the Old World and the New, as 'exotic'. It is yet another example of how vague the boundaries between the exotic and the local, between the centre and the periphery, were in the construction of the knowledge of European natural history at the end of the sixteenth and the beginning of the seventeenth century.[45]

Clusius and the Historia medicinal: *Recataloguing the New World in the Old, 1569-1605*

As has already been pointed out, it seems reasonable to rule out a personal encounter between Monardes and Clusius during the latter's stay in Seville at the beginning of 1565, in the light of the absence of any documentary evidence or other clues to support such a hypothesis. Clusius' first contact with the work of Monardes should therefore be dated to 1569, for it was in April and August of that year that Alfonso Pancio, the physician of the Duke of Ferrara, sent him a text containing a synthesis, in Latin, of Part I of the *Historia medicinal*, which Monardes had published in 1565.[46] Four years later, Clusius was still awaiting a copy of the original edition, to judge from the allusion made by his friend Arias Montano in a letter of August 1569.[47] Finally, in September 1571, during his stay in London, Clusius obtained the first two parts of the work (the second had been published in that very year in Seville).[48] It must have been in the course of 1573, as Clusius himself stated in 1605,[49] that he began to translate them into Latin and to prepare them for the edition that was to come off the Plantin presses in Antwerp in September 1574, with a royal privilege granted by Philip II, not by chance the same person to whom Monardes had dedicated Part II of his work three years earlier.

In the first edition, Clusius' intervention consisted of making the Latin translation and of unifying the two parts in a single work, modifying the order

---

[45] B.W. Ogilvie, 'The many books of nature. Renaissance naturalists and information overload', *Journal of the history of ideas*, 64 (2003), 29-40. Even beyond these chronological limits to the supposed foundation of modern botany with Linnaeus; on this see the interesting comments in S. Müller-Wille, 'Joining Lapland and the Topinambes in flourishing Holland: Center and periphery in Linnaean botany', *Science in context*, 16 (2003), 461-488.
[46] López Piñero and López Terrada, *La influencia española*, 89-90, following Hunger, *Charles de l'Ecluse*, vol. I, 106.
[47] J.L. Barona and X. Gómez, *La correspondencia de Carolus Clusius con los científicos españoles* (Valencia, 1998), 58-59.
[48] Hunger, *Charles de l'Ecluse*, vol. I, 118.
[49] 'Carolus Clusius candido Lectori', in Clusius, *Exoticorum libri decem*, 296: 'Istas duas partes anno Christi millesimo quingentesimo septuagesimo tertio nanciscebar, & ex Hispanico idiomate, quo descriptae erant, in Latinum sermonem convertebam.'

of various chapters, considerably changing the iconographic apparatus of the original edition, and adding complementary comments and annotations of his own to certain chapters. These changes were essentially very similar to the way in which he had treated the *Coloquios* of Garcia da Orta in 1567, which were now reprinted together with the edition of Monardes, as Clusius himself mentioned in the dedication to Thomas Rediger.[50] The thirty-seven sections or chapters of the *Historia medicinal* were transformed into forty-three, mainly due to the fact that Clusius created separate chapters for the various balsamic resins discussed by Monardes, as well as the *hierba de Juan Infante* and the stones found in sharks and caymans, which had not been assigned separate chapters by the physician of Seville.[51] More drastic was the profound reorganisation and arrangement of the chapters compared with the original; after having decided to combine Parts I and II in a single work, Clusius was obliged to recatalogue the materials, although he tried to respect the initial classificatory criterium of Monardes, for if we analyse Clusius' arrangement closely, we find that the organisational criterium is still that of the medicinal use of the remedies: aromatic resins first, followed by cures for *morbus gallicus*, medicinal 'woods' and 'stones' for various ailments, especially antidotes, and finally the important section on purgatives.[52] In some way, these complex migrations from the texts of Monardes to an arrangement that Clusius offers over the years (1574, 1582, 1593, 1605) indicate a crescendo in the decataloguing of the American materials discussed by the physician of Seville, in order to mark them with a taxonomy of Clusius' own devising, which considerably increases their distance from the natural world of the Americas. This distance was already present in Monardes, but in Clusius' Latin version it is consolidated as an unfathomable distance from the natural world of the Americas, which is now no more than a distant, vague horizon, the source of fragments of plants, pieces of stones, animal viscera, seeds that come to fruition with difficulty in European soil, and names of uncertain orthography.

---

[50] 'Et cum Plantinus noster Aromatum historiam recudere vellet, eam auctiorem, & locupletioribus annotationibus, iconibusque insuper illustratam (quoniam eiusdem sunt argumenti) huic coniungendam tradidi': N. Monardes, *De simplicibus medicamentis*, 4. The new edition of the Portuguese author: Garcia ab Orta, *Aromatum et simplicium aliquot medicamentorum apud Indos nascentium historia [...] nunc vero Latino sermone in epitomem contracta, et iconibus ad vivum expressis, locupletioribusque annotatiunculis illustrata a Carolo Clusio* (Antwerp, 1574).

[51] 'Resina ab legna' and 'Resina Carthaginensis'; 'Herba Ioannis Infantis'; and 'Lapis Tiburonum' and 'Lapis Caymanum': Monardes, *De simplicibus medicamentis*, 20, 28, 52 and 53, respectively.

[52] In this respect, the classification of tobacco could appear the most problematic, but this was inevitable given that, in the work of Monardes, it is a remedy for a multitude of ailments and its uses resemble almost all the other categories of products. Monardes placed it at the beginning of Part II, but Clusius decided to move it to the end of the section on aromatic resins and balsams, right before the cures for *morbus gallicus*; Monardes, *De simplicibus medicamentis*, 21-28.

As far as the ilustrations are concerned, Clusius' translation is even more radical in its distancing effect. In fact, Clusius decided to make a clean break with the images used by Monardes: of the ten engravings that he included in his first edition, only that of the armadillo and of the *pimienta luenga*[53] have their counterparts among the twelve engravings in Part II of *Historia medicinal*.[54] The other eight are connected with Clusius' commentary and have no counterparts in the corresponding chapters of Monardes.

The fundamental aspects of Clusius' intervention were already established in the first edition of Monardes' materials (1574), beginning with the Latin translation – which was not to receive any substantial modifications, but only minor corrections – and ending with the criteria for the rearrangement and reorganisation of the chapters, including the form and style of the annotations and images. Certainly, the subsequent editions of Parts I and II increase the number of notes, as well as their length in some cases, but do not reveal any change of orientation in the general criteria adopted by Clusius for editing Monardes in Latin.[55]

The main novelty of the edition of 1593 was to bring together the translation of all three parts of the *Historia medicinal* for the first time, for when Clusius had finally edited Part III in 1582, he did so without the other two parts.[56] This time, however he did no more than combine a reprint of that edition (with a new frontispiece and continuous pagination)[57] with a new edition of Parts I and II, which had been published in 1574 and reprinted in 1579. However, the novel features of the edition of 1593 went further, for Clusius modified seven of his annotations of 1574 (considerably enlarging them in some cases), as well as including for the first time the two well-known engraving of the tobacco plant and a new engraving to his commentary on purgative beans.[58]

---

[53] Monardes, *Historia medicinal*, 66r and 70v; and Monardes, *De simplicibus medicamentis*, 55 and 71, respectively.

[54] Part I does not have any illustrations; Part III has only a small engraving of the receptacle containing the bezoar stones from Peru, which Clusius did not include until 1605; Clusius, *Exoticorum libri decem*, 327.

[55] N. Monardes, *Simplicium medicamentorum ex Novo Orbe delatorum* (Antwerp, 1579) is considered to be a simple reprint of the edition of 1574. On the other hand, the following should be considered to be new editions: N. Monardes, *Simplicium medicamentorum ex Novo Orbe India nascentium liber* (Antwerp, 1593), and C. Clusius, 'Exoticorum liber decimus: sive simplicium medicamentorum ex novo orbe delatorum, quorum in Medicina usus est, Historia', in *Exoticorum libri decem* (Leiden, 1605), 295-354.

[56] N. Monardes, *Simplicium medicamentorum ex Novo Orbe delatorum [...] historiae liber tertius* (Antwerp, 1582).

[57] Ibid., 410-456.

[58] Ibid., 337-338 and 377, respectively.

The number of chapters with annotations in Part III is sixteen,[59] bringing the total number of products on which Clusius provided commentaries or annotations up to thirty-three, almost twice as many as in 1574, although in ten cases the commentary had not changed during the interval of nineteen years.

When Clusius was preparing the *Exoticorum* – the major intellectual project on exotic flora and fauna that he began towards the end of his life – he devoted most of his effort to a satisfactory integration of Part III of the *Historia medicinal* in the orderly arrangement into which he had always wanted to convert his edition of Monardes' material.[60] In fact, the major novelty of the edition, which was included as Book X of the *Exoticorum*, was this unification, in seventy-seven chapters. The chapters were now numbered for the first time, as a sign that Clusius wanted to draw attention to the novelty of his organisational activity. These were not the only innovations. Clusius wrote seven new notes, bringing the number of annotated chapters up to forty, covering more than half of the products dealt with by Monardes. Certainly, three of these new notes were very dry, as in the case of the *lignum aromaticum*.[61] This continuous effort is even clearer in the ten notes that were expanded in 1605, including the inclusion of three new engravings: the *guayaci ramulus*, a fragmentary branch to which the extensive enlargement of his commentary on the guajacum is devoted,[62] and the *lapis tiburonum* and *lapis bezar*; the latter was one of the products that attracted Clusius' most intense interest throughout his career, as can be seen from the references to it in the correspondence with his friends, the progressive history of the annotations, and the additions, including illustrations, that he made to his successive editions of the *Historia medicinal*.[63]

The images in Clusius' edition are always related to his commentaries and the essential criterium by which they are chosen and which tends to make them independent of the *Historia medicinal*. It is an original appropriation – albeit partial and fragmentary – dictated by the objective of showing his readers his own 'experience' with exotic materials. This experience was determined by the peculiar conditions of access to those materials and to certain criteria of proof and demonstration of that experience which seem to be based on the

---

[59] Plus a few small notes in the letter from Pedro de Osma to Monardes, which concluded this version of the Clusian recataloguing; Monardes, *Simplicium medicamentorum*, 394-402.

[60] The text that he placed at the beginning of Book X, to explain to the 'candido lectori' the full gestation of the *Historia medicinal* and of his own version, is significant in this respect; Clusius, *Exoticorum libri decem*, 296.

[61] Clusius, *Exoticorum libri decem*, 324.

[62] Ibid., 314.

[63] Monardes, *De simplicibus medicamentis*, 52-54, without illustrations; Monardes, *Simplicium medicamentorum*, 363-365 and 447-454, with two new engravings; Clusius, *Exoticorum libri decem*, 326-330, with a third new engraving.

'authority' that is conferred on it by is erudite readings and by the wide network of 'eminentissimi ac illustrissimi viri' that Clusius had managed to create around him and which, time and again, he conjures up before the eyes of his readers.

This can be made clear by the example of the sassafras tree, a key product in Part II of the *Historia medicinal*, second only to tobacco, and to which a full-page illustration was dedicated.[64] Monardes wrote that the 'wood and root' of this tree were a medicinal product 'of great properties', introduced by the Amerindians to the French who had tried to settle in the peninsula of Florida. It had reached Monardes through a French intermediary three years earlier (i.e. in 1566-67).[65] From that moment on, the physician of Seville tried out the 'marvellous effects' of the sassafras on his patients; ten pages of his work are devoted to the details of these experiments. Clusius, following his usual practice, translated the entire chapter faithfully enough, although he decided to get rid of the image of the sassafras tree.[66] It is in his commentary, however, that his intervention takes a completely different turn from Monardes. In the first of the five paragraphs, certainly, he states how 'recently' (the year of writing is 1573) he had received from 'Francisco Zennig Pharmacopola Bruxellensi diligentissimo, mihique amicissimo' a wood whose scent and flavour corresponded to Monardes' description, but the rest of the annotation has nothing to do with the sassafras. The scent of that wood – which is taken to correspond to the one described by Monardes – reminds Clusius of that of the *molle* tree that his friend Jean de Brancion ('splendidissimo illustriss. viro') had managed to cultivate in his garden in Mechelen. This leads Clusius to expatiate on this other tree from Peru and to reproduce a full-page image of one of its branches.[67] Without a break, we have passed from the anonymous 'French' of Florida and the witnesses to Monardes' successful experiments with the sassafras in Seville, to a pharmacist from Brussels and the charming garden of his friend Brancion, and to a discussion of the 'history' of a tree, having recourse to his scholarly erudition to illustrate the 'properties' of the 'wine' that is extracted from it in a gloss on what Clusius was able to find in Pedro Cieza de León's *Chronica del Perú*.[68] In 1593, Clusius substantially modified the paragraph of his commentary on how the sassafras came into his hands: logically enough,

---

[64] Monardes, *Historia medicinal*, 39v-49v; the engraving is on 39v.
[65] 'The Frenchman who had been in those parts' probably reached Seville after the expedition of Menéndez de Avilés in 1565 to dislodge the French from the settlement that they had established one year earlier in Florida.
[66] Monardes, *De simplicibus medicamentis*, 44-47.
[67] Ibid., 48, the engraving is on 49.
[68] P. Cieza de León, *Primera parte de la chrónica del Perú* (Seville, 1553), 111-117.

the now obsolete 'recently' has been removed, while the reference to the apothecary from Brussels has now been expanded with the words 'proximis his anniis Londino ab aliis etiam amicissimis viris C. V. Richardo Garth, Hugone Morgano pharmacipoea Regio & Iacobo Gareto iuniore mihi Viennam missi magna eaque libralia fragmenta', unfolding before the reader's eyes a fragment of the map of Clusian geography formed by contacts, readings and erudite notes, friends and important persons.[69] Finally, in 1605 he adds that his friend James Garet, apothecary in London, had sent him a fragment of sassafras in 1600, but he still cannot take his mind off the *molle* tree and adds new information that widens even further that radically European, radically Old World personal map.[70]

Clusius' decisive intervention in the text of Monardes thus lies in the commentaries and related images. These commentaries are of two kinds. On the one hand, Clusius establishes a body of bibliographical references from which he selects for his readers some item of discussion related to the plant and either its novelty or its shared identity with an Old World plant, forming a sort of Clusian library. On the other hand, Clusius provides an extensive and varied list of persons who have provided him with information, drawings, and plants or parts of them. He considers it necessary to make their testimony explicit to the reader, most of the time to lend authority to his own account. In both cases – the bibliographical references and the personal references – their function is above all to provide the authority that the commentator appropriates to establish what is in need of commentary (and if so, of what kind), and what is not in the text that he is translating.

The library on which Clusius drew to comment on Monardes consists of a group of a few works that he uses relatively frequently, and another group of works on which he draws only sporadically. The first group consists, essentially, of works by Francisco López de Gómara, Pedro Cieza de León (each used in nine commentaries),[71] Gonzalo Fernández de Oviedo (used on seven

---

[69] N. Monardes, *Simplicium medicamentorum*, 359-361.

[70] C. Clusius, 'Exoticorum liber decimus: sive simplicium medicamentorum ex novo orbe delatorum, quorum in Medicina usus est, Historia', in *Exoticorum libri decem*, 322. The references to his friends and his fragmentary posssessions of branches, roots and seeds of the *molle* are now extended to include the physicians Simón de Tovar and Everardus Vorstius, because the former wrote from Seville in 1593 on the 'grapes' of the *molle*, and the latter told Clusius that he had managed to cultivate an exemplar in Rome. Finally, he relates his own experience with a *molle* seed in England and offers as his ultimate and final authority on the subject 'C. V. Matthia de Lobel', whose drawing appears in his commentary on balsam, which had not yet been published. All of this, it should be recalled, is to be found in a note commenting on Monardes' chapter on the sassafras, from whose wood a 'water' was extracted that the *Indios* of Florida used to cure 'their ailments'.

[71] F. López de Gómara, *Primera y segunda parte de la historia general de las Indias, con todo el descubrimiento y cosas notables que han acaecido desde que se ganaron asta el año de 1551. Con la conquista de México y de la nueva España* (Medina del Campo, 1553) and P. Cieza de León, *Primera parte*.

occasions),[72] and Juan Fragoso (whose name appears five times).[73] Among the second group we find a total of seventeen authors, four of whom are cited on three occasions (Agustin de Zárate,[74] Jean de Léry,[75] André Thevet,[76] and Garcia da Orta),[77] while the others are only cited once or twice. Clusius also refers to his own works: three times to other parts of the *Exoticorum*,[78] and once to his *Per Hispanias*.[79] Moreover, two of the three references to Garcia da Orta are actually references to Clusius' scholia, not to the text of the Portuguese writer.

Clusius' temporally and spatially extensive re-reading thus leads to a decisive and radical modification of the original work, in sharp contrast to his relative fidelity with regard to the translation of Monardes' text and his criteria for grouping and classifying the products.

By way of conclusion

Clusius did everything possible to integrate the work of Monardes in his own *Exoticorum* project. To that end, he turned his attention to the botanical materials deriving from the New World, to which his attitude was primarily scholarly, since his attempts to gain knowledge based on his own experience were conducted with fragments, pieces of branches or roots, seeds that did not always grow, rotten fruit, dried herbs, etc.

---

[72] G. Fernández de Oviedo, *Primera parte de la historia natural y general de las Indias, yslas y tierra firme del mar oceano* (Seville, 1535). Clusius also knew Oviedo's earlier work that anticipated the *Historia natural y general*, his *Sumario de la natural historia de las Indias* (Toledo, 1526), quoting from it in the 1593 edition: see N. Monardes, *Simplicium medicamentorum*, 440.

[73] J. Fragoso, *Discursos de las cosas Aromáticas, árboles y frutales, y de otras muchas medicinas simples que se traen de la India Oriental, y sirven al uso de la medicina* (Madrid, 1572).

[74] A. de Zárate, *Historia del descubrimiento y conquista del Perú, con las cosas naturales que señaladamente allí se hallan, y los sucesos que ha avido* (Antwerp, 1555).

[75] J. de Léry, *Histoire d'un voyage fait en la terre du Bresil, autrement dite Amerique* (La Rochelle, 1578).

[76] A. Thevet, *Les singularitéz de la France antarctique, autrement nommée Amérique & de plusieurs terres & isles decouvertes de nostre temps* (Paris, 1557).

[77] Garcia da Orta, *Coloquios dos simples e drogas e cousas medicinais da India* (Goa, 1563). Clusius translated this work into Latin long before he came into contact with Monardes' work: D. Garcia ab Orta, *Aromatum et simplicium aliquot medicamentorum apud Indos nascentium historia* (Antwerp, 1567). After this first edition, Clusius' translation and commentaries to the *Coloquios* of Garcia da Orta accompanied the editorial adventure of the Clusian translation of Monardes in the successive editions of 1574, 1579, 1593; of course, the work was also included in the *Exoticorum libri decem* of 1605.

[78] Clusius, *Exoticorum libri decem*, 324 (*lignum aromaticum*, reference to lib. 4, cap. 11), 326 (on *lapis tiburonum*, reference to lib. 6, cap. 18) and 330 (on *armadillo*, reference to lib. 5, cap. 15).

[79] C. Clusius, *Rariorum aliquot stirpium per Hispanias observatarum historia, libris duobus expressa* (Antwerp, 1576). It concerns Clusius' commentary on the chapter that Monardes devoted to the *Caçavi*; the commentary first appeared in Clusius, *Simplicium medicamentorum*, 440.

It is obvious that Monardes' approach was a particular one, since he never set foot on the other side of the Atlantic either and his primary interest was in collecting information that would enable him to know the potential of the American plants as efficacious medicinal remedies within the conceptions of the Galenic medicine in use, often as succedanea for the exotic medicinal products that did not come from the Americas and which he presented as more expensive, less abundant, and less effective.

In fact, as the examples quoted have been intended to show, both Monardes and Clusius present us less with gazes of the New World than with rapid glimpses of what was brought by ships (Monardes) or sent by friends (Clusius), before rapidly refocusing on that Old World that enveloped them and called for their genuine attention.

Monardes already presupposes a consolidated distancing of the other, but Clusius accentuates this process to the point of rendering the other almost invisible, like a distant horizon that fades into the background. Clusius faithfully translates the passages in which Monardes presents the knowledge of the Amerindians about their plants, but in doing so he codifies them in such a way that the Latin reader whom he is addressing receives a sort of fossilised version of the text of Monardes. This fossilisation is primarily due to Clusius' manifest lack of interest in all that Monardes conveys concerning the knowledge and practices of the Amerindians, or concerning his experiments with his patients and the effects of the remedies. This explains why his notes never add anything on these points, except when his published source has something to say, as in the case, for example, of the names in Náhuatl that he adds thanks to his reading of López de Gómara, or the Tupi words that he includes thanks to his reading of Jean de Léry. Clusius' exposure of the texts of Monardes to his readings and to what his friends have sent him certainly separates the Clusian translation from that fondness for 'experience' lacking in erudition that Monardes almost always imposed on himself.

At the end of the sixteenth century, the natural world of the Americas seems to have been relegated by European intellectuals to the status of a quarry (a quarry that they had never visited, in most cases) whose only function was to supply new succedanea, variations on or exceptions to the flora and fauna of the Old World.

Instead of referring to the perception of the other, rapidly ignored for their incapacity to confer meaning on the knowledge that has to be extracted from the plants, we can speak of an authentic expropriation of the other by way of an apparent interest in or respect for their names, plants, ailments and remedies. The expropriation is already perceptible in Monardes, and much more so when Clusius translates him for his readers. What is conveyed to them is thus a definitively fragmented American nature, converted into dozens of products, some of them succedanea for other known products, that are only useful for the construction of European knowledge if they are tested by the authorised experience of the only ones who can give them meaning. In the last resort, it is the names of those *clarissimi eruditissimique viri* that have passed into the irreversible construction of the history of European natural science thanks, among other things, to the Clusian annotations.

To gaze at the other – to collect this or that of their names, their plants, their healing practices – is only a form of self-reflection, never a way of thinking about the other. Alterity is the excuse to return to oneself and to insist that the only gaze that deserves to be reflexive and eternal is the one turned on oneself; the others are just glimpses, rapid and furtive glances by those who only want to return rapidly to the mirror that reflects time after time their own image, though without understanding anything of what the mirror offers or conceals.

Americana in the *Exoticorum libri decem* of Charles de l'Écluse

Peter Mason

Well, there are many worthy creatures waiting in the wings – aardvarks, armadillos, penguins, and more.[1]

Images lead lives of their own, irrespective of the intentions of those who produce them, and these lives may be very long indeed.[2] William B. Ashworth Jr has devoted several articles to charting the long lives of many Renaissance illustrations of animals,[3] although he unfortunately tends to ignore the evidence of non-printed sources.[4] Two of the candidates that he mentions as deserving further study – the armadillo and the penguin – are both illustrated in the *Exoticorum libri decem* (hereafter *Exoticorum*) of Charles de l'Écluse (Carolus Clusius) (Ill. 20), who was the first to name and describe for the scientific community the Magellanic penguin. What the armadillo and the Magellanic penguin also have in common is that they both derive from the American continent. Indeed, the armadillo is so closely associated with that continent that it features regularly in visual allegories or personifications of America as one of the four continents.[5] And what they also have in common is that Clusius was receiving

---

[1] W.B. Ashworth Jr, 'The persistent beast: Recurring images in early zoological illustration', in A. Ellenius (ed.), *The natural sciences and the arts* (Stockholm, 1985) [Acta universitatis Upsaliensis, Figura nova series, 22], 46-66, here 65.
[2] P. Mason, *The lives of images* (London, 2001).
[3] W.B. Ashworth Jr, 'Remarkable humans and singular beasts', in J. Kenseth (ed.), *The age of the marvelous* (Hanover [N.H.], 1991) [Exhibition catalogue, Hood Museum of Art, Dartmouth College], 113-144; 'Natural history and the emblematic world view', in D.C. Lindberg and R.S. Westman (eds), *Reappraisals of the scientific revolution* (Cambridge, 1990), 303-332; 'Emblematic natural history of the Renaissance', in N. Jardine, J.A. Secord and E.C. Spary (eds.), *Cultures of natural history* (Cambridge, 1996), 17-37.
[4] For instance, his discussion of the iconography of the marmoset in the *Exoticorum* and later ('The persistent beast', 54) fails to refer to what is probably the earliest European representation of a marmoset: the platyrrhine monkey in a portrait of Cardinal Antonio del Monte (National Gallery of Ireland, Dublin) by Sebastian del Piombo that may date from as early as 1516. See M. Donattini, 'Orizzonti geografici dell'editoria italiana (1493-1560)', in A. Prosperi and W. Reinhard (eds.), *Il Nuovo Mondo nella coscienza italiana e tedesca del Cinquecento* (Bologna, 1992), 79-154, esp. 116-117.
[5] If the artist of the armadillo added to a manuscript of Pietro Candido Decembrio's *De omnium animalium naturis atque formis* (Cod. Urb. Lat. 276, Biblioteca Apostolica Vaticana, Rome) was a disciple

Ill. 20. Carolus Clusius. Painting, dated 1606 by Filippo Paladini after a woodcut portrait of Clusius at the age of 75 by Jacques de Gheyn II, published as a frontispiece to Clusius' *Rariorum plantarum historia* (Antwerp, 1601).

the latest information about them in the period immediately prior to the completion of the *Exoticorum*. Since both information about *americana* and the American items themselves were arriving in the Netherlands in the years immediately after the return of the first Dutch expeditions to the South Atlantic,[6] Clusius' treatment of them offers us a fascinating glimpse of the 'kitchen' in which he was putting the finishing touches to his *magnum opus*.

The sources of Clusius' information about Americana

Among the *Americana* that we find described in the *Exoticorum* are fruits, different kinds of wood, birds, geese, armadillos, serpents, lizards, cacti, beans, potatoes, sloths, monkeys, and the manati. Not all of these representatives of the animal and vegetable worlds were or are equally well known. But when it comes to examining the sources of the information presented by Clusius in his *Exoticorum*, the least known can turn out to be the most interesting, and vice versa.

This list is not exhaustive, nor could it ever be. In a number of cases, Clusius himself is uncertain whether the item in question came from the Americas or not, so that it is impossible to compile a full list of *Americana* on the basis of his descriptions alone.[7] One could, it might be argued, bring the insights of modern science to bear on the question to facilitate more reliable identifications. There are, however, two objections to such a practice. First, the descriptions provided by Clusius are not always sufficient to provide a modern zoologist or botanist with enough material for a firm identification. Second, and more important, if we aim to get closer to an understanding of Clusius' own methods and practice, we have to bracket such later insights as irrelevant, if not confusing. For it is with the categories of his own day that we must start if we are to avoid the twin perils of anachronism and scientific triumphalism.

---

or imitator of Raphael, this would make his rendering one of the earliest illustrations of this creature; see D. Franchini et al., *La scienza a corte. Collezionismo eclettico, natura e immagine a Mantova fra Rinascimento e Manierismo* (Rome, 1979), 65-68. On the early iconography of the armadillo – a very popular creature in the European *Kunst- und Wunderkammern* – see also F. Egmond and P. Mason, 'Armadillos in unlikely places. Some unpublished sixteenth-century sources for New World *Rezeptionsgeschichte* in Northern Europe', *Ibero-Amerikanisches Archiv*, 20/1-2 (1994), 3-52.

[6] On the presence of *Americana* in European collections, see C.F. Feest, 'European collecting of American Indian artefacts and art', *Journal of the history of collections*, 5/1 (1993), 1-11, P. Mason, 'From presentation to representation: *Americana* in Europe', *Journal of the history of collections*, 6/1 (1994), 1-20, idem, 'Faithful to the context? The presentation and representation of American objects in European collections', *Anuário antropológico*, 98 (2002), 51-95; E. Bujok, *Neue Welten in europäischen Sammlungen. Africana und Americana in Kunstkammern bis 1670* (Berlin, 2004).

[7] This problem is basic to any encyclopaedic attempt to chart the extent of *Americana* in textual or visual sources; see the remarks in P. Mason, 'Escritura fragmentaria: aproximaciones al otro', in G.H. Gossen, J.J. Klor de Alva, M. Gutiérrez Estévez and M. León-Portilla (eds.), *De Palabra y Obra en el Nuevo Mundo*, vol. III: *La formación del otro* (Madrid, 1993), 395-430.

Clusius' interest in and contact with *Americana* go back at least to the year 1564, when he visited Sebald Linz in Lisbon during his travels through the Iberian peninsula, which were to result in the *Rariorum aliquot stirpium per Hispanias observatarum historia* (Antwerp, 1576).[8] Sebald's son Roderic had sailed to the Portuguese port from Pernambuco (in Brazil) with some parrots, monkeys and a marmoset, whom he fed on maize, another American novelty.[9] Only one marmoset survived the voyage. It was still alive when Clusius left Lisbon for Spain in January 1565. At the time of writing the Appendix to the *Exoticorum*, Clusius could remember its shape, and subsequently obtained a coloured image of it (Ill. 21).[10] It is also probably to his Iberian travels that we can attribute the connection with Simón de Tovar, who had two botanical gardens in Seville and at some point in time sent Clusius a Peruvian bean.[11]

Clusius was so impressed by the dragon tree from the eastern side of the Atlantic which he saw in Lisbon that it was the first plant to be described and illustrated in his *Rariorum aliquot stirpium per Hispanias observatarum historia*, published by Clusius' friend Christoffel Plantijn in Antwerp in 1576.[12] Besides descriptions of the flora of Spain and Portugal based on this journey, however, the *Rariorum [...] historia* also included accounts of items from the other side of the Atlantic, such as the American avocado tree that Clusius saw in a monastery in Valencia, the sweet potato, the thuya, the agave,[13] and a type of American cane.[14] The publishing history of this work is revealing of Clusius' eagerness to incorporate new data in his publications, a characteristic that, as we shall see, applies to the publication history of the *Exoticorum* as well. Though

---

[8] On Clusius' Iberian journey see J.L. Barona, this volume.

[9] *Exoticorum libri decem*, 364. Whether the earliest European representation of American maize should be attributed to Hans Burgkmair for his *Triumph of Maximilian* (H. Honour, *L'Amérique vue par l'Europe* [Paris, 1976] [Exhibition catalogue, Grand Palais, Paris], 11) or to Giovanni da Udine (for his floral and vegetable decorations to the Villa Chigi 'detta Farnesina' in Rome [G. Caneva, *Il mondo di Cerere nella Loggia di Psiche* (Rome, 1992)]), the motif certainly existed by the end of the second decade of the sixteenth century.

[10] *Exoticorum libri decem*, 372.

[11] *Exoticorum libri decem*, 69.

[12] See P. Mason, 'A dragon tree in the Garden of Eden. A case study of the mobility of objects and their images in early modern Europe', *Journal of the history of collections*, 18/2 (2006), 169-185.

[13] On the agave, which features under the same – Catalonian – name of *filagul* not only in the *Rariorum aliquot stirpium [...]* but also in the herbarium that Pier'Antonio Michiel compiled between the beginning of the 1570s and his death in August 1576, see J. Pardo Tomás, 'Tra 'oppinioni' e 'dispareri': la flora americana nell'erbario di Pier'Antonio Michiel (1510-1576)', in G. Olmi and G. Papagno (eds.), *La natura e il corpo. Studi alla memoria di Attilio Zanca* (Florence, 2006), 73-100, esp. 91-92.

[14] See L. Ramón-Laca Menéndez de Luarca, 'Charles de l'Écluse y la flora ibérica', in idem and R. Morales Valverde (eds.), *Charles de l'Écluse de Arras, Descripción de algunas plantas raras encontradas en España y Portugal* (Castilla y León, 2005), 22.

Ill. 21. Cercopithecus sagouin. From Clusius' *Exoticorum libri decem* (Leiden, 1605), 372.

he probably contacted Plantijn with a proposal for the publication soon after returning from the Iberian peninsula, political troubles in the Netherlands, followed by Clusius' departure for Austria, both delayed the appearance of the publication. Characteristically, Clusius turned this delay to his advantage: besides descriptions of the flora of Spain and Portugal, the work incorporated not only the results of Clusius' earlier observations of the flora of Southern France (carried out during his stay in Montpellier in 1551-1554), but also an appendix on the plants of European Turkey, including the famous tulip, which had been sent to him by the imperial representative in Constantinople, Ogier van Busbeck.[15] Clusius was in Vienna from 1573 to 1578, and again from 1581 to 1588. It was during this period that he received several American objects connected with British voyages to the New World, such as the Brazilian bean purchased by the British apothecary Richard Garth and sent to Clusius in

---

[15] An exemplar of the *Rariorum aliquot stirpium [...]* preserved in the Koninklijke Bibliotheek Albert I in Brussels (VH 6773 A LP), was a gift by Clusius to Ogier van Busbeck. See E. Cockx-Indestege and F. de Nave (eds.), *Christoffel Plantijn en de exacte wetenschappen in zijn tijd* (Ghent, 1989) [Exhibition catalogue, Museum Plantin-Moretus Antwerp], no. 32.

Vienna in 1585,[16] and two other types of Brazilian bean that Clusius received from another British apothecary, James Garet Jr.[17]

Clusius moved from Vienna to Frankfurt in 1588. Three years later Garth sent him a branch of a Brazilian 'Junipappeeywa',[18] and in the same year James Garet sent him a Virginian Macoqwer, a kind of gourd. Clusius' account of it in the *Exoticorum* includes a surprising amount of ethnographic detail: the Virginian Indians first emptied it, then filled it with small stones and attached it to a stick as a rattle, but without feather ornaments[19] (unlike the Brazilian rattles, which were well known to the educated European public through the description of them in Michel de Montaigne's essay *Des cannibales* of 1580[20]). Two years after the publication of Walter Raleigh's *The discoverie of the large, rich and bewtiful empyre of Guiana* in 1596, Garet sent Clusius a 'scaly fruit' (*squamosus fructus*) which had been brought to London by Raleigh's expedition.[21] It was also during his years in Frankfurt that Clusius received the famous image of the potato.[22]

In 1593 Clusius moved to Leiden in the Netherlands to take up his post as praefectus of the botanical garden. He was joined there after a few years by a manati calf: brought back from the Atlantic by Dutch sailors in 1600, it was hung from the gateway of the Leiden *hortus botanicus* after Clusius had had the opportunity to observe it in Amsterdam and to have an illustration of it made (Ill. 22).[23]

---

[16] *Exoticorum libri decem*, 69.
[17] *Exoticorum libri decem*, 60-61.
[18] *Exoticorum libri decem*, 10.
[19] *Exoticorum libri decem*, 23.
[20] M. de Montaigne, *Oeuvres complètes*, ed. A. Thibaudet and M. Rat, introduction and notes M. Rat (Paris, 1962), 206.
[21] *Exoticorum libri decem*, 25.
[22] F.W.T. Hunger, *Charles de l'Ecluse. Carolus Clusius. Nederlandsch kruidkundige 1526-1609*, 2 vols. (The Hague, 1927-43), vol. I, 175.
[23] The other manatis brought back by the sailors did not achieve such an illustrious posthumous fame: a male manati had been stuffed with straw and hung from the beam of the ship with the calf on its back, but all that was left of the female manati was her ribs with a bit of flesh still clinging to them; *Exoticorum libri decem*, 132-135. On the curiosities on show in the Leiden hortus see E. de Jong, 'Nature and art. The Leiden Hortus as "Musaeum"', in idem and L.A. Tjon Sie Fat (eds.), *The authentic garden. A symposium on gardens* (Leiden, 1991), 37-60. Among the inventories published by De Jong, the first 25 items in the list of contents drawn up in 1617 must date from between 1600 – when the old gallery was in operation – and 1610-1612 – when the new gallery was opened. The 'two Indian hanging rope beds' mentioned there must be hammocks (the 1659 inventory uses the word 'hamack'), but these are the only objects in that list that can be clearly identified as American. Once again, Michel de Montaigne had helped to disseminate knowledge about the Amerindian hammock in his essay *Des cannibales*; see note 20.

Ill. 22. Manati. From *Exoticorum libri decem*, 133.

Clusius' interest in both the East and West Indies probably grew after his move to Leiden. He was in contact with Jan Huygen van Linschoten and Bernardus Paludanus in Enkhuizen, and with the merchants Johannes de Weely, David Sinapius and Emanuel Sweert in Amsterdam, and Simon Parduyn in Middelburg.[24] Stephen Jan Scharm, an apothecary in Amsterdam, sent him a bean from Haiti in 1600.[25]

An indication of the speed at which access to information about the exotic world was taking place can be seen from the fact that the sailing instructions contained in Linschoten's *Itinerario* were rushed through the press in 1595 in order to be given to the first Dutch fleet to sail to the East Indies in April of that year. Cornelis de Houtman, who was in command of that voyage, Jacob van Neck, who commanded the second Dutch voyage to the East Indies (1598-1599/1600), and his vice-admiral Wijbrant van Warwijck, are all mentioned in the pages of the *Exoticorum* as suppliers of *exotica*. As Nicolás Monardes had enjoyed his privileged position in the port of Seville,[26] so Clusius took advantage of his proximity to the main Dutch port of Amsterdam. Clusius himself mentions that he went to Amsterdam in August 1601 'to see whether the ships returning from Java and the Moluccas had brought back any exotica'.[27] The United Dutch East Indian Company (VOC) was established in the following year.

---

[24] Hunger, *Charles de l'Ecluse*, vol. I, 266-268.
[25] *Exoticorum libri decem*, 62.
[26] See J. Pardo Tomás, this volume.
[27] *Exoticorum libri decem*, 8.

Dutch voyages to the South Pacific followed the same pattern. An account of the catastrophic expedition of Jacques Mahu and Simon de Cordes that left Rotterdam in 1598 and passed through the Strait of Magellan to harry the Spanish on the Chilean coast, during which more than a hundred men, including both commanders, lost their lives, was published (with illustrations) in 1600.[28] Likewise, Olivier van Noort's expedition of 1598-1601, which operated in the waters of the Strait simultaneously with the fleet of Mahu and De Cordes, was described in an account published in both Amsterdam (by Cornelis Claesz) and Rotterdam (by Jan van Waesberghe) in 1602.[29] Jan de Maes, a relative of Clusius, mentions having read Van Noort's account with great interest (in a letter to Clusius dated 15 April, 1602).[30] Incidentally, the chronology of these voyages itself could furnish Clusius with a clue to the provenance of an item. In discussing a 'Serpens peregrinus', Clusius deduced that it must be from America because there was no voyage to the East Indies in the year in which it was brought back.[31]

By the time of writing the *Exoticorum*, then, Clusius had been intermittently exposed to *Americana* for almost forty years, and the pace at which such objects became available to him was accelerating soon after his move to Leiden. He had connections with local traders in exotica, access to scholars like Paludanus who were involved in collecting and describing them, was on friendly terms with several of the persons who led the overseas expeditions, and could draw on a network of individuals abroad, such as the British apothecaries, to supply him with objects themselves, or with verbal and/or visual representations of exotic objects.

The weak link in this knowledge network was an epistemological one. It was the gap between first-hand knowledge of the object in question, and having to rely on second-hand verbal or visual information in the absence of the object, that was to create the greatest difficulty for Clusius in the task of compiling the *Exoticorum*, for Clusius repeatedly stressed that if an opinion was to be authoritative, it had to be based – and to be shown to be based – on accurate, first-hand observation. This emerges clearly from the case of what is

---

[28] Barent Jansz Potgieter, *Wijdtloopigh verhael van tgene de vijf schepen (die int jaer 1598 tot Rotterdam toegherust werden, om door de Straet Magellana haren handel te drijven) wedervaren is [...]* (Amsterdam, 1600). See also W. Klooster, *The Dutch in the Americas 1600-1800* (Providence, 1997), 11.

[29] O. van Noort, *Beschryvinghe vande voyagie, om den geheelen werelt cloot, ghedaen door Olivier van Noort [...]* (Amsterdam, 1602). A summary account of the voyage, *Extract oft kort verhaal wt het groote journael*, was published by Jan van Waesberghe within a month of Van Noort's return in August 1601.

[30] Leiden University Library, VUL 101 (= CLUY202). There are eighteen letters in French from Jan de Maes to Clusius covering the period from 1596 to 1607.

[31] *Exoticorum libri decem*, 113-114.

probably the most famous of Clusius' American exotica: the drawing of a potato.[32] The watercolour now in the Plantin Museum in Antwerp is inscribed with a note by Clusius: 'Taratoufli à Philippo de Sivry acceptum Viennae 26 Januarii 1588'. It is not entirely clear from this statement alone whether what Clusius received from Philippe de Sivry (Lord of Walhain and governor of Mons) was the potato itself or the drawing of it. We do know that James Garet and Caspar Bauhin also supplied him with drawings of potatoes. The *verbal* description of the potato in Clusius' *Rariorum plantarum historia* of 1601 follows the earlier description given by Pedro Cieza de León in the *Parte primera dela chronica del Perú* (Seville, 1553). For the visual representation of the potato in the *Rariorum plantarum historia*, however, Clusius did not use any of these drawings, but commissioned new ones.

The potato was not exactly a novelty in Europe at this time. After its introduction to Spain, the potato spread rapidly through Europe, so the difficulties in obtaining first-hand observation were less. But when it came to less common exotica, Clusius ran up against considerable obstacles, as we shall now see.

Object, report, image

If I never have, can, must or shall see a white bear alive, have I ever seen the skin of one? Did I ever see one painted? – described? Have I never dreamed of one?[33]

After 1557, there was no need to dream of the Brazilian sloth. The first European illustration of a Brazilian sloth appeared as one of the woodcuts in *Singularitez de la France antarctique* by André Thevet, published in that year, who claimed to have kept a wounded sloth for twenty-six days.[34] And by the time of the publication of Clusius' *Exoticorum* in 1605, two more images of the sloth had been added to the repertoire.[35] Clusius obtained the first of these

---

[32] J. Balis, *Hortus Belgicus* (Brussels, 1962) [Exhibition catalogue, Bibliothèque Albert I, Brussels], 33-34; Honour, *L'Amérique vue par l'Europe*, no. 35; P. Vandenbroeck (ed.), *America bride of the sun* (Antwerp, 1992) [Exhibition catalogue, Royal Museum of Fine Arts, Antwerp], no. 206.

[33] Lawrence Sterne, *The life and opinions of Tristram Shandy, gentleman*, vol. V, chap. 43.

[34] André Thevet, *Les singularitez de la France Antarctique, autrement nommée Amerique* (Paris, 1557), f. 99v. For a modern edition, see *Le Brésil d'André Thevet. Les singularités de la France Antarctique (1557)*, ed. with commentary by F. Lestringant (Paris, 1997), 200.

[35] To the intervening period can be dated the image of a sloth walking upside down along the branches of a tree on folio 64 of a 134-leaf manuscript illustrating the fauna, flora and people of America; the manuscript has been attributed to an anonymous French Huguenot who served under Sir Francis Drake in the last years of the sixteenth century by F. Lestringant, *L'Expérience huguenote au nouveau monde (XVI*[e] *siècle)* (Geneva, 1996), 265-290. Since this manuscript was not published until a facsimile edition appeared in 1996 as *Histoire naturelle des Indes: The Drake manuscript in the Pierpont Morgan Library* (New York), this image of a sloth based on direct observation failed to have any

Ill. 23. Sloth. From *Exoticorum libri decem*, 111.

(Ill. 23) via one of his correspondents, Dietrich Clemensz Coornhert, who arranged for a drawing to be made of the preserved sloth that was in the collection of Rutger Jansz in Amsterdam.[36] The second came into Clusius' hands after the first six books of the *Exoticorum libri decem* had already been printed, and with them the engraving of the preserved sloth. This time it was the famous floriculturist and merchant in rare and curious specimens Emanuel Sweerts who told Clusius that he had just acquired a sloth in Amsterdam which had died on the voyage from America only a few days before. Sweerts obligingly dispatched the sloth to Clusius, who included a woodcut of this rather fierce-looking creature and a description of it in the 21-page appendix to the work (Ill. 24).[37]

---

further repercussions. For a fuller discussion of the iconography of the sloth, which corrects W.B. Ashworth's account in 'The persistent beast' on a number of points, see P. Mason, 'Il contributo dei Libri Picturati A. 32-38 alla comprensione dell'iconografia del Brasile olandese nei dipinti di Albert Eckhout e di Frans Post', in G. Olmi and G. Papagno (eds.), *La natura e il corpo. Studi alla memoria di Attilio Zanca* (Florence, 2006), 101-120.

[36] *Exoticorum libri decem*, 111.

[37] *Exoticorum libri decem*, 373. Although Rudolf II had a stuffed sloth in his collection in Prague, his court physician, Anselmus de Boodt, based his drawing of a sloth, signed 'A.D.B', on the image in Clusius' appendix. See A. Balis, 'Naar de natuur en naar model', in M.-C. Maelis, A. Balis and R.H. Marijnissen, *De albums van Anselmus de Boodt (1550-1632). Geschilderde natuurobservatie aan het Hof van Rudolf II te Praag* (Tielt, 1989), 63 and note 52. The drawing is on folio 35 of the second volume of the De Boodt albums.

Ill. 24. Sloth. From *Exoticorum libri decem*, 373.

The Appendix to the *Exoticorum* opens with a discussion of the bird of paradise. Sweerts had sent a bird of paradise to Clusius along with the sloth, and he also sent Clusius a Rex bird of paradise, which Clusius described and had illustrated (Ill. 25).[38] But the bird of paradise and the sloth were not just related in the mind of Emanuel Sweerts:

L'accordéon de la mappemonde, à une époque où la longitude ne peut être fixée avec précision […] permet de considérer dans le manucodiata, le fabuleux oiseau de Paradis des îles orientales, et le bradype du Brésil, aux moeurs nocturnes, deux créatures placées dans un rapport de symétrie inverse, diamétralement opposées sur le parallèle.[39]

Antonio Pigafetta, in his account of the first circumnavigation of the world by Magellan, had mentioned the gift to the King of Spain by one of the kings of the Moluccas of two

very beautiful dead birds, which are as thick as stock-doves, with small head and long beak, and legs a palm in length and as thin as a feather. They have no wings, but instead long feathers of diverse colours like large plumes. The tail is as long as that of a

---

[38] *Exoticorum libri decem*, 362.
[39] F. Lestringant, *Écrire le monde à la Renaissance* (Caen, 1993), 323 (= 'Le déclin d'un savoir. La crise de la cosmographie à la fin de la Renaissance', *Annales E.S.C.*, mars-avril 1991, no. 2, 243).

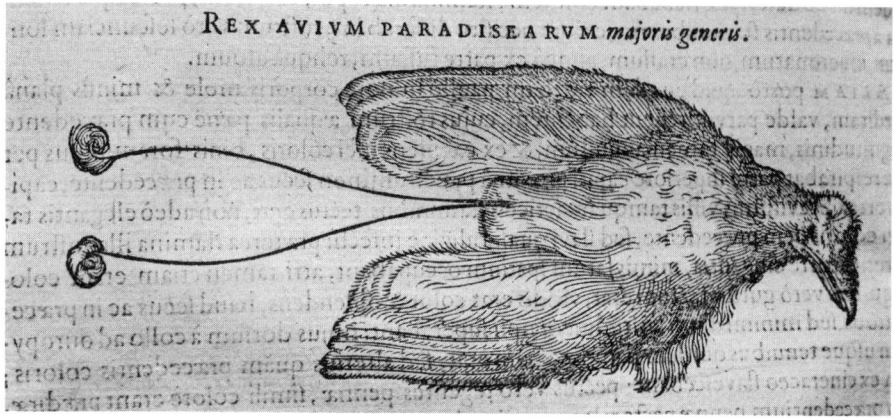

Ill. 25. Rex bird of paradise. From *Exoticorum libri decem*, 362.

stock-dove, and all the other feathers except the wings are of a tawny colour, and they never fly except when there is wind. We were told that those birds came from the earthly paradise, and were called *Bolon diuata*, that is to say, birds of God.[40]

A few years earlier than Pigafetta's account is a letter by Maximilian Transylvanus, a secretary to the Emperor Charles V, to the Archbishop of Salzburg on 5 October 1522. Transylvanus reports that the expedition was presented with no less than five *manucodiatas*, one of which he managed to obtain from the captain of the vessel for the Archbishop (along with some cinnamon, nutmeg, mace and cloves). His letter may be the earliest mention of the bird of paradise in European literature. He notes:

They hold these manucodiatas to be celestial, and even when they are dead they never corrupt or smell. Their plumage is of diverse and very beautiful colours, they are the size of turtle-doves, and have a very long tail, and if one of their feathers is plucked, another grows, even when they are dead. The kings take them into battle, and believe that if they have them with them they are safe and invincible in battle.[41]

One of the earliest birds of paradise to appear in a European collection is the one recorded in the 1523-4 inventory of the collection of Margaret of Austria in Malines, who kept a stuffed bird of paradise wrapped in taffeta in a small wooden casket in her library.[42] Among the early collectors of birds of

---

[40] J. Sebastián de Elcano, A. Pigafetta, M. Transilvano, F. Albo, G. de Mafra et al., *La primera vuelta al mundo* (Madrid, 2003), 300-301.
[41] J. Sebastián de Elcano et al., *La primera vuelta al mundo*, 62.
[42] 'Item, ung oyseau mort, appellé oyseau de paradis, envelopé de taffetaf, mis en ung petit coffret de bois'; see D. Eichberger, 'Dürer's nature drawings and early collecting', in D. Eichberger and

paradise we can also mention Conrad Peutinger in Augsburg and Johann Kramer in Nuremberg.[43]

From the Appendix to the *Exoticorum*, we know that Clusius had heard reports that the birds of paradise had feet, but he was unable to confirm them because the ones with feet taken to Amsterdam from the East were sold and transferred to Rudolf II in Prague before he had had a chance to see them and to have them illustrated.[44] So the woodcut of the bird of paradise in the Appendix to the *Exoticorum* is footless (Ill. 26).[45] The image is of a bird of paradise in the collection of Pieter Paaw in Leiden. In the Appendix, Clusius quotes Jehan de Weely, who sold the bird with feet to Rudolf, for confirmation of the fact that it did have feet. De Weely's letter to Clusius (in Dutch, dated 13 June 1605) has been preserved:

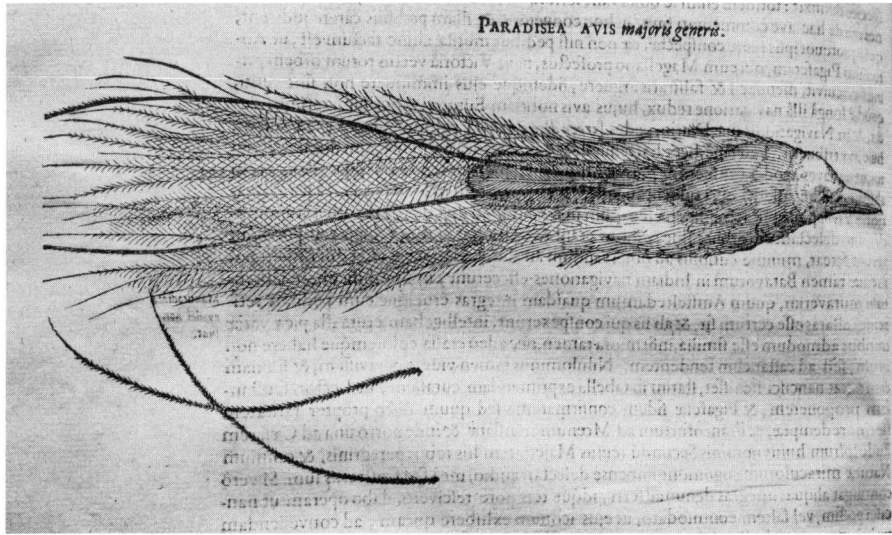

Ill. 26. Bird of paradise. From *Exoticorum libri decem*, 360.

C. Zika (eds.), *Dürer and his culture* (Cambridge, 1998), 26, and eadem, *Leben mit Kunst, Wirken durch Kunst. Sammelwesen und Hofkunst unter Margarete von Österreich, Regentin der Niederlande* (Turnhout, 2002), 185 (where the author mistakenly states that the bird of paradise came from Central America and not from the Moluccas).

[43] F. Koreny, *Albrecht Dürer and the animal and plant Studies of the Renaissance* (Boston, 1988) [Exhibition catalogue, Albertina, Vienna], 100.
[44] *Exoticorum libri decem*, 359ff.
[45] *Exoticorum libri decem*, 360.

The bird of paradise was in every respect like the vulgar sort, somewhat flat, not of the round kind that they call papaulben, but the other sort with its [snawen?] that they call the feet. It had two feet like a sparrow hawk or harrier, that looked unseemly and ugly, being pressed flat against the belly so that little more than the claws could be seen. The leg was dried and looked ugly too, so that the Indians very sensibly cut off the feet together with the leg, for it is the ugliest part of the bird, and in my opinion they all have similar feet.[46]

In fact, the 1607 inventory of the collection of Rudolf II mentions a large number of birds of paradise, variously without feet or wings, without feet but with wings, with feet but without wings, and one with both feet and wings.[47]

The problem here is that many of the items that reached Clusius did so in a more or less fragmentary state.[48] This was a problem which plagued many collectors of curiosities.[49] The ideal state in which in Clusius hoped to receive objects is exemplified by the *Echinomelocactus* that he bought in Holland in 1601. Having purchased it in perfect condition, he was able to dissect it and thus to arrive at authoritative knowledge based on first-hand observation (Ill. 27).[50] The other extreme is exemplified by the scaly horn that Johannes van Hoghelande bought from Dutch sailors in 1601 and lent to Clusius so that the latter could have an illustration made (Ill. 28). Clusius was given no indication of its provenance; he did not even know whether the creature it came from had one or two horns.[51]

It was not just the objects themselves, but also visual representations of them that could be fragmentary. Jacques Plateau sent Clusius a picture of the head and beak of a bird, but not of the whole bird (Ill. 29). Clusius wondered whether it might be the bird *alcatraz* described by the chronicler Gonzalo Fernández de Oviedo y Valdés, who had spent about thirty years in America and whose *Historia general y natural de las Indias* had earned him the reputation of the Pliny of the New World.[52] But without the full picture it was impossible for him to decide.

Clusius endeavoured to fill in the gaps in his information through an appeal to Oviedo in connection with several other items. For instance, the description

---

[46] Leiden University Library, VUL 101.
[47] A. Schnapper, *Le géant, La licorne, La tulipe. Collections françaises au XVIIe siècle* (Paris, 1988), 81.
[48] See the astute remarks on the Venetian patrician Pier'Antonio Michiel, who faced similar problems in the compilation of his herbarium, which included illustrations of and comments on more than forty American plants, in J. Pardo Tomás, 'Tra 'oppinioni' e 'dispareri': la flora americana nell'erbario di Pier'Antonio Michiel (1510-1576)'.
[49] P. Mason, *Infelicities. Representations of the exotic* (Baltimore/London, 1998), 70.
[50] *Exoticorum libri decem*, 92.
[51] *Exoticorum libri decem*, 109. The object in question looks as though it may have belonged to a pangolin. I am grateful to Espen Waehle for further information on this identification.
[52] *Exoticorum libri decem*, 106.

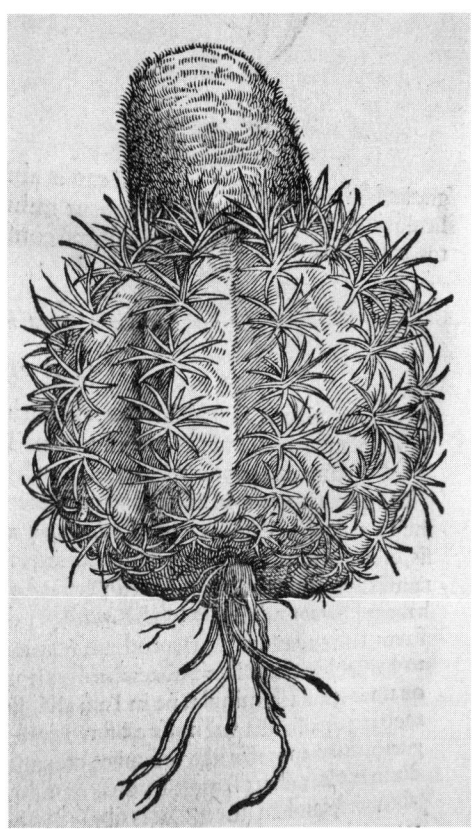

Ill. 27. Echinomelocactus. From *Exoticorum libri decem*, 92.

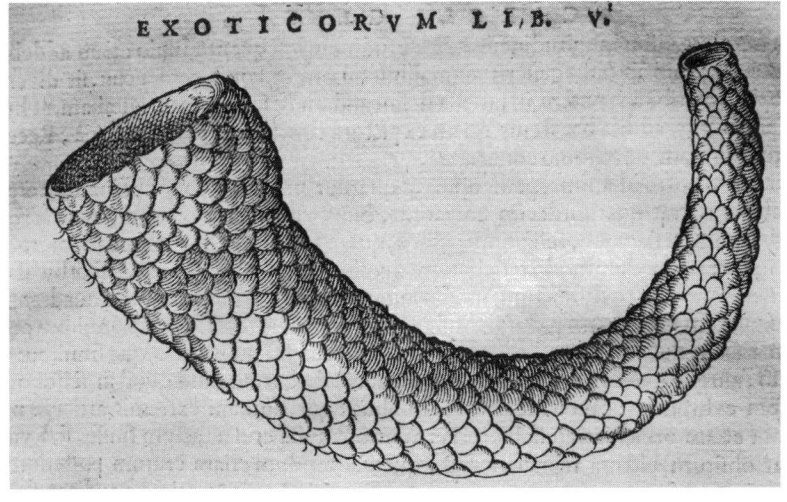

Ill. 28. Scaly horn. From *Exoticorum libri decem*, 109.

Ill. 29. Bird (alcatraz?). From *Exoticorum libri decem*, 106.

and woodcut of the 'Yvana', a type of lizard, are based on the animal in the collection of Pieter Paaw, but Clusius also refers to Oviedo in this connection.[53] Although the 'Ficus Indica' was not American, Clusius suggested a connection with an American tree, the mangrove (*mangle*), as described by Oviedo.[54] And after Paludanus had lent Clusius the branch of a tree in 1595 for him to describe and illustrate, Clusius wondered whether it might by the 'Gaguey' described by Oviedo.[55]

The quantity of information about America expanded considerably after 1600, in the years immediately prior to the completion of the *Exoticorum*. One can mention the *Bixa orellana* seed sent to Clusius by Juan de Castañeda in September 1602,[56] followed by a *Bixa orellana* branch sent by Peter Garet a month later. The woodcut in the *Exoticorum* (Ill. 30) is based on the branch, and Clusius recorded the use of *Bixa orellana* for body painting in America –

---

[53] *Exoticorum libri decem*, 116.
[54] *Exoticorum libri decem*, 3.
[55] *Exoticorum libri decem*, 11. On Oviedo's illustrations of New World plants see J. Pardo Tomás, 'Le immagini delle piante americane nell' opera di Gonzalo Fernández de Oviedo (1478-1557)', in G. Olmi, L. Tongiorgi Tomasi and A. Zanca (eds.), *Natura-cultura. L'interpretazione del mondo fisico nei testi e nelle immagini* (Florence, 2000), 163-188.
[56] There are fourteen extant letters from Juan de Castañeda to Clusius for the period from September 1600 to February 1604. See J.L. Barona in this volume.

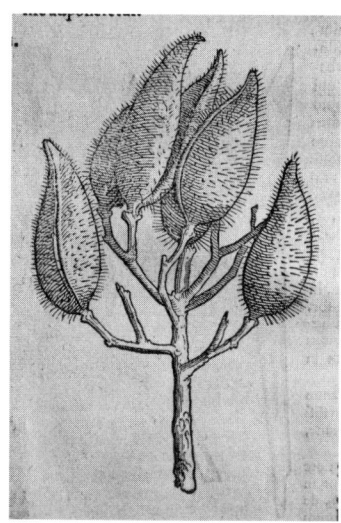

Ill. 30. Bixa orellana. From *Exoticorum libri decem*, 74.

evidence of his continued ethnographic interest.[57] In 1601 Peter Garet sent him a 'Lignum exoticum' from America.[58] Another piece of wood that Clusius received from Peter Garet arrived without a provenance; Clusius learnt that it had come from the Strait of Magellan from someone who had received a similar piece from Sebald de Weert, captain on the Mahu and De Cordes voyage, whose ship had been unable to pass through the Strait of Magellan and who had returned to Europe in 1600.[59]

Clusius cites the diaries of the Dutch expedition to the Strait of Magellan as the source for information about and an illustration of a bird that he was the first to describe for the scientific community: the Magellanic penguin, or *Anser Magellanicus*.[60] The illustration in the *Exoticorum* (Ill. 31) is indeed taken from an engraving of Dutch sailors killing penguins for food in the *Wijdtloopigh*

---

[57] *Exoticorum libri decem*, 74. The earliest European representation of the use of *Bixa orellana* in the context of American body painting is an anonymous French painting on parchment (Bibliothèque municipale, Rouen, XIR 188745) representing the Brazilian Indians who took part in the official entry of Henri II of France and his wife, Catherine de'Medici, in the port of Rouen on 1 October, 1550. The work was presumably completed soon after the event it portrays. For an illustration in colour see the exhibition catalogue *Americas lost*, ed. D. Levine, Musée de l'Homme (Paris, 1992), 117.
[58] *Exoticorum libri decem*, 7.
[59] *Exoticorum libri decem*, 8.
[60] *Exoticorum libri decem*, 101.

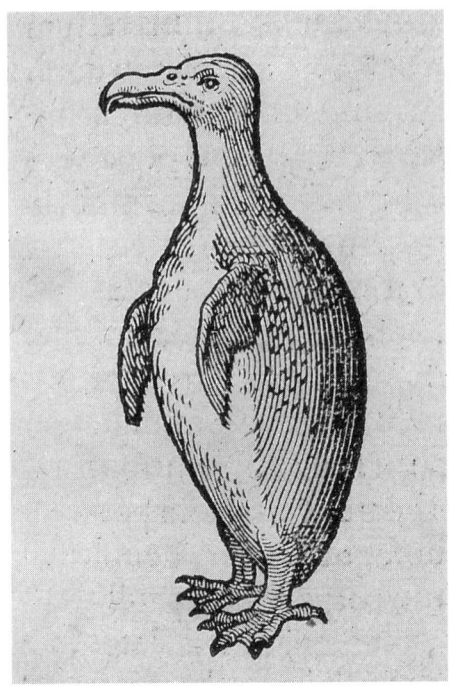

Ill. 31. Anser Magellanicus. From *Exoticorum libri decem*, 101.

Ill. 32. Anser Magellanicus. From Barent Jansz Potgieter, *Wijdtloopigh verhael van tgene de vijf schepen (die int jaer 1598 tot Rotterdam toegherust werden, om door de Straet Magellana haren handel te drijven) wedervaren is [...]* (Amsterdam, 1600).

verhael [...], an account of the voyage of Jacques Mahu and Simon de Cordes published in Amsterdam by Zacharias Heyns in 1600 (Ill. 32).[61] Clusius thus had ample time to include its findings in his work.

In the following year the same publisher issued a book with almost 150 woodcuts of individuals in national costumes, each accompanied by a descriptive quatrain, the *Dracht-Thoneel*, which included versions of the native peoples of Tierra del Fuego, based on, though not identical to, the woodcuts contained in the *Wijdtloopigh verhael* [...]. Though hardly relevant to a collection of national dress (intended as a guide for stage productions), the volume also contained a woodcut of shells (*klipklevers*) from the Strait of Magellan, and one of the Magellanic penguin (Ill. 33), based on the one in the *Wijdtloopigh verhael* [...], though with webbed feet like those of a duck. Another version of the image, but with more claw-like feet, appeared in the account of the same voyage included in Part 2 of the ninth volume of the encyclopaedic work *America*, published in a German-language version by the De Bry brothers in Frankfurt in the same year of 1601 (Ills. 34 and 35).[62] Incidentally, The presence of a male and a female Magellanic penguin like the one illustrated in *Exoticorum*, though lacking its rear toe, among the South American birds illustrated by various artists for Marcus zum Lamm, a Protestant cleric and jurist at the court of the Prince-Electors of the Palatinate in Heidelberg and compiler of a 33-volume *Thesaurus Pictarum*, leads one to wonder whether there was any personal contact between zum Lamm and Clusius prior to the former's death in 1606.[63]

Ill. 33. Anser Magellanicus. From Zacharias Heyns, *Dracht-thoneel* (Amsterdam, 1601), 145.

---

[61] B.J. Potgieter, *Wijdtloopigh verhael van tgene de vijf schepen (die int jaer 1598 tot Rotterdam toegherust werden / om door de Straet Magellana haren handel te drijven) wedervaren is* [...] (Amsterdam, 1600).

[62] There is one extant letter from Johan Theodore and Johan Israel De Bry to Clusius, dating from 1604.

[63] See R.K. Kinzelbach and J. Hölzinger (eds), *Marcus zum Lamm (1544-1606). Die Vogelbücher aus dem Thesaurus Picturarum* (Stuttgart, 2000), 59-60. The authors accuse Clusius of improving on the woodcut by Potgieter by adding a rear toe to the penguin. This is not true: Potgieter's illustration already has the (biologically accurate) rear toe.

Ill. 34. Dutch penguin-hunting. From De Bry, *America*, vol. IX, part II (Frankfurt, 1601).

One of those who supplied Clusius and others with both *Americana* and images of them during these years was Jacques Plateau, who had a private museum in Doornik.[64] In 1603, for instance, he sent Clusius the coloured image of a Brazilian bird with bright colours,[65] but the most celebrated of the fauna with which the names of both Plateau and Clusius are connected is probably the armadillo, which had been first described by Oviedo. In a letter to Clusius dated November 1602, Plateau mentioned three types of armadillo. The first was the type that had already been described by Clusius in his annotations to Monardes, which were published as volume X of the *Exoticorum*. It is the only animal to be illustrated in those annotations (Ill. 36); the other illustrations are of (parts of) plants and of two animal products (the *lapis tiburonum*

---

[64] See J. Balis, *Van diverse pluimage. Tien eeuwen vogelboeken* (Antwerp, 1968) [Exhibition catalogue, Antwerp, The Hague and Brussels], 27.
[65] *Exoticorum libri decem*, 96. Clusius adds that Everard Vorstius claimed to have seen a similar one from Mexico in the collection of Cardinal Paleotti's secretary.

Ill. 35. Anser Magellanicus, detail of Ill. 25.

and the bezoar).[66] While Monardes had noted the medicinal use of the armadillo's tail, Clusius' annotations provided a fuller description of the animal, drawing on André Thevet, Hans Staden and Jean de Léry, who had all been in Brazil, as well as Pierre Belon, who had not.[67] Plateau's second type was a smaller version of the first type. The third type, however, was very different, and Clusius claimed to be the first to describe and illustrate it. Jacques Plateau sent Clusius a coloured image of this type of armadillo, which is the model for the woodcut (Ill. 37), but failed to include any dimensions. When Clusius insisted on the need for the creature's vital statistics, Plateau obligingly sent them on later.[68] Incidentally, although most of today's tourists to Rome are probably ignorant of the fact, Clusius' first type of armadillo was immortalised in the

---

[66] On Clusius' drastic changes to the illustrative material of Monardes, see J. Pardo Tomás, this volume.
[67] *Exoticorum libri decem*, 330.
[68] *Exoticorum libri decem*, 109.

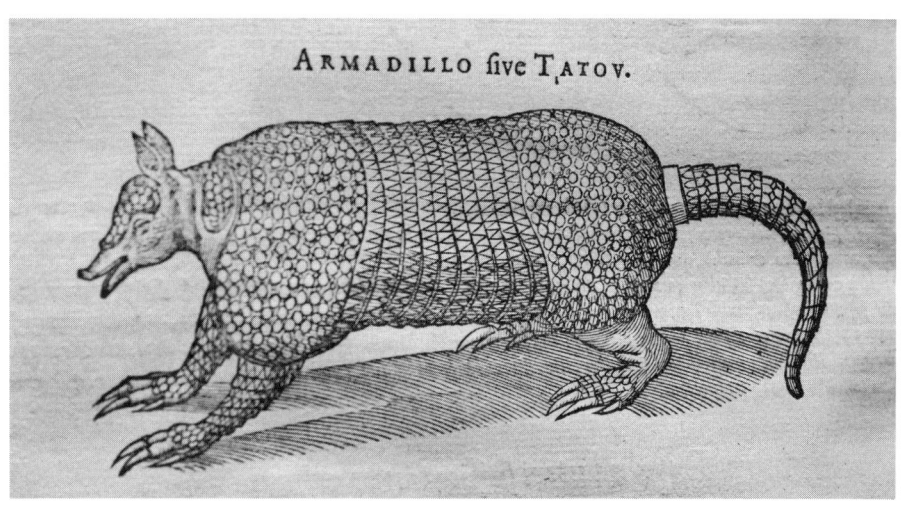

Ill. 36. Armadillo. From *Exoticorum libri decem*, 330.

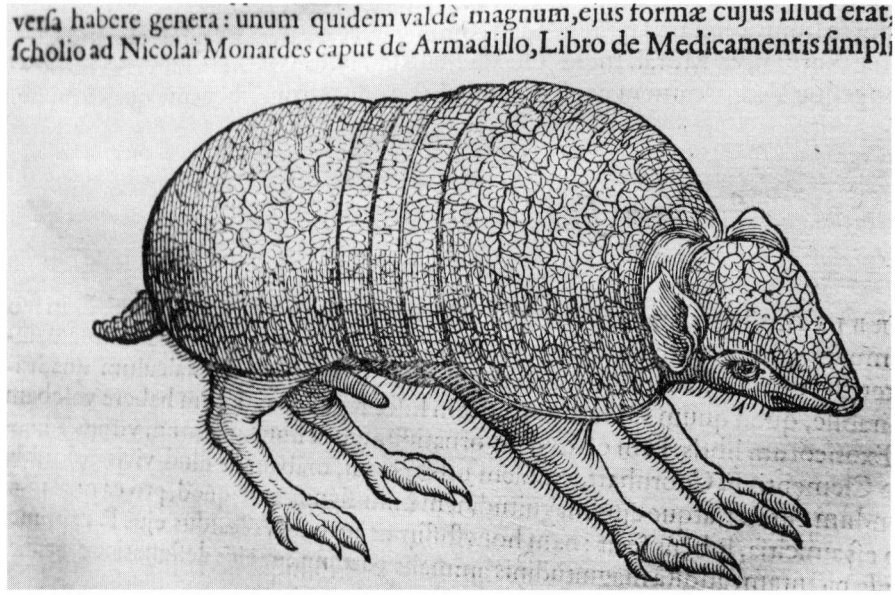

Ill. 37. Armadillo. From *Exoticorum libri decem*, 109.

Ill. 38. Gian Lorenzo Bernini, Fontana dei Quattro Fiumi, 1651. Piazza Navona, Rome.

group of American attributes (including an opuntia cactus) of the Río della Plata in Gian Lorenzo Bernini's *Fontana dei Quattro Fiumi* in the Piazza Navona, completed in 1651 (Ill. 38).[69]

This was not the only occasion on which Clusius criticised Plateau for failing to provide full details. For instance, Plateau sent Clusius a 'mus aquaticus' in 1604, but failed to indicate its provenance.[70] (The same problem arose in connection with the scaly lizard that the Leiden apothecary Christian Porret showed to Clusius in 1604: its provenance was unknown.[71]) The woodcut of the 'Cercopithecus' in the Appendix to the *Exoticorum* is based on an image sent by Plateau, but Clusius complains that he had received no indication of the size of the creature. Clusius' artist had to improve on the crude original image.[72] We might compare the stress on the importance of exact measurements in the correspondence between the Italian jurist and collector Cassiano dal Pozzo and the French antiquarian Nicolas Fabri de Peiresc – with whom Clusius also corresponded –,[73] although this was clearly of even greater

---

[69] An armadillo can be seen hanging from the ceiling of Athanasius Kircher's Roman College Museum in the frontispiece to the 1678 catalogue of that museum: see E. Capanna, 'Zoologia Kircheriana', in E. Lo Sardo (ed.), *Athanasius Kircher. Il Museo del Mondo* (Rome, 2001) [Exhibition catalogue, Palazzo di Venezia], 172.

[70] *Exoticorum libri decem*, 375.

[71] *Exoticorum libri decem*, 374.

[72] *Exoticorum libri decem*, 370.

[73] There are eight extant letters from Peiresc to Clusius, dating from between 1602 and 1606.

importance and relevance to Peiresc since he was writing a treatise on the measures of antiquity.[74] In the case of a 'Mergus Americanus', an image of which was sent to Clusius by Plateau, Clusius thought that it was probably the first description of this bird, but regretted that all he could offer was a description based on the image he had received and expressed the hope that the artist had got it right.[75]

If Clusius sometimes appears ungrateful, we should bear in mind that many of the images Clusius received from Jacques Plateau reached him when time was running out. He could not wait for further, more detailed information. Literally stop press news was the picture of a 'Psittacus elegans' that Clusius received from Plateau in 1605, the very year of publication of the *Exoticorum*.[76] The accelerating pace at which a growing number of *Americana* reached Clusius between 1600 and 1605 coincides with the increasing pressure he was under to complete the *Exoticorum*. In fact, many of those who supplied him with information felt the same pressure: in a letter of 21 August 1601, Juan de Castañeda stated: 'I would like to send you the drawings of so many new plants before the completion of the printing of your book.'[77]

Two years after the publication date of the *Exoticorum*, Clusius was shown a book of plants and animals from various regions by a certain Johannes van Uffele, who had just returned from Brazil. Clusius begged him for copies of the drawings, which were incorporated in the posthumous *Curae posteriores, seu plurimarum non ante cognitarum, aut descriptarum stirpium, pegerinorumque aliquot animalium novae descriptiones* published in Antwerp in 1611.[78] The Capuchin Friar Gregorio da Reggio, five of whose letters to Clusius have been preserved, sent him not only dried plants from near Innsbruck, but also a brief treatise – the only work of the friar's to be published – on the American pepper, which was likewise incorporated in the *Curae posteriores*.[79]

The fact that Clusius included a 21-page Appendix to the *Exoticorum* already bears witness to his concern to be abreast of the latest news in the world of natural history. An edition of Clusius' *Historia Plantarum & Exoticorum*

---

[74] N.F. de Peiresc, *Lettres à Cassiano dal Pozzo (1626-1637)*, ed. and comm. J.-F. Lhote and D. Joyal (Clermont-Ferrand, 1989).
[75] *Exoticorum libri decem*, 103.
[76] *Exoticorum libri decem*, 364.
[77] J.L. Barona and X. Gómez Font, *La correspondencia de Carolus Clusius con los científicos españoles* (València, 1998), 84.
[78] P.J.P. Whitehead, 'Georg Markgraf and Brazilian zoology', in E. van den Boogaart (ed.), *Johan Maurits van Nassau-Siegen 1604-1679. A humanist prince in Europe and Brazil* (The Hague, 1979), 437.
[79] G. Olmi, 'Lettere di Fra Gregorio da Reggio, cappuccino e botanico del tardo rinascimento', in M. Beretta, P. Galluzzi and C. Triarico (eds.), *Musa Musaei. Studies on scientific instruments and collections in honour of Mara Miniati* (Florence, 2003), 117-139, esp. 121.

bound in one volume, now in the Leiden University Library,[80] takes us one stage further in documenting the scholar's relentless desire to update his work in the light of the new discoveries that were being made almost every day in the New World. The title page, *Aucta omnia et recognita partim ope exemplaris [...] Clusii emendati, partim ex praescripto scedulae ab ipso paucis ante obitum suum septimanis Iusto Raphelengio commisse. Suis videlicet locis ubique accomodatis, quae ex Appendicibus Auctariisve, necnon Curis Posterioribus, dictisque exemplari emendato ac scedula, addi vel mutari Autor voluerat. Interserta etiam alicubi nonnulla ab eodem Raphelengio, quae diversitate characterum scholii instar distinguuntur*, indicates the author's desire to incorporate all of the latest discoveries in the published edition of his work. Although it includes many handwritten additions on plants that had been published in the *Historia Plantarum*, the main changes to the *Exoticorum libri decem* are connected with the incorporation of the Appendix of that work into the body of the text of this projected second edition. Weeks before his death, Clusius was still working, Sisyphus-style, to ensure that the finished garment would be seamless.

---

[80] Leiden University Library 755 A 3. Hunger, *Charles de l'Écluse*, vol. I, 298, n.1 refers to this work.

Uses of pictures in printed books:
The case of Clusius' *Exoticorum libri decem*

Sachiko Kusukawa

As Ivins noted many years ago, exactly replicable pictures became a viable means of communicating visual information for the first time after the advent of the movable-type press.[1] It does not follow straightforwardly, however, that the increasing use of pictures, say, in printed books about the *materia medica* in the early modern period proves a dramatic change in the visual practice of science.[2] Printers, for instance, played a significant role in determining the presence, quality and quantity of pictorial matter in a printed book. Carolus Clusius wrote and published in a world where printed books and their pictures had their advantages as well as limitations as conveyers of knowledge. My aim in this chapter is to discuss an attempt that he made, in the *Exoticorum libri decem*, to use pictures effectively in order to create knowledge about the nature of the exotic that was credible and 'legitimate'. In order to appreciate what we may or may not be able to infer from the pictures in printed illustrated books, however, I shall first discuss the world of printed books in which Clusius lived as an author.

The world of printers, books and woodcuts[3]

In 1542, Michael Isengrin at Basel published Leonhard Fuchs' *De historia stirpium commentarii insignes*, with 511 figures with no repeats. The figures were mostly of folio size, had minimum shading, were intended to be coloured, and

---

I would like to thank Florike Egmond, Robert Visser, Paul Hoftijzer and Kasper van Ommen for their hospitality at the Scaliger Institute in September 2004, where an early version of this paper was delivered; I am especially indebted to Florike Egmond for sharing her research material with me.

[1] W.M. Ivins Jr, *Prints and visual communications* (London, 1953).

[2] A more fruitful way to look at this topic is in terms of the relationship between text and image, as exemplified in B.W. Ogilvie, 'Image and text in natural history, 1500-1700', in Wolfgang Lefèvre, Jürgen Renn and Urs Schoepflin (eds.), *The power of images in early modern science* (Basle/Boston/Berlin, 2003), 141-166.

[3] As will be clear from my notes, for this section, I am greatly indebted to the scholarship by Leon Voet, Francine de Nave and Dirk Imhof.

functioned as an 'Idealbild', to represent an object as universal as possible.[4] These woodcuts were used again, with additional cuts, in the following year in Fuchs' *New Kreüterbuch*. This German edition was not an exact translation of the Latin edition which sought to recover the pristine knowledge of medicinal herbs. Instead, it abridged some of the Latin arguments, and with the new index of diseases that could be treated by the plants described, it became more a reference manual for medical treatment. Some copies of these folio editions appear to have been sold coloured.[5] In 1545, Isengrin had these pictures copied and re-cut to a smaller size (octavo), printing two pictorial editions, one in German and another in Latin, with minimum text. Fuchs described them as being made for the students of plants ('studosis herbariae rei') to take with them on trips or walks into the country and compare them with plants found in the country.[6] Although in modern bibliographic terms, the German and the pictorial versions might be described as editions of Fuchs' *De historia stirpium*, these editions address slightly different (though not necessarily mutually exclusive) audiences, and thus different sectors of the book-buying market. In this way, the same, exactly repeatable pictures could be part of a printer's strategy of diversifying his product in order to maximise its market appeal.

In 1552, the Antwerp printer Jan van der Loe procured a privilege to revise and augment Fuchs' *New Kreüterbuch*, for which he solicited the help of Rembert Dodoens (Dodonaeus).[7] Van der Loe spread the cost of the project by publishing illustrated herbals in a smaller format (octavo), and in stages: *De frugum historia liber unus* (1552); *Trium priorum de stirpium historia* (1553); *Trium*

---

[4] For the universalising tendency of Fuchs' figures, see S. Kusukawa, 'The uses of pictures in the formation of learned knowledge: The cases of Leonhard Fuchs and Andreas Vesalius', in eadem and I. Maclean (eds.), *Transmitting knowledge: Words, images, and instruments in early modern Europe* (Oxford, 2006), 73-96. The reasons for Fuchs' insistence on a one-to-one match between text and image is discussed in S. Kusukawa, 'Leonhart Fuchs on the importance of pictures', *Journal of the history of ideas*, 58/3 (1997), 403-427.

[5] A. Arber, 'The colouring of sixteenth-century herbals', in her *Herbals: Their origin and evolution: A chapter in the history of botany 1470-1670*, ed. W.T. Stearn (Cambridge, 1990), 315-318. For the problems of identifying current copies that were coloured originally, see F.G. Meyer, with E.E. Trueblood, and J.L. Heller (eds.), *The great herbal of Leonhart Fuchs:* De historia stirpium commentarii insignes, *1542*, 2 vols. (Stanford, 1999), vol. I, 119f.

[6] 'Caeterum cum eius operas propter suam molem ac magnitudinem, non nisi domi usus esse possit, de ratione aliqua mihi cogitandum fuit, qua efficerem ut herbariae rei studiosis ita consulerem, ut peregrinantem etiam ac deambulantes haberent, quibus cum nativas herbas rure inventas conferrent. Neque enim ulla via ad recte cognoscendas stirpes expeditior, quam illa nativarum ad pictures diligens collatio.' L. Fuchs, *Primi de stirpium historia commentariorum tomi vivae imagines, in exiguam angustioremque formam contractae* (Basle, 1545), A2r.

[7] The imperial privilege is dated 27 May, 1551, R. Dodoens, *De frugum historia* (Antwerp, 1552), A1v.

posteriorum de stirpium historia (1554).[8] The complete Dutch version of Fuchs' *New Kreüterbuch* by Dodoens, the *Cruijdt-boeck*, was then printed in 1554 in folio, containing 710 illustrations (500 cuts copied from Fuchs, and 210 newly cut). Van der Loe proceeded to print a French edition, translated by Clusius, in 1557, with an additional 108 woodblocks.[9] Like Isengrin, Van der Loe then produced a pictorial, octavo version in two volumes in 1559. By the year of Van der Loe's death (1563), when a revised edition was printed, the *Cruijdt-boeck* had 861 woodcuts.

Plantin emulated van der Loe in the way he published works on plants: by printing in stages to increase gradually the stock of woodcuts, and by maximising the use of the woodcuts in diverse publications. In 1566, he first printed Dodoens' *Frumentorum, leguminum, palustrium et aquatilium herbarum [...] historia*, which was a revised edition of his *De frugum historia* (1552), and would eventually become part of the *Stirpium historiae pemptades sex* (1583). The *Frumentorum [...] historia* contained 280 octavo pages with 81 illustrations.[10] 800 copies were printed, for which paper cost 35 fl. 12 st., and the setting, printing, and compiling of an index cost 27 fl., a total of 62 fl. 12 st.[11] Although Plantin did not include the cost of the illustrations when setting the price of this book, we do know that Peeter vander Borcht was paid 5 patavars per drawing copied from Van der Loe's *Cruijdt-boeck*.[12] The drawings, in turn, were cut by Cornelius Muller, Gerard Janssen van Kampen or Arnold Nicolai, who were paid 8 stuivers for each wood block.[13] Thus the cost for illustrations would have come around to 54 fl., just under 50% of the total cost of printing.[14] This amount of outlay itself was negligible within the context of Plantin's annual running cost of 13,041 fl. for that year,[15] but it is still significant that newly cut woodcuts

---

[8] F. de Nave and D. Imhof (eds.), *Botany in the Low Countries (end of the 15th century – ca. 1650)* (Antwerp, 1993) [Exhibition catalogue, Plantin-Moretus Museum, Antwerp], 98-100.

[9] Ibid., 101.

[10] I.e. 78 woodblocks with repeats, C. Depauw, 'Peeter vander Borcht (1535/40-1608): The artist as inventor or creator of botanical illustrations?', in De Nave and Imhof (eds.), *Botany in the Low Countries*, 49.

[11] L. Voet, 'Christopher Plantin as a promoter of the science of botany', in De Nave and Imhof (eds), *Botany in the Low Countries*, 44.

[12] De Nave and Imhof (eds.), *Botany in the Low Countries*, 103. Cf. Dodoens' claim: 'De ijs [imaginibus] autem, quae huic historiae additae sunt, affirmare possumus, eas ex vivarum herbarum imitatione depictas.' R. Dodoens, *Frumentorum, leguminum, palustrium et aquatilium herbarum [...] historia* (Antwerp, 1566), 19. My interpolation. See also Depauw, 'Peeter vander Borcht', 49f.

[13] In his accounting, Plantin used the Carolus guilder (abbreviated as 'fl.'), divided into 20 stuivers or patavars (abbreviated as 'st.'); L. Voet, *The Golden Compasses: A history and evaluation of the printing and publishing activities of the Officiana Plantiniana at Antwerp*, 2 vols. (Amsterdam/London, 1969-72), vol. I, 440.

[14] Voet, 'Plantin as a promoter', 44.

[15] Voet, *The Golden Compasses*, vol. II, 227.

cost more than the paper, which was normally the largest cost involved in printing. Plantin would then spend 82 guilders for another new set of 108 woodblocks, for Dodoens' *Florum et Coronarum odoratorumque [...] historia*, printed in 1568 (and used again in the *Stirpium historiae pemptades sex*).[16] In 1574, Dodoens' *Purgantium aliarumque [...] historia libri III*, containing 221 illustrations was published.[17] In 1578 Plantin lent 37 blocks from Dodoens' works to Hendrik van der Loe, who used them, together with woodblocks from his father's (Jan van der Loe) stock, to print an English translation (by Henry Lyte) of Dodoens' *Cruijdt-boeck*.[18]

It is as if there was a division of labour between Dodoens and Clusius for the Plantin press, since Clusius first published with Plantin on medicinal plants not commonly found in the comprehensive herbals of Fuchs and Dodoens.[19] Clusius, in turn, believed that the study of plants was so immense that nobody so far had managed to publish a 'complete (absoluta) historia', though some had brought to light plants hitherto unknown.[20] Clusius persuaded Plantin to publish books on New-World flora and fauna which had largely been issued in the vernacular. He helped Plantin expand the market for these books by translating into Latin the vernacular works on the subject already published, thus turning Garcia ab Orta's *Coloquios dos simples* (Goa, 1563) into the *Aromatum et simplicium aliquot medicamentorum apud Indos nascentium historia* (Antwerp, 1567); Nicolas Monardes' *De las drogas de las Indias* (Seville, 1565 and 1571) into the *De simplicibus medicamentis ex Occidentali India delatis quorum in medicina usus est* (Antwerp, 1574); Christophorus a Costa's *Tradado de las drogas medicinas y plantas de las Indias Orientales* (Burgos, 1578) into the *Aromatum et medicamentorum in Orientali India nascentium liber* (Antwerp, 1582). Each title underwent revisions and all were published together in 1593. It is important to note the traditional, medicinal orientation of these publications. For instance, in *De simplicibus medicamentis* Clusius described the 'Lapis tiburonum', stones found in the head of sharks ('tiburones'), which had been described by Monardes as white, large, heavy, and when ground into a powder, very effective for 'nephriticis', urinary

---

[16] Voet, 'Plantin as a promoter', 41; De Nave and Imhof (eds.), *Botany in the Low Countries*, 105.

[17] L. Voet, *The Plantin press (1555-1589): A bibliography of the works printed and published by Christopher Plantin at Antwerp and Leiden*, 6 vols. (Amsterdam, 1980-83), no. 1099.

[18] De Nave and Imhof (eds.), *Botany in the Low Countries*, 105.

[19] Cf. Fuchs' new world plants, see Meyer, *The great herbal*, vol. I, 133. For the tobacco plant in Dodoens' work, see De Nave and Imhof (eds.), *Botany in the Low Countries*, 110f.

[20] 'Rei vero herbariae studium adeo immensum est, ut absolutam plantarum historiam nemo hactenus evulgarit, sed aliqui duntaxat ignotas nobis plantas interdum in lucem proferunt.' N. Monardes, *De simplicibus medicamentis ex occidentali India delatis, quorum in medicina usus est*, tr. C. Clusius (Antwerp, 1574), A2r.

difficulties and stones in kidneys.[21] As was often the case, Clusius then provided a note and quoted a description from Francisco López de Gómara's *Historia general de las Indias* (1551) which suggested that a fish known by the name of 'manatus' had much in common with the 'tiburones'.[22]

It is also worth nothing the modest outlay for the illustrations in these works. Ab Orta's *Aromatum et simplicium aliquot medicamentorum apud Indos nascentium historia* (1567) contained 264 octavo pages and 16 illustrations (drawn by Vander Borcht and cut by Nicolai), and was issued in 1250 copies, for which the total cost of printing was 91 fl. 2 st., including 47 fl. 6 st. for paper and 10 fl. 10 st. for illustrations; thus the woodcuts cost just over 10% of the total cost, but the paper over 50%.[23] Plantin gradually increased the number of illustrations (and the sales price) in successive editions, but the increases were modest: the second edition in 1574 contained 27 illustrations and the third edition of 1579 had 28 illustrations.[24] This was similarly the case with the work of Monardes: *De simplicibus medicamentis ex Occidentali India delatis* (1574) contained 10 illustrations and the second edition in 1579 had just three more.[25] Clusius was critical of the pictures in A Costa's original edition, and the first edition of the *Aromatum et medicamentorum [...] liber* (1582) contained only two illustrations, only one of which had been copied from the original edition.[26] This single picture, in fact, was included in order to show how false and unreliable A Costa's description of the 'Caryophyllus' was. Clusius referred the reader instead to the proper ('legitima') picture of the clove tree in Ab Orta's *Aromatum et simplicium aliquot medicamentorum apud Indos nascentium historia* (1567).[27] The

---

[21] Ibid. 52.

[22] Ibid. 52f. Cf. F. López de Gómara, *Historia general de las Indias*, ed. P. Guibelalde and E.M. Aguilera, 2 vols. (Barcelona, 1954), vol. I, 54f.

[23] Voet, *The Plantin press*, no. 1838.

[24] Voet, *The Plantin press*, nos. 1839-1840. The *Aromatum et simplicium [...] historia*, first edition (1567) was sold for 4 st., the second edition (1574), 4½ st., and the third edition (1579) 6 st.; we only know the price for the first edition of Monardes' *De simplicibus medicamentis [...]* (1574): 1½ st; and of the first edition of A Costa's *Aromatum et medicamentorum [...] liber* (1582) at 1 st. Voet, 'Plantin as a promoter', 42.

[25] Voet, *The Plantin press*, nos. 1710-1712.

[26] 'Icones praeterea, quas ad vivum expressisse passim gloriatus, suis locis insperserat, reieci, quoniam plane ineptae essent, et nihil minus, quam legitimas stirpes referrent: uti ex unica Caryphyllorum arborum effigie (quam idcirco intuli, ut cum legitima, Garciae adiecta, conferre liceat) quilibet iudicare poterit.' C. A Costa, *Aromatum et medicamentorum in Orientali India nascentium liber*, tr. C. Clusius (Antwerp, 1582), 5. The other picture is that of a fragment of the 'Lignum Colubrinum' given to Clusius by Hector Nunez, ibid., 52f.

[27] '[...] ipsius verba, lectori proponenda censui, ut animadvertere queat quam parum fidei interdum huic auctori sit adhibendum, qui veritatis assertorem se gloriatur, et plantas ad vivum expressisse asserit, cum tamen ipsius icones nullius stirpis vivam effigiem imitentur, praesertim earum quas hactenus nobis videre licuit. Caryophyllorum certe legitimam iconem in Aromatum Garciae historia

second edition of A Costa's *Aromatum [...]* (1593) retained only the figure of the 'Caryophyllus' from the 1582 edition, rather than replacing it with another picture. For Clusius, a false picture could be used to distinguish between true and false descriptions by different authors. The modest uses of pictures in Clusius' translations in general suggest that Plantin's strategy for publishing this genre of exotic medicine remained rather conservative, building on existing publications (in the vernacular), turning them into Latin to expand into an international market on rare and exotic medicines, but limiting the financial risk by creating only a small amount of woodcuts and keeping the cost per title down.

This may partly explain why Clusius' own work on the rare and the exotic suffered somewhat at the hands of Plantin. Clusius had collected plants on a trip in 1566 to Spain and Portugal, accompanying Jacob Fugger.[28] In the summer of 1568, he carefully supervised Peeter vander Borcht to draw pictures of plants from dried specimens.[29] The pictures were ready by 1569, but Plantin used them first for Dodoens' *Purgantium [...] herbarum historiae libri IIII* (1574).[30] Clusius' own work was published in 1576 (the privilege is dated 1575), as the *Rariorum aliquot stirpium per Hispanias observationum historia*, with 544 octavo pages and 233 figures, which in turn included woodcuts from Dodoens' *Purgantium [...] herbarum historia*. Despite his claim of including the rare and the unknown varieties, Clusius conceded that some would notice that the pictures had appeared in other books before. He explained, however, that he had let Dodoens freely use his pictures for the *Purgantium [...] herbarum historia*, because the bonds of true friendship ('vincula amicitiae verae') are such that possessions should be freely shared rather than printed as one's own.[31] Such a friendship in

---

exhibui, [...]' A Costa, *Aromatum [...]* (1582), 33. Voet, 'Plantin as a promoter', 52f. De Nave and Imhof (eds.), *Botany in the Low Countries*, 117. Garcia ab Orta, *Aromatum et simplicium aliquot medicamentorum apud Indos nascentium historia* (Antwerp, 1567), 102 for the picture of the 'Garyophillus'.

[28] 'In ista peregrinatione plurimarum formam, natales, et nomina memoriae causa adscripsi, nonnullarum etiam effigies ipse carbone aut rubria delineavi, atque omnes fere inde rediens exsiccates detuli; aut earum semina, vel ipsas etiam plantas, quae videlicet vecturae tradidatem ferre potuerunt (quales sunt bulbosae et tuberosae) amicis inde misi.' C. Clusius, *Rariorum aliquot stirpium per Hispanias* (Antwerp, 1576), 7.

[29] 'Eaque adeo de causa, biennio post, industrium et diligentem pictorem nactus, stirpium icones in tabellis ligneis depingendas curavi, et plerunque etiam ipsi pictori adstiti, ut de his quae in siccarum plantarum forma exprimenda diligentius erant observanda, commonefacerem.' Clusius, *Rariorum aliquot stirpium per Hispanias*, 8.

[30] Voet, *The Plantin press*, no. 1099; Dodoens' work included 33 woodcuts from Clusius' Spanish flora, and Clusius borrowed 6 from Dodoens' work, making the overlap between the two works 39 woodcuts.

[31] '...nec etiam cuiquam novum videatur, si plerasque stirpium effigies in hoc libello conspexerit, quas apud alios, qui suas lucubrationes ante me ediderunt, viderit. Ea enim sunt verae amicitiae vincula, ut illam nihil peculiare, nihil sibi proprium habere putem: sed quaecunque habent amici,

sharing pictures was made possible because both Clusius and Dodoens were publishing with the same printer, Plantin; from the point of the printer, such a sharing of woodcuts was financially necessary. This is in marked contrast to the way Fuchs, for instance, felt proprietary about his own woodblocks.[32]

In 1576, Plantin printed a 'hybrid' work by Matthias Lobelius, the *Plantarum seu stirpium historia* (1576),[33] with 782 woodblocks cut anew at the cost of 523 fl.[34] Of the total 1790 illustrations contained in this work, over 700 were those previously used in the works of Dodoens and of Clusius. As the second part of this work, Plantin tacked on Lobelius' *Stirpium adversaria nova*, composed with Pierre Pena and published in London by Thomas Purfoot in 1571. Plantin had bought 800 copies of Purfoot's edition of *Stirpium adversaria nova* for 1,200 fl., and around 250 of the woodblocks (the bulk of which arrived in Antwerp in 1580), at the extra expense of 120 fl.[35] The *Plantarum seu stirpium historia* dealt with the *materia medica*, especially of Dioscorides; in it Lobelius defended the knowledge of the ancients against claims of 'new medicines' promoted by the works of Paracelsus and his followers,[36] and he included a list of formulas for medicines by Guillaume Rondelet.[37]

In addition to the Purfoot woodblocks, Plantin acquired some 1000 of the plant woodcuts from Jan van der Loe's widow in 1581.[38] These were put to good use in the same year, in Lobelius' Flemish edition of the *Plantarum seu stirpium historia*, the *Cruijdt-boeck*, containing 1408 folio pages with 2187 illustrations, and sold for 6 fl. 10 st a copy.[39] In the same year, at the behest of Severinus Gobelius, physician to the Elector of Brandenburg, Plantin also issued a pictorial album, the *Plantarum seu stirpium icones*, in a horizontal quarto

---

liberaliter inter se communicare debere. Inde factum est, ut clarissimus vir Rembertus Dodonaues, nunc Caesareus medicus, veteri amicitia mihi conjunctus, quas ex meis iconibus voluerit, libere in suam Purgantium historiam intulerit.' Clusius, *Rariorum aliquot stirpium per Hispanias*, 9f.

[32] Kusukawa, 'Fuchs on the importance of pictures'; see also Fuchs' rejection of Conrad Gessner's request to borrow the woodcuts of the *De historia stirpium*, J.L. Heller and F.G. Meyer, 'Conrad Gessner to Leonhart Fuchs, October 18, 1556', *Huntia*, 5/1 (1983), 67, 69, 75.

[33] 'Hybrid' is Voet's expression in his 'Plantin as a promoter', 42.

[34] Voet, *The Plantin press*, no. 1578.

[35] '1580. Adi 4ᵉ de may ils ont livré les figures de Londres dont sont d'accord avec C. Plantin qu'ils auront 20 livres, val. fl. 120', *Bibliotheca Belgica: Bibliographie générale des Pays-Bas*, ed. F. van der Haeghen, re-ed. M.-T. Lenger, 6 vols. (Brussels, 1979), vol. III, 1129-1130.

[36] 'Sunt enim recentiorum complures qui profitentur, summeque gloriantur nova se invenisse remedia, novasque praeparandorum medicamentorum formulas, e quorum numero tenebricosis suis scriptis est Paracelsus; eiusque asseclae, cum tamen in Dioscoridis, aliisque veterum auctorum monumentis, huiuscemodi extrahendarum facultatum herbarum, praeparandorumque medicamentorum rartionem posteris traditam legamus.' Lobelius, *Plantarum seu stirpium historia* (Antwerp, 1576), 4.

[37] *Formulae aliquot remediorum Guillielmi Rondelletii [...]*.

[38] Voet 'Plantin as a promoter', 41.

[39] Voet, *The Plantin press*, no. 1579.

format, with 1140 pages and 2176 illustrations; Gobelius bought 150 copies at 36 st. each.⁴⁰ This pictorial album had minimal text – just the plant-names and references to Lobelius' *Plantarum seu stirpium historia* and *Cruydt-boeck*. The index of plant names in Dutch, French, German, Italian, Spanish and English, made the album saleable across Europe. Gobelius also asked for a coloured copy of the *Icones*, to which Plantin replied that he did not have a coloured copy, and that furthermore, it would take at least three months, as it would be a laborious task. Instead, Plantin offered him one of the three coloured copies of Lobelius' *Cruijdt-boeck* that he had ready, and this is what Gobelius purchased. Plantin charged Gobelius one stuiver each for colouring 2,100 illustrations and hence this coloured herbal cost an extra 105 fl. on top of the 8 fl. for the book itself.⁴¹ This kind of charge was exceptional, but it does suggest that Plantin had some ready-coloured copies of the *Cruijdt-boeck*. Nevertheless, it does not appear to have been a regular practice for Plantin to offer coloured herbals in the regular way that he offered coloured maps.⁴² Nor is it clear whether the three coloured copies now in the Plantin Moretus museum were archetypes from which other coloured copies were made.⁴³ In 1575, Plantin had offered his customers the choice between woodcut or engraved illustrations for a breviary, priced at 3 fl. and 4 fl. respectively.⁴⁴ Plantin did not offer this option with his books on plants, not even with the grander herbals of Lobelius or Dodoens. Perhaps the market for these books was deemed not large enough to allow for such differentiation, or building up a large stock of engravings was too expensive: a copper engraving print would have been ten to twelve times more expensive than a woodcut print of an equivalent size.⁴⁵

In 1583, Dodoens' *Stirpium pemptades sex*, in 808 folio pages with 1306 illustrations, was published for 6 florins.⁴⁶ As the title suggests, this work was divided into thirty parts, and contained 1306 woodcut illustrations.⁴⁷ Some of

---

⁴⁰ Voet, *The Plantin press*, no. 1580.
⁴¹ Plantin's letter to Gobelius is reproduced in Voet, *The Golden Compasses*, no. 1579. Voet believes that Gobelius' coloured copy was intended for the Duke of Prussia, Gobelius' patron. Voet, 'Plantin as a promoter', 44.
⁴² Note that many of Plantin's woodcuts for plants have heavy shading, implying that they were not originally designed to be coloured. For Plantin's colouring practice, see Voet, *The Golden Compasses*, vol. II, 242f.
⁴³ For the three coloured copies, see De Nave and Imhof (eds.), *Botany in the Low Countries*, 65f., 122. For the practice of colouring after an archetype, see Arber, 'Colouring'.
⁴⁴ Voet, *The Golden Compasses*, vol. II, 381.
⁴⁵ I use the estimate in D. Woodward, *Maps as prints in the Italian Renaissance: Makers, distributors and consumers* (London, 1996), 33. But see Plantin's use of copper engraving, Voet, *The Golden Compasses*, vol. II, 194-305.
⁴⁶ Voet, 'Plantin as a promoter', 43.
⁴⁷ Voet, *The Plantin press*, no. 1101.

Van der Loe's woodblocks were re-used and the book also included copies of pictures from the Juliana Codex of Dioscorides that had recently arrived in Vienna.⁴⁸ Clusius felt that the codex contained few pictures which referred to a true image of a plant, while Dodoens appears to have been more casual about their veracity.⁴⁹ In the preface to the *Pemptades*, Dodoens described the sharing of woodblocks with the works of Lobelius and Clusius as saving expenses for Plantin.⁵⁰

By the time Plantin died in 1589, he had accumulated 7,530 woodblocks valued at 3,726 fl. and 2,391 copperplates worth 4,384 fl. 14 st.⁵¹ The woodblocks and copperplates were then divided between the 'officinae' in Antwerp (Jan I Moretus) and Leiden (Franciscus I Raphelengius), but the sharing of woodblocks continued.⁵² Although I have stressed Plantin's financial considerations in the inclusion of woodcuts in his printed herbals, it would be misleading to see him as entirely profit-driven. For apart from financial exigencies, it is also possible that printers felt more than justified in the repeated use of plants in different books. The arguments used, inter alia, by the Frankfurt printer Christian Egenolff in his defense against the charge of plagiarism (of the pictures in Otto Brunfels' *Vivae eicones herbarum*) brought to the Reichskammergericht by the printer Hans Schott, certainly point in this direction. They included the statement that pictures of plants may resemble each other because one cannot draw or copy a picture of a rosemary, a daffodil, or a

---

⁴⁸ De Nave and Imhof (eds.), *Botany in the Low Countries*, 106; e.g. R. Dodoens, *Stirpium pemptades sex* (Antwerp, 1583), 436: 'Aconitum Lycoctonon ex Cod. Caes.' Arber, *Herbals*, 239. For the Juliana Codex, see M. Collins, *Medieval herbals: The illustrative tradition* (London, 2000), 39-50.

⁴⁹ 'Caesareum codicem aliquando conspexi: sed paucae istic inerant icones veram stirpium effigiem referentes: sed Dodonaeum novi, qui an verae sint non magnopere curat, modo suo argumento deserviant. Memini enim illum admonere cum fictitiam illam Eriophori imaginem incidi curaret, quae ut nomen referret lanuginoso flore expressa fuit a Cortuso et nobis missa, ne eam in suum opus inferret, suspectam etenim maxime mihi esse, et ad phantasiam expressam; at ille, quid mea, inquit, refert? Cortuso acceptam referam.' Clusius to Camerarius, 1583, F.W.T. Hunger, *Charles de l'Escluse: Carolus Clusius, Nederlandsch kruidkundige 1526-1609*, 2 vols. (The Hague, 1927-1943), vol. II, 394.

⁵⁰ 'Icones autem plurimas quidem nostra opera et cura iam olim delineatas fuisse facile agnoverit, qui frumentorum, florum et coronariarum, purgantiumque historias cum appendice prius habuerunt seu viderunt. His accesserunt non paucae (praeter nonnullas novas et ante non editas) et me quidem procurante supra aliquot annos expressae, quae in vernaculis ac Gallicis de stirpium historia commentariis a Ioanne Loëo quondam editis extant: reliquae partim ex Caroli Clusii, sed plures ex M. Lobelij observationibus accesserunt… Non existimavi enim easdem (nisi forte non satis recte expressas) iterum depingendas, ac duplici sumptu Christophorum Plantinum typographum diligentissimum gravandum, qui olim nostras de floribus, purgantibus, frumentisque historias, ac deinde Caroli Clusij et M. Lobelij observationes suis typis in publicum dedit.' R. Dodoens, *Stirpium pemptades sex* (Antwerp, 1583), ††2v.

⁵¹ His whole printing asset was valued at 22,607 fl. Voet, *The Golden Compasses*, vol. II, 231.

⁵² Voet, 'Plantin as a promoter', 43.

borage plant in any other form than it is;⁵³ nor, as he argued, did the privileges granted to Dürer or Jacopo de' Barbari imply that no other painter might paint the same subjects, such as Adam and Eve.⁵⁴ Egenolff's point was that privileges over pictorial matter did not cover the subjects of the pictures, only their forms. But even the likeness of forms could not be prohibited in the case of pictures of plants, because plants have to be depicted the way they are. One daffodil was going to look similar to another daffodil; thus depictions of plants would necessarily converge in form. An assumption along these lines could well justify copying from printed pictures, which in turn may have been copied from elsewhere, so long as at some point up the chain of copying, the pictures had been drawn from nature. Since they represented objects in nature, pictures of plants could thus be repeatedly used by printers in different publications.⁵⁵ In the second half of the sixteenth century, Plantin cornered the market for illustrated herbals, but not just out of financial acumen. Although there is no evidence to suggest that Plantin – like the printer Paul Arnold in Amsterdam who acted for Emmanuel Sweert – sold plants alongside his books, Lobelius certainly regarded Plantin as a central figure in upholding a republic of letters in the matter of plants, as he (Lobelius) urged others to send in new plants to the 'Plantinian garden'.⁵⁶

Plantin's case highlights how the copying and re-use of woodcuts was common practice in illustrated printed books in the second half of the sixteenth

---

⁵³ 'Und wan gleich die Kreuther unter einander sich ein wenig vergleichen, so wolle doch Ew. Gnaden erwegen, daß man Rosmarin Affodilis oder ein ander Krauth nie kann in einer anderen formb oder gestalt mallen oder conterfeyen, dann es an im selbst ist.' Altona, 'Aus den Akten des Reichskammergerichts', *Zeitschrift für die gesamte Strafrechtswissenschaft*, (1892), 901. Egenolff's argument is summarised in H. Grotefend, *Christian Egenolff, der erste ständige Buchdrucker zu Frankfurt a. M.* (Frankfurt, 1881), 16f.

⁵⁴ 'Es wäre ja ein absurdum, daß Kay. Privileg also sollte verstanden werden, daß dieweyl Hannes Schott hatte das Kreutherbuch getruckt, daß derhalben man müßte ein Krauth, das kleine schmahle blettlein hatt, mit langen breiten Blettern und contra drucken wider Arth gestalt formb und natur der kreuther; etwas unförmblichs nit gesehen, denn wiewol Albrecht Dürer Jacob Meller zu Wittenberg und andere Privilegien haben, das niemandt ihre gemälte nachmallen darff, so folgt doch derhalben nit, daß dieweyllen dieselben einen Adam et Evam Acteonem Achillem pinxissent, daß derohalb khein andere maller auch dergleichen fabbeln nit malen dürfft.' Altona 'Reichskammergerichts', 901. 'Jacob Meller' is identified as Barbari in Peter Parshall, '*Imago contrafacta*: Images and facts in the Northern Renaissance', *Art history*, 16 (1994), 569.

⁵⁵ This did not necessarily prevent printers from obtaining privileges to cover pictures in order to protect the cost and labour that had gone into producing illustrative figures.

⁵⁶ E. Sweertius, *Florilegium [...]* (Frankfurt am M., 1612), verso of title page. Also: 'Quare omnes rogatos velim qui in hoc studium incumbunt, ut si quid praeter has novarum plantarum, aut quidpiam aliud nova Naturae foetu exortum reperiant, in has Plantinani horti areolas liberaliter conferant, cum omnes homines adniti debeant ut rempublicam literarim pro sui ingenii facultate et viribus iuvent et exornent.' Lobelius, *Plantarum seu stirpium historia*, 457.

century; but it also shows a printer's willingness to invest in producing pictures for printed books, although these were not necessarily drawn from life each time. As in the case of Clusius' supervision of vander Borcht, authorial control could be exercised in the drawing the pictures, but this was not always the case. The origin of pictures, their placement in the text, and their quantity varied according to printers' practices. This suggests, then, that historians ought be cautious in interpreting the pictures in printed books on plants.

The Exoticorum libri decem

In 1605, Raphelengius published the *Exoticorum libri decem*, which included Clusius' own studies on foreign plants (the first six books), besides new editions of the works of Ab Orta, A Costa, Monardes, and of Petrus Bellonius' *Plurimarum singularium memorabilium rerum [...] observationes* and *De neglecta stirpium cultura*.[57]

British Library, shelfmark C.60.e.5. is Clusius' own copy of Bellonius' *Observationes* (1589), *De neglecta stirpium cultura* (1589), and the 1593 edition of the works of Monardes, Ab Orta and A Costa. The books are annotated throughout in a single, fine, neat hand in brown ink. Judging from the corrections on the title-pages, these annotations were the basis of a revision that became part of the *Exoticorum libri decem* of 1605. There are extensive revisions and additions to the text, frequently with additional paste-ins. These textual annotations are all in Latin. Some, though not all, of the pictures also receive some comment, either in Dutch or in Latin.

For instance, on page 390 of Bellonius' *Observationes* (1589), against the picture of the 'abies' (Ill. 39), Clusius noted that the picture should be taken out because it was inept ('inepta') and that the second picture on page 231 of the 'second volume of pictures' ought to be placed in its stead.[58] This refers to the pictorial album (1581/1591), the *Plantarum seu stirpium icones*, in the second volume of which, on page 231, there is indeed a picture of the 'abies' on the right and a picture of the 'picea' on the left ((Ill. 40). Some of Clusius' comments are thus instructions to replace a picture with another from within the Plantinian stock of woodcuts.[59] It is somewhat odd that the picture of the

---

[57] Bellonius' *Observationes* was originally published by Plantin in 1555, but the woodcuts had to be re-cut, as Plantin could not recover the woodblocks that were sold in 1562; Clusius' Latin translation first appeared in 1589, as did that of the *De neglecta stirpium cultura*. De Nave and Imhof (eds.), *Botany in the Low Countries*, 115f.
[58] 'haec icon tollenda, nam inepta, et eius loco 2a pagina 291 Tomi II iconum reponenda.'
[59] See also Clusius' annotations in the British Library copy, P. Bellonius, *Plurimarum singularium and memorabilium rerum [...] observationes*, tr. C. Clusius (Antwerp, 1589), 93, 24, 187.

390 P. BELLONII OBSERVATIONVM
Abietis Icon.

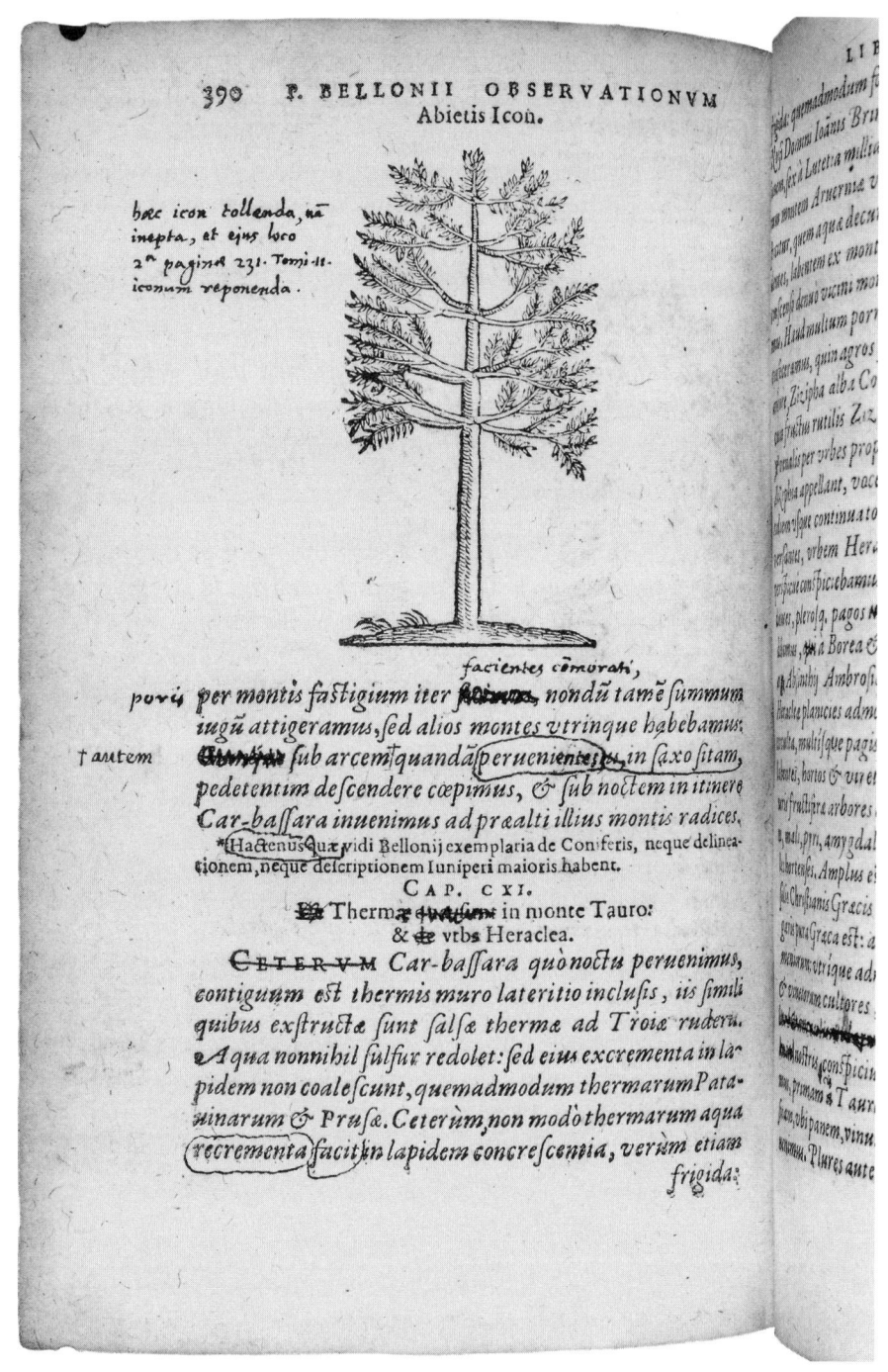

hæc icon tollenda, n̄ā
inepta, et eius loco
2ᵃ paginā 231. Tomi II
iconum reponenda.

facientes cōmorati,
porq̀ *per montis fastigium iter* ~~facimus~~, *nondū tamē summum*
iugū attigeramus, sed alios montes vtrinque habebamus:
✝ autem ~~Quum~~ sub arcem quandā peruenientes, *in saxo sitam,*
pedetentim descendere cœpimus, & sub noctem in itinere
Car-bassara inuenimus ad præalti illius montis radices.
Hactenus quæ vidi Bellonij exemplaria de Coniferis, neque delinea-
tionem, neque descriptionem Iuniperi maioris habent.
CAP. CXI.
Thermæ ~~quædam~~ in monte Tauro:
& vrbs Heraclea.

CETERVM *Car-bassara quò noctu peruenimus,*
contiguum est thermis muro lateritio inclusis, iis simili
quibus exstructæ sunt salsæ thermæ ad Troiæ rudera.
Aqua nonnihil sulfur redolet: sed eius excrementa in la-
pidem non coalescunt, quemadmodum thermarum Pata-
uinarum & Prusæ. Ceterùm, non modò thermarum aqua
recrementa facit *in lapidem concrescentia, verùm etiam*
frigidæ

Ill. 39. P. Bellonius, *Plurimarum singularium and memorabilium rerum [...] observationes*, tr. C. Clusius (Antwerp, 1589), 390.

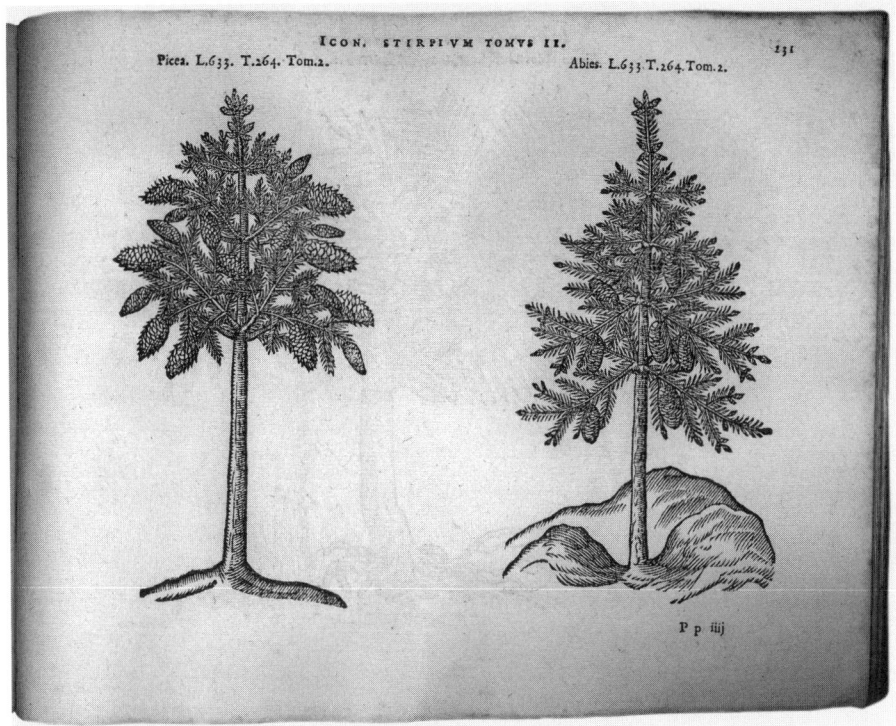

Ill. 40. M. Lobelius, *Plantarum seu stirpium icones* (Antwerp, 1581), vol. II, 231.

'abies' was indeed replaced in the 1605 edition (page 163) ((Ill. 41), but not by the woodcut on the specified page of the *Icones*.[60] It is not clear whether this had to do with the division of the woodblocks between Raphelengius and Moretus, or some other reason.

In another case – the picture of the 'civet' cat on page 220 of the *Observationes* ((Ill. 42) – Clusius' correction would have required cutting a new woodblock as he explained that the ears of the cat in the picture had to be rounder, and not so pointed.[61] However, in the *Exoticorum libri decem* (page 94) ((Ill. 43), the old woodcut was retained, perhaps for financial reasons or because the change appeared so slight. The cat did finally acquire its new ears in Clusius' posthumous publication, the *Curae posteriores* (1611) ((Ill. 44).

---

[60] Note that the 'abies' of Lobelius' *Icones* (vol. II, 231) was reproduced as 'picea' in Bellonius' *Observationes*, 93.
[61] 'hujus iconis aures rotundiores esse debent, ut emendavi, non mucronatae'. For the vexing identity of the 'civet', see S. de Renzi, 'Writing and talking of exotic animals', in M. Frasca-Spada and N. Jardine (eds.), *Books and the sciences in history* (Cambridge, 2000), 157f.

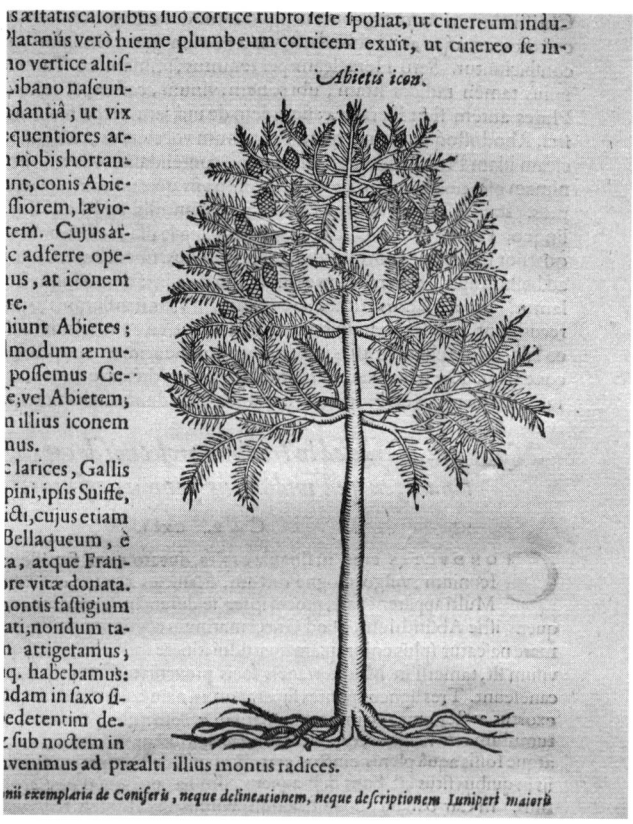

Ill. 41. P. Bellonius' *Observationes* in C. Clusius, *Exoticorum libri decem* (Antwerp, 1605), 163.

Not all pictorial corrections were ignored in the *Exoticorum libri decem*, however. In the case of the 'Lapis tiburonum' in the 1593 Monardes edition (pp. 364-365), Clusius pasted on the page a brown-ink sketch of the stone found by Francis Drake's ship, which had been drawn to scale by James Garet and sent to Clusius at Frankfurt in 1593 ((Ill. 45). This was indeed cut anew for the *Exoticorum libri decem*, page 326 ((Ill. 46). Clusius' textual additions to the section of the 'Lapis tiburonum' in the British Library copy were also reproduced faithfully in the *Exoticorum libri decem*. What had changed, however, was the inscription on the picture: from 'Lapis tiburonum' to 'Believed to be the 'Lapis tiburonum', but more truly the 'Lapis manati" in the 1605 edition.[62]

---

[62] 'Lapis Tiburonum creditus, sed verius Manati.' C. Clusius, *Exoticorum libri decem* (Leiden, 1605), 326.

220 P. BELLONII OBSERVATIONVM

barbæ pilos illi similes, oculis est splēdentibus & subrubentibus: sub quibus binis maculis nigris insignita est. Eius aures orbiculares sunt, ad ~~Taxi~~ *Melis* auriū formam accedentes. Corpus habet maculis aspersum nigris, cùm alias eius color albicet: ~~eius~~ crura & pedes pilo nigro obsiti sunt, quemadmodum in Ichneumone: cauda oblonga, superius nigra, infima verò parte albis maculis aspersa. Agilis est, & carne vescitur. Hæc est Ciuette historia, quæ si cum Hyænæ descriptione cōferatur, apparebit ~~Ciuettam non Claeon esse sed~~ veterum Hyænam esse.

hujus iconis aures rotundiores esse debent, ut emendavi, non mucronatæ.

Plurima
~~Decorum videndum~~ in Alexandria conspiciendā, ~~septē~~ obelisci, & magni Aegyptiorum colossi.

CAP. XXI.

POSTRIDIE urbe egressi, præaltā Pompey columnam spectatum iuimus, in exiguo quodam promontorio sitam, octaua milliaris ab urbe parte. ~~Disciplinam~~ ~~ante nobis conspexi~~ *est † dime-* mēsæ, altitudinis, & crassitiē, reliquas ~~quas unquam~~ *dine* ~~hactenus~~ superat. Colūnæ Agrippæ quæ in Pantheo Romano sunt, nihil ad eius crassitiem accedunt. Vniuersa eius materies, cùm epistyliy, eum forma cubica ex Thebaico marmore est; ex eodem videlicet, quo omnes obelisci ex Aegypto delati constant. Eam Cæsaris iussu, ob victoriam aduersus Pompeium partam, istic erectam aiunt. Adeò crassa est hæc columna, ut impossibile iudicetur,

Ill. 42. P. Bellonius, *Plurimarum singularium and memorabilium rerum [...] observationes*, tr. C. Clusius (Antwerp, 1589), 220.

llius etiam nomen quo eam nuncupare solemus, ex Arabibus
licto veteri. Compacto, ut Meles, est corpore, major tamen:
liebrem naturam habeat meatum, ex quo Civetta emanat,
gentes, existimarunt illam esse Melem sive Taxum, vulgo
& Aristoteli Trochus dictus est. In cervice & per dorsi spinam
igit cùm indignatur, non minus quàm sus. Inde factum, ut
uncupata sit. Os ha-
am felis, sed barbæ
lendentibus & sub-
s maculis nigris insi-
lares sunt ad Melis
. Corpus habet ma-
liàs ejus color albi-
obsiti sunt, quem-
cauda oblonga, su-

Ill. 43. Bellonius' *Observationes* in Clusius, *Exoticorum libri decem* (Antwerp, 1605), 94.

Ill. 44. C. Clusius, *Curae posteriores* (Leiden, 1611), 57.

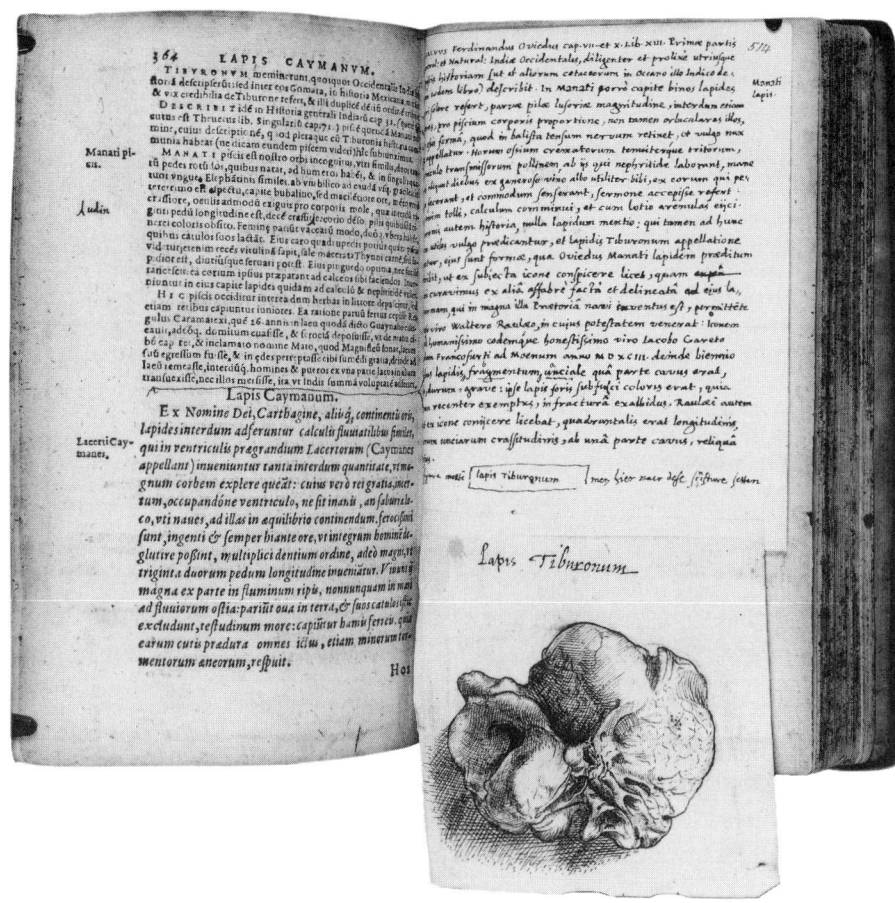

Ill. 45. N. Monardes, *Simplicium medicamentorum ex novo orbe delatorum, quorum in medicina usus est*, tr. C. Clusius (Antwerp, 1593), 364-365.

Clusius had already pointed out in the first edition that Gómara's description of the 'tiburones' and the 'manatus' were very similar. Now Clusius added for the 1605 edition that Oviedo, who discussed both the 'tiburones' and the 'manatus', had given the 'Lapis manati' the features of the 'Lapis tiburonum'.[63] Clusius further explained that he had acquired a fragment of Drake's 'Lapis tiburonum' in 1605.[64]

---

[63] Clusius, *Exoticorum libri decem*, 326. Clusius referred to Fernandez de Oviedo y Valdez, *La historia general de las Indias* (Seville, 1535), civ^v–cv^r (Part 1, book 13, chapter 7) 'De los Tiburones'; cvj^r (Part 1, book 13, chapter 10).

[64] Clusius, *Exoticorum libri decem*, 326.

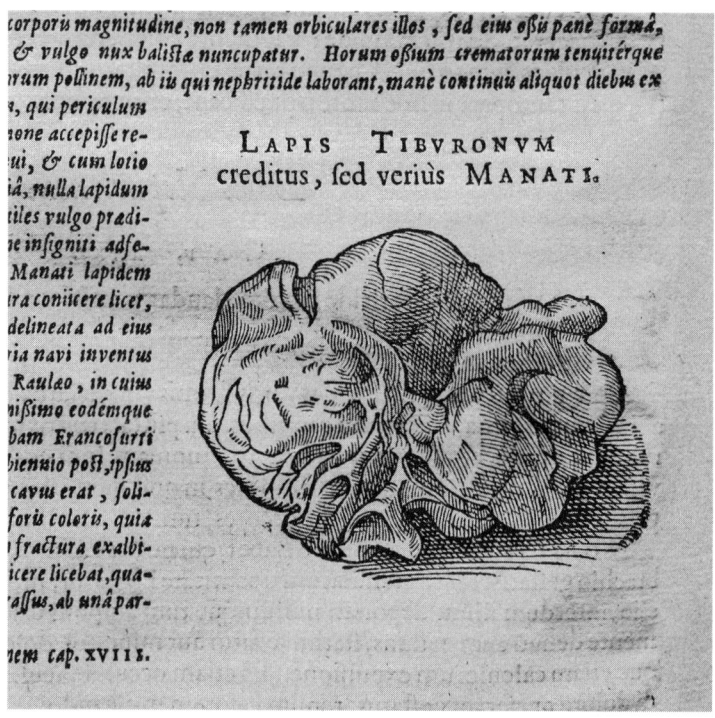

Ill. 46. N. Monardes, *Simplicium medicamentorum ex novo orbe delatorum*, in C. Clusius, *Exoticorum libri decem* (Antwerp, 1605), 326.

The British Library copy is a further good indication that Clusius, as author, was writing in a world of books, with its advantages and limitations. The advantage was that he could obtain knowledge of flora and fauna of lands he had never travelled to through printed books of those who had. The limitation was that the existence, placement and correction of woodcuts and text in his own books were ultimately in the hands of the printer.

Within such limitations Clusius developed several effective uses of images for his work which were in stark contrast to the idealising and universalising tendencies of a Fuchs or a Vesalius.[65] By default, rare and exotic objects are singular and hard to come by; their uniqueness raises questions of authenticity and credibility. It is not surprising, then, that the existence and origins of such alien objects must be established for every case. Therefore, Clusius gives the details of when, from whom, where, and in what condition he had received an

---

[65] For this trend in Fuchs and Vesalius, see Kusukawa, 'Uses of pictures'.

object. Details of the senders or donors – for instance, that Franciscus Roderiguez was a prefect in Java; that the Amsterdam surgeon Johannes Langhe had been to Brazil; that Dr Lambert Hortensius returned from Java in 1601 – all help to establish the credibility of the alien origin of their objects.[66] The particular and individual details, as spelt out in the text, were crucial in persuading the reader that the images depicted an exotic object that truly existed. Clusius often provided details of several sources of an object or its picture. Such an accumulation of details, again, would enhance the veracity of a pictured object.

This kind of strategy was not unique to Clusius. The problem with respect to a rare or unknown plant had already come to the fore in a dispute over the true identity of the 'aconitum' between Pier Andrea Mattioli and Conrad Gessner. The latter thought that the 'aconitum primum' depicted in Mattioli's commentary on Dioscorides was made up in order to fit Dioscorides' description, or otherwise that Mattioli had been duped. However, if Mattioli could show the plant to two or three erudite men and got their testimony, Gessner was prepared to retract his position.[67] Clusius too thought that Mattioli's pictures were not reliable.[68] As printed pictures of plants proliferated without their names remaining stable, the authority of printed images (exactly repeatable though they may be) was becoming highly contested. Especially for pictures of rare and exotic ones, written evidence was thus becoming necessary.

Clusius had never left Europe, and his study of the exotic and the rare necessarily depended on books, on others who had been there, and on his large circle of correspondents. Peter Mason has described in this volume the variety of sources and the diverse quality of information Clusius obtained on Americana. Pictures, of course, had a role to play in the gathering of knowledge. Clusius could not obtain every exotic object himself, and in several cases went to see an object: he often reported on how he was 'shown' an object, of which

---

[66] Clusius, *Exoticorum libri decem*, 12, 15, 20.

[67] 'Matthaeoli Senensis quidem pro aconito primo delineata imago, plane fictitia mihi videtur: sive ipse ad Dioscoridis descriptionem confinxerit, sive ab alio deceptus acceperit [...] Quod si herbam ipsam quam pingit, duobus aut tribus eruditis viris demonstret, illorumque testimonio eam nobis vel publice approbet, palinodiam facile meditabor, et insuper gratias agam.' Conrad Gessner, *De raris et admirandis herbis, quae sive quod noctu luceant, sive alias ob causas, Lunariae nominantur* (Zurich, 1555), 40. The 'aconitum primum' is depicted in P.A. Mattioli (ed.), *Commentarii in libros sex Pedacii Dioscoridis [...] de materia medica* (Venice, 1554), 479. For the Mattioli-Gessner exchange, see now Candice Delisle, 'The letter: Private text or public place? The Mattioli-Gesner controversy about the *aconitum primum*', *Gesnerus* 61 (2004), 161-176.

[68] 'Velim tamen multis Matthioli fictitiis iconibus abstineas, quae si non animi malignitate, animo certe parum ingenuo in ejus Commentarios sunt illatae: ob quam adeo causam an venia dignus sit, multum ambigo.' Clusius to Camerarius, 1584, Hunger, *Charles de l'Escluse*, vol. II, 402.

he was allowed to make a drawing.[69] Several collectors who could or would not part with their prized rare objects sent Clusius pictures instead, sometimes coloured ones.[70] In one instance, working from the picture alone, Clusius conceded, was less accurate than seeing the object directly.[71] Conversation with others was also a source of knowledge, as Clusius often reported, but there is only one case in the first six books of the *Exoticorum libri decem* (namely the case with which the book opens) where a picture of an exotic tree was drawn on the basis of a conversation. But this was a conversation with a courtier, Fabricius Mordentius Salernitanus, of Maximilian II, whose word was presumably sufficient in terms of credibility.[72] This means, of course, that the pictures printed in the *Exoticorum libri decem* were not always the result of Clusius' own firsthand observation.[73]

Nor were the pictures in the *Exoticorum libri decem* always an exact representation of the object in question. This could partly be the craftsman's fault, if he drew the surface of a fruit smooth when it should have been rough, or (without consultation) filled in the eye of a fish, which in the original dried specimen was just a cavity.[74] Another reason is that some of the original specimens were not live: Clusius explains how he soaked a dried plant in water for several hours before having it drawn by an artist.[75] Indeed, as was the case with vander Borcht over the illustrations for the Spanish flora, Clusius was keenly aware of the difficulty of deducing a proper image or 'historia' of a plant from a dried sample, especially if one had not seen it live and growing.[76] However

---

[69] Clusius, *Exoticorum libri decem*, 28, 29, 63, 71, 80, 86, 90, 108, 124.

[70] Clusius, *Exoticorum libri decem*, 63 (Franciscus le Clerc), 103 (Jacob Plateau), 137 (Volcardus Coornhart).

[71] This was the picture of the sloth; Clusius, *Exoticorum libri decem*, 110f. For the identity of the sloth, see W.B. Ashworth Jr, 'The persistent beast: recurring images in early zoological illustrations', in A. Ellenius (ed.), *The natural sciences and the arts* (Uppsala, 1985), 46-66. See further Peter Mason's chapter in this volume.

[72] 'Huius arboris iconem, quam concinne fieri potuit, ex Fabricij narratione, accedente etiam ipsius iudicio, adumbrari iussimus, eamque in illorum qui hoc studio delectantur gratiam, hic subijcimus.' Clusius, *Exoticorum libri decem*, 2. For this case, see B. Ogilvie, *The science of describing*, forthcoming, ch. 5, section: The *ficus indica*: reliable witnessing. I thank Prof. Ogilvie for allowing me to read a draft of this chapter. Mordentius was also versed in mathematics (*Exoticorum libri decem*, 184). Cf. the case of Paullus Choartus Buzenvallus, *Exoticorum libri decem*, 24.

[73] For the limits of Clusius' pictures for classifying objects, see C. Swan, 'From blowfish to flower still life paintings: classification and its images circa 1600', in P.H. Smith and P. Findlen (eds.), *Merchants and marvels: Commerce, science and art in early modern Europe* (New York/London, 2002), 114-122.

[74] *Exoticorum libri decem*, 44, 142. Cf. also praises for the 'perito pictore', ibid., 88, 92.

[75] Clusius, *Exoticorum libri decem*, 87, 90.

[76] 'Quam vero difficile sit ex siccis plantis genuinas earum icones exprimere, nisi pictori adsit rei herbariae non vulgariter peritus, ipse in Hispanicis expertus sum, qui tamen pictorem in exprimendis earum iconibus versatissimum nactus fui: ex siccis praeterea stirpibus earum historiam (nisi

naturalistically executed, pictures of exotic plants and animals printed in the *Exoticorum libri decem* do not guarantee, therefore, that Clusius had directly observed a live specimen. In some cases, the pictures showed what an object would or should have looked like. The net effect of the printed pictures was, however, to show in a uniformly vivid state objects that Clusius got to know originally in various ways – live, dried, whole, partial, pictured, reported by an eye-witness or by hearsay.[77]

The kind of people with whom Clusius was in contact concerning rare and exotic naturalia included fellow physicians, such as Stephan Backerus, Ioannes de Castaneda, Jacobus Colius, Henricus Hoierus, Lambertus Hortensius, Franciscus le Clerc, Bernardus Paludanus, Peter Paaw, Tobias Roelsius, Simon de Tovar, and Aelius Everhardus Vorstius; apothecaries such as the Garet brothers (James in London,[78] Pieter in Amsterdam), Hugh Morgan, the Royal Apothecary in London,[79] Giovanni Pona in Verona, Christianus Porretus in Leiden, Johannes Scharm and Walichius Syvertz in Amsterdam; merchants such as Ioannes Gorvertz van der Aer (Amsterdam), Hendricus Tilmannus, Volcardus Coornhardt (Amsterdam), and the company of nine merchants (the forerunner of the Dutch East India Company) who had organised a fleet to Java (Hendricus Hudde, Reynerus Paaw, Petrus Hasselaer, Ioannes Ioannis F. Caerl, Ioannes Popper, Henricus Buyck, Theodoricus ab Os, Sylvertus Pietersen, Arnoldus Grotenhuys).[80] There were also other citizens, such as the Chancery clerk Richard Garth,[81] Simon Parduynus, councillor of Middelburg; and Emmanuel Sweert, a citizen of Amsterdam who traded in rare flowers.[82]

---

nascentes videris) describere, non levem laborem esse mecum judicare potes.' Clusius to Camerarius, 1584, Hunger, *Charles de l'Escluse*, vol. II, 403.

[77] For the various media by which information on new-world flora and fauna can be obtained, see De Renzi, 'Writing and talking of exotic animals.'

[78] James Garet, an apothecary of London, died in 1610, leaving one estranged son, Fernando, from his first marriage, and two daughters, Elizabeth and Mary, by his second wife Jacomina. According to his will, it appears that Garet did not have a museum. PCC will prob/11/115, Image ref: 295. Jacomina's will PCC will prob/11/182, image ref. 387 (http://www.nationalarchives.gov.uk/).

[79] C.A. Bradford, *Hugh Morgan: Queen Elizabeth's apothecary* (London, 1939).

[80] The gifts from this company were given on return of the first fleet (1597); Clusius, *Exoticorum libri decem*, 26, 36, 88. H. Terpstra, 'De Nederlandsche voorcompagnieën', in F.W. Stapel (ed.), *Geschiedenis van Nederlandsch Indië*, 5 vols. (Amsterdam, 1938-40), vol. II, 275-475.

[81] Richard Garth (d. 1597), President of the Chancery, who was related by marriage to the Savilian family. His sister must have married Sir John Savile, son of Henry, the warden of Merton and founder of the Savilian chairs: Garth names John Savile's children, Henry, Jane and Elizabeth Jackson his 'nephew' and 'nieces'. PCC will prob/11/92/145r-146r. For objects sent by Garth, see *Exoticorum libri decem*, 6, 31, 69, 78. Garth's widow, Joanna Busher, was credited with a recipe for Merlin's potion, M. Lobelius, with P. Pena, *Stirpium nova adversaria* (London, 1605), 473.

[82] Sweertius, *Florilegium*. Cf. Hunger, *Charles de l'Escluse*, vol. I, 267-269.

Physicians and apothecaries formed a natural group of correspondents for Clusius, given their shared interest in flora and fauna as *materia medica*. This was the age when the composition of medicines was gradually being codified by way of pharmacopoeias. As early as 1561, Clusius had translated a Florentine pharmacopoeia into Latin: the *Antidotarium, sive de exacta componendorum miscendorum medicamentorum ratione*. The city magistrates of Nuremberg, Augsburg and Amsterdam each established their pharmacopoeia to be used by practitioners in this period.[83] At the behest of Plantin, the Antwerp apothecary and botanist Pieter van Coudenberghe revised and corrected Cordus' *Dispensatorium* (1568), which was further modified by Lobelius (1580) and eventually adopted by the Antwerp city magistrates in 1659.[84] In the preface to his *Pemptades*, Dodoens declared that an apothecary's mistake in substituting certain components of a medication should rest squarely on the physician ignorant of the *materia medica*, since the apothecary derived his authority from the physician. The problem was, in fact, not just ignorance, but also the work of fraudsters and impostors who peddled adulterated medicines.[85] The import of foreign medicines into the European market accelerated during this period, and was accompanied by an increasing sense of danger about relatively unknown drugs and the possibility of adulterated ones.[86] The English physician Timothy Bright was therefore not alone in insisting that in the

---

[83] Valerius Cordus, *Dispensatorium sive antidotarium* (Nuremberg, 1592); *Pharmacopoeia seu medicamentarium pro Rep. Augustana* (Augsburg, 1580); *Pharmacopoeia Amstelredamensis (1636)*, facs. edition by D.A. Wittop Koning (Nieuwkoop, 1961).

[84] De Nave and Imhof (eds.), *Botany in the Low Countries*, 95-97. See also Lobelius' tabulation of medicines, Swan, 'Blowfish', 122-128.

[85] 'Sit pro exemplo Electarium Diamargariton calidum ab Avicenna descriptum Canonis tertij, fen xxi. Tract. Ii, cap. 11 uterum gestantibus convenitus in cuius compositionem Seitaragi Arabibus dictum, venit admiscendum. Huius autem loco indocti Pharmacopoei Turbith appellatum accipiunt: radicem valide purgantem, et corpus insigniter commoventem. Ita salutare medicamanetum in noxium commutant. Quis hic venit culpandus? Cui imputandus error? Pharmacopoeo ne an medico? Pharmacopoeus fortassis se alicuius Medici auctoritate tuebitur: culpa idcirco in Medicum recidit imperitum, et simplicis materiae medicae ignarum. Si etenim sciret Seitaragi lignosum quoddam esse, tenue, garyophyllis simile (qualia sunt lignosa sarmenta quae garyophyllis inferuntur) ut Avicenna testatur. Nec Hali Abbas repugnat; haudquaquam Turbith eius loco substitui permisisset [...]. Non lubet autem hic referre, quam multis modis imposturas ac fraudes moliantur: vel dum compositiones et vetustate exoletis aut situ corruptis parant: aut easdem depravant, quaedam omittentes, alia addentes: vel cum spuria, factitia, adulterataque pro legitimis exquisitisque venum exponunt.' Dodoens, *Pemptades*, 2.

[86] This was certainly the case for England: see R.S. Roberts, 'The early history of the import of drugs into Britain', in F.N.L. Poynter (ed.), *The evolution of pharmacy in Britain* (London, 1965), 165-185. Cf. the case and fear of being duped by fake objects, in P. Findlen, 'Inventing nature: commerce, art and science in the early modern cabinet of curiosities', in Smith and Findlen (eds.), *Merchants and marvels*, 303f.

face of such threats, local (English) medicine was sufficient to cure all diseases.[87] Clusius' attempt at proper identification of new world *materia medica* in the works of Ab Orta, Monardes and A Costa thus could be seen in this context to have serious practical implications. The juxtaposition in the *Exoticorum libri decem* of the 'legitima' and 'spuria' pictures of the clove tree ((Ill. 47) could not only clarify the veracity of authors' claims, but also help readers distinguish between true and false objects. 'Legitima' is an adjective that Clusius frequently used, for knowledge as well as pictures. In the preface

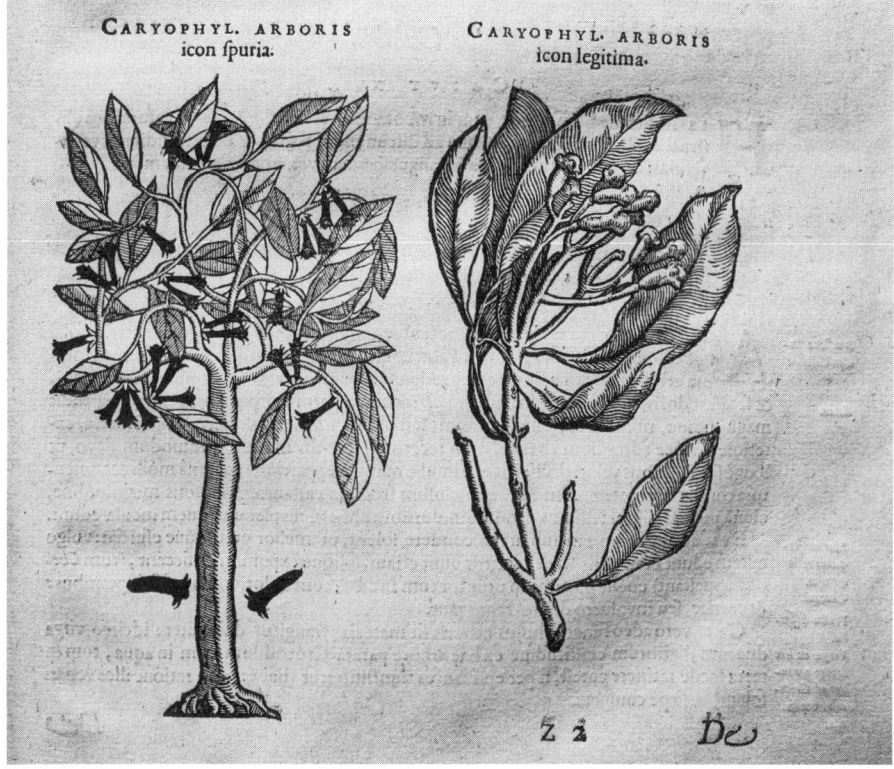

Ill. 47. C. A Costa, *Aromatum et medicamentorum in Orientali India nascentium liber* in C. Clusius, *Exoticorum libri decem* (Antwerp, 1605), 267.

---

[87] Timothy Bright, *A treatise, wherein is declared the sufficiencie of English medicines for the cure of all diseases cured with medicine* (London, 1580). For the suspicion of foreign drugs, and sufficiency of local drugs, see Andrew Wear, *Knowledge and practice in English medicine, 1550-1680* (Cambridge, 2000), 66-78. For the professional implications of foreign and native medicines in London in this period, see D.E. Harkness, "'Strange' ideas and 'English' knowledge. Natural science exchange in Elizabethan London', in Smith and Findlen (eds), *Merchants and marvels*, 137-160.

of the *Exoticorum libri decem* he explained, for instance, that he had included only 'true' and 'legitimate' things in the *Rariorum plantarum historia*, and that now he presented his study of exotic things with 'legitimate' pictures.[88] For Clusius, then, a 'legitimate' picture also helped make his own knowledge 'legitimate'. Clusius took special care with pictures, as he echoed the language of pharmacy – an adulterated picture could detract much from the authority of 'legitimate' ones.[89]

Some of the physicians among his contacts, such as Tobias Roelsius,[90] were collectors, as were some apothecaries: Giovanni Pona, who is perhaps less well known than his fellow Veronese apothecary Francesco Calzolari, also had a museum, which was described in the *Index multarum rerum quae repositorio suo adversantur* (1601).[91] The Leiden apothecary Christian Porret had a collection as well.[92] Rare and exotic objects displayed in apothecaries' shops attracted customers to come in, buy their medicines, and exchange gossip.[93] Merchants too, such as Joannes Rutgerus, had a museum, which was not unusual given the example of the Fuggers.[94] Moreover, other respectable citizens like Garth and Parduyn appear to have had a keen interest in collecting exotic objects, if not building up a museum. This also meant that profit could be made by supplying rare and exotic objects to these people – something which worried Clusius. He pointed out that there were rich people prepared to spend

---

[88] 'Proximis his annis rariorum plantarum quas in variis meis peregrinationibus observavi historiam publici iuris feci, cujus lectionem non inutilem fuisse, sed fructum aliquem rei herbariae studiosis attulisse, mihi persuadeo: summopere enim curavi, ut nihil nisi quod *verum et legitimum* esset, in ea traderem. [...] et amicorum quorundam diligenda, et mea sedulitate effectum est, ut nonnullas adquisiverim, quarum descriptionem in suas classes distinctum et sex libris comprehensam, in lucem profero, confidens aliquam etiam utilitatem studiosae iuventuti allaturam: nam in illa pleraque Aromata diligenter descript, et *legitimis iconibus* ad vivum expressis (non modicum, meo iudicio, momentum ad eorum cognitionem adipiscendam allaturis) illustrate reperiat.' *Libri exoticorum decem,* †6v. My emphasis.
[89] 'Ego vero contraria plane sum in sententia, nullam fictitiam aut suspectam meis admiscere sciens velim: possent enim hujusmodi adulterinae legitimarum aliarum authoritati multum adimere.' Clusius to Camerarius, 1583, Hunger, *Charles de l'Escluse,* vol. II, 394, in the context of the veracity of the pictures in the Juliana codex, see also n. 49 above.
[90] *Exoticorum libri decem,* 65.
[91] P. Findlen, *Possessing nature: Museums, collecting, and scientific culture in early modern Italy* (Berkeley, 1994), 180.
[92] A. Goldgar, 'Nature as art: the case of tulips', in Smith and Findlen (eds.), *Merchants and marvels,* 332.
[93] Findlen, 'Inventing nature', 307; for apothecaries' and barbers' shops as sites for exchanging political gossip, see Filippo de Vivo, 'Wars of papers: Communication and polemic in early seventeenth-century Venice' (PhD dissertation, Cambridge, 2003).
[94] Clusius, *Exoticorum libri decem,* 110f. For merchants and their collections, see M.A. Meadow, 'Merchants and marvels: Hans Jacob Fugger and the origins of the Wunderkammer', in Smith and Findlen (eds.), *Merchants and marvels,* 184.

a lot of money to acquire a rare plant so that possessing it would bring glory to them; and enticed by the prospect of profit from such people, there were merchants, tailors, craftsmen and contemptible conmen who wanted to have dealings with the study of plants; this in his view could make the very study of plants contemptible.[95]

In the case of the nine merchants who formed a company to send ships to the East Indies, their act of giving exotic gifts was presumably seen as an important gesture of generosity and respectability.[96] Jacob van Neck, whose triumphant return from the second voyage (1599) sparked off the Dutch rush to the East Indies, and Dr Hortensius, the physician appointed to the company's fleet, respectively brought back some exotic nuts and pepper for Clusius.[97] Hortensius was also physician for the 1602 sailing to the Far East. For this sailing Clusius asked Theodoricus ab Os to give the ships' apothecary Clusius' own memorandum which listed what type of objects and information to bring back (which included, if possible, their medicinal effects).[98] The Dutch rush to the East Indies thus was another route by which Clusius obtained objects, and another reason why merchants were becoming interested in knowledge of the exotic.

In the *Exoticorum libri decem*, we catch a glimpse of Clusius' interaction with several of his correspondents. It is in marked contrast to the way Mattioli behaved towards others with expertise of plants. As Paula Findlen has shown, the inclusion, exclusion and criticism of individuals in his publications was a method that Mattioli perfected in the successive editions of his commentary on Dioscorides. This helped not only to define and order the republic of botanists, but also fashioned Mattioli himself as the pinnacle and centre of that republic.[99] So far as I can see, Clusius did not aspire to such megalomania when referring to individuals, and he certainly disapproved of Mattioli's arrogance.[100] Instead, Clusius describes himself as receiving letters, pictures

---

[95] 'Vile tandem fiet istud studium, mi Lipsi, quia et mercatores, imo sartores et cerdones, aliique viles artificies, id tractare volunt, spe quaestus illecti: nam vident opulentos istos pecuniam interdum profundere, ut plantulam aliquam redimant, quae raritatis nomine commendetur; ut gloriari apud suos possint, se illam possidere. Clusius to Lipsius, 1594, Hunger, *Charles de l'Escluse*, vol. II, 251. n. 1, as pointed out in Goldgar, 'Nature as art', 346, n. 82.
[96] Clusius, *Exoticorum libri decem*, 36, 88.
[97] *Exoticorum libri decem*, 53 (van Neck), 20, 78 (Hortensius).
[98] Hunger, *Charles de l'Escluse*, vol. I, 266f., and Clusius, *Exoticorum libri decem*, †2v.
[99] P. Findlen, 'The formation of a scientific community: natural history in sixteenth-century Italy', in A. Grafton and N. Siraisi (eds.) *Natural particulars: Natural philosophy and the disciplines in early modern Europe* (Cambridge [Mass.], 1999), 369-400.
[100] 'Fuit Matthiolus, dum vixit, arroganti ut apparet ingenio praeditus, aliosque reprehendendi summa prurigine (ut nosti) laboravit: itaque non mirum, si quidam suborti sunt qui ejus vestigiis insistentes illum egregie exagitarunt.' Clusius to Camerarius, 1584, Hunger, *Charles de l'Escluse*, vol. II, 402.

and objects in the major cities of Europe at the time: Antwerp, Amsterdam, London, Vienna and Frankfurt. These cities were rapidly becoming major centres of commercial exchange, and thus also of information exchange, since news about shipments from the New World could easily have commercial value.[101] Clusius thus described himself as placed in centres of exchanging objects and knowledge of naturalia and exotica. Although Clusius mentions the fact that he (or a friend) occasionally purchased objects, he also explains how many objects and pictures were sent and given to him out of friendship or generosity of the donor. Perhaps it was important to him that he was describing part of an exchange system that would mark out the connoisseurs from the peddlers.[102] It goes without saying that neither Clusius nor others interested in rare and exotic naturalia were the only scholars who made use of the communication networks converging on commercial or diplomatic centres in Europe. Those with antiquarian interests also exchanged letters, objects (such as coins and fragments), and pictures for their study of ancient monuments.[103] As Clusius' own pursuits testify, antiquities and nature were not mutually exclusive interests, however. It may well be that there was much common ground (perhaps also including their use of pictures) in the way scholars approached objects from the distant past and from distant lands. For Clusius, pictures were important in making knowledge of objects from distant lands 'legitimate', though the authority of the picture itself would often rely on the text it accompanied.

---

[101] W.D. Smith, 'The function of commercial centres in the modernization of European capitalism: Amsterdam as an information exchange in the seventeenth century', *The journal of economic history*, 44/4 (1984), 985-1005; P. O'Brien et al. (eds.), *Urban achievement in early modern Europe: Golden Ages in Antwerp, Amsterdam and London* (Cambridge, 2001); J.J. McCusker and C. Gravestijn, *The beginnings of commercial and financial journalism. The commodity price currents, exchange rate currents and money currents in early modern Europe* (Amsterdam, 1991) for publication of commodity price-lists. For more on information exchange, see Florike Egmond's chapter in this volume.

[102] Goldgar, 'Nature as art', 337.

[103] J. Papy, 'An antiquarian scholar between text and image? Justus Lipsius, humanist education and the visualisation of ancient Rome', *Sixteenth century journal,* 35 (2004), 97-131. I thank Prof. I. Maclean for drawing my attention to this piece. Cf. also J. de Landtsheer, 'Justus Lipsius and Carolus Clusius: A flourishing friendship', in M. Laureys (ed.), *The world of Justus Lipsius: A contribution towards his intellectual biography* (Brussels, 1998), 273-295.

PART IV

Ideas and influence of Clusius

The influence of Clusius in Italy.
Federico Cesi and the Accademia dei Lincei

Irene Baldriga

Introduction

As one of the main characters in the early modern European scientific community, Carolus Clusius established important relationships with a wide variety of Italian natural philosophers. Among these were the Bolognese Ulisse Aldrovandi, the Neapolitan Ferrante Imperato, and the famous Veronese collector Francesco Calzolari. The failed attempt by the founders of the Roman Accademia dei Lincei, mainly Johannes Eckius and Federico Cesi, to start a long-lasting correspondence with the famous Northern botanist is far less known. The aim of this contribution is not only to offer an interpretation of such an uncanny and surprising failure, but also to raise some further questions concerning certain methodological changes that took place within the *République des Lettres* at the beginning of the seventeenth century.

A considerable corpus of the Italian correspondence collected by Clusius was published by De Toni in 1911.[1] The majority of these letters relate to the purchase of botanical specimens, bulbs ('cipolle') and seeds, but in some cases they can offer more informative clues about the nature of scientific relationships at the time. Once a contact had been established and assured, letters mainly became containers of specific requests for new species to grow and collect in private and academic gardens, or could include complex questions concerning watering and exposure.[2] As a general rule, we can observe that two main aspects should be considered in the analysis of these documents. First, this kind of correspondence completely lacks digressions on philosophical problems, so that it is often necessary to read between the lines in order to identify them. Second, the constant, stubborn and apparently blind insistence

---

* I would like to thank Florike Egmond and Peter Mason for their precious help and thoughtful advice.
[1] G.B. De Toni, 'Il carteggio degli italiani col botanico Carlo Clusio nella Biblioteca Leidense', *Memorie Regia Accademia, Scienze, Lettere ed Arti, Modena*, 10 (1911), 1-147.
[2] On this, see G. Olmi, 'Molti amici in varii luoghi: Studio della natura e rapporti epistolari nel secolo XVI', *Nuncius. Annali di storia della scienza*, 6 (1991), 3-31.

on practical affairs displayed by the correspondents should be interpreted in the perspective of a modern empiricism, which was to be sustained more through experimental observations than through the use of literary sources and erudite quotations.

The present contribution proceeds by way of a comparison between two very different kinds of correspondence which involved Clusius and some Italian scientific scholars: the scientific-practical correspondence undertaken by the Tuscan botanist Caccini; and the philosophical-theoretical correspondence attempted, in vain, by the Lincean Academy.

A failed overture: Clusius and the Lincei

Between September 1606 and March 1609, Carolus Clusius – who by this time was an old man and was suffering seriously from bad health – exchanged a constant correspondence with a 33-year-old Florentine botanist called Matteo Caccini. The letters sent by the great scientist were discovered and published in 1939 by Piero Ginori Conti,[3] but they are worth further consideration, especially with regard to the nature of some relationships established by Clusius at that time.

Caccini, who might be considered a dilettante in botany, approached Clusius via the friar Gregorio da Reggio, who had been corresponding with the scientist from Bologna since 1602. In his first letter to Caccini, Clusius not only enthusiastically agrees to reply to his Italian admirer, but also adopts an informal and quite friendly tone, apologizing for his poor command of the Italian language, refusing any title, and asserting that he had always done his best to please those who like devoting themselves to this honest activity and exercise ('honesto passatiempo et exercitio').[4]

Nevertheless, it must be stressed that, despite such a generous declaration, Clusius cannot help himself from specifying that the new relationship will certainly be fruitful in terms of mutual material exchange (bulbs, seeds, images of flowers and blooms). He also appears to be well informed about the good acquaintances made by Caccini in Tuscany and elsewhere among flower growers. The material aspect of the correspondence between Clusius and Caccini is clearly testified by the text of the letters themselves: they mainly discuss problems of cultivation connected with the weather conditions, watering and manuring. It was Caccini, more than his new friend, who was to send precious

---

[3] The letters from Clusius to Caccini were published in P. Ginori Conti, *Lettere inedite di Charles De L'Ecluse a Matteo Caccini floricultore fiorentino* (Florence, 1939).

[4] Leiden University Library, VUL 101, letter from Clusius in Leiden to Caccini in Florence, 30 September 1606.

bulbs and seeds, fighting against the many difficulties that beset travel in seventeenth-century Europe, not to mention the risks that such a long trip entailed for those delicate samples. The iron boxes used to carry the beloved goods did not always assure their survival, and often gave the anxious correspondent the unpleasant surprise of a rotten specimen. Clusius himself was extremely sceptical about merchants and intermediaries, especially those living in the Southern Netherlands:

[...] everyone in Brabant and in these provinces wants to be a dealer in plants, even the most humble and wretched [...].[5]

[...] it is true that in Brussels there are some humble and uncultivated persons who trade in flowers to make a profit and who name the plants in their own way [...].[6]

I would like to suggest that it was advantageous for Clusius to establish a brand-new correspondence with an unknown and quite young Italian botanist like Caccini, to whom he nevertheless displayed sincere gratitude for his precious contribution to his own research. Many species sent by Caccini to Leiden were later included by Clusius in his last work, in which he mentioned the good offices of his Florentine correspondent.

However, if Clusius certainly benefited from Caccini's zealous friendship, we should also wonder about the opposite: why was Caccini so eager to obtain the favour of the famous scientist? Although he had never visited Italy, Clusius enjoyed a high reputation among Italian scholars, as Andrea Ubrizsy-Savoia has clearly pointed out.[7] Thanks to his highly praised books, mostly the *Rariorum plantarum historia*, and to the wide web of relationships he had established through his correspondence, Clusius could count on a wide popularity, as the case of Caccini also demonstrates.

More striking in this respect is the story of the overture made by the Accademia dei Lincei to obtain Clusius' approval and support.[8] In 1604 Johannes Eckius, co-founder of the Academy together with Federico Cesi, Anastasio de Filiis and Francesco Stelluti, sent two letters to the prestigious

---

[5] Leiden University Library, VUL 101, letter from Clusius in Leiden to Caccini in Florence, 8 December 1608: 'ognuno in Brabante et in queste provincie vol essere mercatore de piante, fin a vilissimi guanapani e gente mesquina.'

[6] Leiden University Library, VUL 101, letter from Clusius in Leiden to Caccini in Florence, 30 October 1608: 'E ben vero che in Bruxelles sono alcuni vili et mechanice persone che fanno mercantia di fiori et con ganancia, li quali battisano le piante a modo loro.'

[7] A. Ubrizsy Savoia, 'I rapporti tra Carolus Clusius ed i naturalisti italiani del suo tempo', *Physis*, 20 (1978), 4-69.

[8] I. Baldriga, '"La fatiga di pigliar i disegni delle piante": Federico Cesi, la pittura filosofica e la riproduzione del mondo vegetale', in *Federico Cesi un principe naturalista, Atti del Convegno Internazionale di Studi, Acquasparta, September 29-30, 2003, Accademia Nazionale dei Lincei* (forthcoming), 505-525.

director of the Leiden Botanical Garden, informing him about the nature and goals of the new association and asking for his participation.⁹ Strangely enough, it seems that Clusius did not pay any attention to Eckius' request, probably ignoring even the core of the ambitious project that Cesi and his friends were dreaming of. The apologetic historical approach which has been devoted to the Lincean Academy over the past two hundred years has completely neglected the absolute failure of this first chapter of Cesi's adventure. Not only Clusius, but also Bauhin and Lobel were invited in vain by Eckius to join his circle.

It might be intriguing to ask why Clusius actually refused the Lincean overture. In his letters, Eckius appears quite bold in his description of the Roman circle: in referring to a wide, universal web of contacts and relationships, he was counting his chickens before they were hatched. Probably inspired by Cesi himself, who – among the founders of the Academy – was certainly the one most interested in botany, Eckius proposed to Clusius to share observations and discoveries in the field, but his requests apparently fell on deaf ears. It is very likely that the Linceans appeared quite vague in their proposals to Clusius: their letters did not present any kind of organic scientific plan, and the theoretical approach they seemed to embrace was likely to arouse Clusius' scepticism. Most of all, Eckius included among the goals of the new Academy the investigation of the secret and arcane aspects of the universe. Could it be this esoteric aspect of the Lincean research that induced Clusius to reject Eckius' invitation? It should be stressed that Cesi's interest in the esoteric, alchemy and arcane knowledge has been, deliberately or not, often disregarded by the vast

---

⁹ G. Gabrieli, *Il carteggio linceo* (Rome, 1996), nos. 6 and 11. The bold approach adopted by Eckius is clearly testified by the letter of April 1604:

Escellens D.S.P.,

Miraberis forsan, vir doctissime, has inconsuetas totius Ill.mae Lyncae Academiae literas: mirari desines, ubi te ipsum tuamque in stirpium differentiis disciplinam consideraveris, et famam, quam de tua habemus sapientia, speculatus fueris: ea enim est causa literarum, hi studiorum fructus duntaxat gloria.

Et ea est quod discendi causa ad te venimus, tuaeque petimus disciplinae commercium de plantarum seminum aliorumve differentiis similium, novis et arcana quaedam vel generata natura rara: si non ipsa, saltem eorum descriptiones.

Caeterum unicum hoc ut tuam scientiam nobis communices epistolis, ut ita etiam tecum alloquamur, qui universi terrarum orbi[s] doctissimis quibuscumque viris confabulamur, ut omnibus nostrae Academiae satisfaciamus professoribus, et precipue Principi Ill.mo Marchioni Caesio.

Apud quem curatum habebimus, ut quascumque [velis] plantarum varietates ad te mittantur. Sic etiam si quod nostrir rarum videbitur partibus adfuerit, ut mittas petimus.

Vale interim et salve. Roma XIII. Calendas Aprilis 1604.

studiosissimus

Joannes Heckius.

majority of scholars, who were mainly concerned to sing the praises of the institution that could embrace Galileo Galilei. Even recent contributions seem to insist on the relationship between Galileo and the Academy, and in so doing they ignore the many pieces of evidence for a knowledge which was based on late-Renaissance epistemology.

The result of Caccini's approach to Clusius was very different. In considering the reasons for his success, we should consider not only Fra Gregorio's introduction, but also the offer of specimens and seeds. It was like showing carrots to a donkey. In the letter sent by Gregorio da Reggio to introduce Caccini to Clusius, the request for an erudite correspondence is clearly compensated by the offering of botanical tokens: 'You should know that I am in contact with a gentleman from Florence who begs me [...] to recommend him to you [...]; you will be well satisfied with the new friendship and the new things he is growing in his garden [...].'[10]

Even more important is what Fra Gregorio says about Caccini's intellectual approach: 'Do not mind his youthfulness and lack of expertise in the field, because at any rate he has a strong desire to be practical and also to be useful to you.'[11] The reference to a 'strong desire to be practical' might be interpreted not only as a clear reference to Caccini's desire to obtain a deeper knowledge of botanical matters, but also as a comment intended to reassure Clusius about the objective nature of Caccini's interests and to exclude any suspicions of pedantic speculations.

With regard to the correspondence with the Linceans that never seems to have got off the ground, Eckius' visit to Clusius in Leiden seems particularly striking. In his *Fructus itineris ad Septentrionem*, a sort of a diary of his 1605 peregrination through Northern Europe, Eckius reports the following: 'Visited Scaliger and Clusius in Leiden and made them both friends of the Lyncei.'[12] Thus, a real contact between our scientist and the most adventurous member of the Academy did happen; had he replied to Eckius' letters or not, in the end Clusius made his acquaintance and was certainly informed about Cesi's projects.

The fictitious nature of such a relationship between Clusius and the Linceans is attested by the absence of any known letters between them in the archives and confirmed by certain facts of his scientific life. A confrontation with Caccini's correspondence is once again revealing. In the letters sent to Caccini, Clusius mentions a fact which clearly demonstrates his lack of contacts with the city of

---

[10] De Toni, 'Il carteggio degli italiani col botanico Carlo Clusio nella Biblioteca Leidense', 147.
[11] Ibid.: 'né guardi V.S. ch'egli in questo sij giovine e non molto esperto della professione, perché ad ogni modo egli ha un vivo et ardente desio, sì di farsi pratico, come anco di servir V.S.'
[12] 'Lugduni visitavit Scaligerum et Clusium, utrumque amicos Lynceaorum fecit.' G. Gabrieli, *Contributi alla storia della Accademia dei Lincei*, 2 vols. (Rome, 1989), vol. II, 1110.

Rome: this is related to his deep desire to receive the *Minus cognitarum stirpium* written by the well-known Neapolitan botanist Fabio Colonna.[13] Having lost any contact with the author and believing that the book had already been published in Rome, Clusius asks Caccini to purchase a copy, and even suggests contacting the Dutch apothecary Henricus Corvinus, who was living in Rome and practised his profession in his workshop 'The imperial eagle'.[14] It is a curious coincidence that both Colonna and Corvinus would later be in close contact with the Lincean Academy; what this episode seems to show is that, despite his eagerness to receive the book, Clusius did not contact Cesi or his fellow Linceans, but found it more convenient to involve the Florentine Caccini. In brief, there are enough hints to raise the hypothesis that Clusius' decision to reject the Lincean Academy was deliberate. Not even when a direct relationship with Cesi might be attractive for personal reasons was Clusius disposed to ask for the Prince's help.

The story of the failed relationship between Clusius and the Linceans is the mirror of a failed dream. The international glamour pursued by the young Linceans, especially during the first phase of their research, never materialised. Cesi and his young fellows were not able to establish the solid web of contacts they needed in order to achieve the official recognition of the *République des Lettres*. Clusius' refusal to include them in the vast community of his *contubernales* was not an isolated case: it came along with Bauhin's, Lobel's, and later also Kepler's and Bacon's. I am deeply convinced that this still fuzzy chapter of the Lincean history should be investigated more attentively and that it could help to explain the accusation of heresy which hit Eckius and, indirectly, the other founders of the Academy. On the basis of these considerations, it is clear that if a relation between Clusius and the Lincei did exist, it had a one-way direction and mainly involved the Roman circle founded by Federico Cesi.

Some aspects of Clusius' influence on the botanical research of Cesi and the Lincei: Empiricism versus sublimation

Beyond any doubt, the investigation of the vegetable world soon became a very important part of the Lincean project. Cesi himself engaged in the study

---

[13] Fabio Colonna, *Minvs cognitarum rariorvmque nostro coelo orientivm stirpivm Ekphrasis. Qua non paucae ab antiquioribus Theophrasto, Dioscoride, Plinio, Galeno aliisq[ue] descriptae, praeter illas etiam in Phytobasano editas disquiruntur ac declarantur. Item de aquatilibus aliisque nonnullis animalibus libellus [...] Omnia fideliter ad vivum delineata, atque aeneis-typis expressa cum indice in calce voluminus locupletissimo* (Rome, 1616).

[14] On Henricus Corvinus see G.J. Hoogewerff, 'Henricus Corvinus', *Mededeelingen van het Nederlandsch Instituut te Rome*, 2 (1922), 113-118; I. Baldriga, *L'occhio della lince. I primi lincei tra scienza, arte e collezionismo* (Rome, 2002), 227-233.

of botany more than in any other aspect of the theatre of nature. Cesi was well aware of the importance of acquiring large amounts of visual and written information concerning the object of his research. It is possible that the wideness of Cesi's view and its international profile have not been sufficiently examined. Most of all, insufficient attention has been paid to the important role played by the Netherlands as an example to be taken into consideration in the Lincean adventure. In fact, there are many clues that suggest that the Linceans had a strong interest in the Low Countries which went beyond the Batavian origin of the young Johannes Eckius (known as the 'Illuminated') and which included gardening, collections of curiosities (Paludanus), scientific instruments, and publishing. In the specific context of this contribution, I would like to stress the importance of the attention paid by the Linceans to the species cultivated in the Leiden botanical garden. An investigation of the Lincean library inventories has demonstrated the presence of the catalogue of the Leiden Hortus published by Pieter Paaw in 1601.[15] We do not have enough documents to provide final conclusions about the use Cesi made of such information, but we can certainly assert that the Lincean Prince pursued his projects more through a theoretical route than through the practice of cultivation. I would also suggest that the reason why he basically privileged botany over the other natural sciences was the existence of a huge amount of books and previous experiences on which he could base his assertions.

Cesi was firmly convinced that the creation of a taxonomic system with information derived from the direct observation of plants, bulbs, flowers and seeds could be the key to open the secret door of natural knowledge. In a letter to Johannes Faber, sent on 19 November 1622, Cesi describes his investigation: 'I am now in that great *Chaos* of the methodical distribution of plants and it seems to me to have quite found a solution; this will be an important part of my Mirror of Wisdom and Theatre of Nature.'[16] Besides, Cesi's fidelity to traditional terms like 'Theatre' or 'Mirror', considered as appropriate descriptions of his research, reveals his renewed ambition to provide a fully comprehensive reconstruction of the universe.[17]

---

[15] P. Paaw, *Hortus publicus academiae Lugduno-Batavae, eius ichnographia, descriptio, usus* (Leiden, 1601). This work is mentioned in the inventory of Federico Cesi's private library, next to Jan Jansz Orlers' description of the city of Leiden (*Beschrijvinge der stadt Leyden* [Leiden, 1614]). Cf. Baldriga, *L'occhio della lince*, 75.

[16] Gabrieli, *Il carteggio linceo*, 778. This passage was also commented and stressed by P. Findlen, *Possessing nature: Museums, collecting, and scientific culture in early modern Italy* (Berkeley etc., 1994), 74.

[17] The term 'Theatrum' may refer to Kaspar Bauhin's anatomical and botanical theatres, but had already became quite unusual in the titles of books on nature published after the second decade of seventeenth century. On the other hand, the concept of 'Mirror' was much more popular in the field of moral treatises (e.g. Boaistuau's *Theatrum mundi et speculum vitae humanae*) and esoteric philosophies (e.g. Lindhout's *Speculum astrologiae*; De Villanova's *Speculum alchimiae*).

The fruit of such exhausting work and devotion was the famous *Tabulae phytosophiche*, considered as the first attempt at a systematic description of the vegetable world, examined on the basis of morphological, physiological and pathological patterns. Nevertheless, despite their proclaimed 'modern' character, the *Tabulae* still preserved the symptoms of a certain epistemological orthodoxy. As Paula Findlen has pointed out, Bauhin would have certainly expressed his scepticism towards Cesi's approach, especially with regard to the 'global' view of his discussions that connect observations of nature with the system of the Liberal Arts.[18] Despite the adoption of some extraordinary tools like the microscope, Cesi's investigation hides, under the ambition of his unifying complexity, a mental approach which it would be outrageous to consider as revolutionary. Even the use of tables and graphic illustrations should be connected, as I have tried to demonstrate elsewhere, with Cesi's interest in the ancient – but at his time still appreciated – art of memory.[19] On the other hand, it is still possible that, in elaborating his 'metodo sinottico', Cesi took into consideration the examples of previous botanical publications like those by Clusius and other Northern scholars.[20] As Claudia Swan has suggested, the adoption of schemes might present some striking contradictions with the visual approach generally embraced by late-sixteenth-century scholars like Clusius.[21] In this respect, it is quite interesting to underline a fascinating comment made by the botanist Fabio Colonna on the tabular classification used by Cesi for his botanical observations: 'I have seen your subtlety and how much you have *sublimated* the plants.'[22]

The concept of sublimation, associated by Colonna with the use of a tabular system, clearly defines the distance it involved from common visual observation and suggests the perception of an abstract approach to the field, which – in Cesi's ambitions – would eventually embrace the entire complexity of the theatre of the world: 'I am working on the distinction and division of all things

---

[18] Findlen, *Possessing nature*, 74.

[19] On the interpretation of this issue, compare Baldriga, *L'occhio della lince*, 123-147; and D. Freedberg, *The eye of the lynx. Galileo, his friends, and the beginnings of modern natural history* (Chicago/London, 2002). See also the more recent contribution by A. Ubrizsy Savoia, 'Il metodo sinottico, collante tra la Syntaxis Plantarum di Aldrovandi e le Tavole Fitosofiche di Cesi', in *Federico Cesi un principe naturalista, Atti del Convegno Internazionale di Studi, Acquasparta, September 29-30, 2003, Accademia Nazionale dei Lincei* (forthcoming).

[20] On Clusius' use of tables and schemes, see C. Swan, 'From blowfish to flower still life paintings. Classification and its images, circa 1600', in P.H. Smith and P. Findlen (eds.), *Merchants and marvels. Commerce, science and art in early modern Europe* (New York/London, 2002), 109-136.

[21] Swan, 'From blowfish to flower still life paintings'.

[22] 'Ho veduto la sottigliezza di S. Ecc., et quanto habbi sublimato le piante' (Naples, 20 October 1628, published in Gabrieli, *Il carteggio linceo*, 186). See also Baldriga, 'La fatiga di pigliar i disegni delle piante'.

and although I proceed with certain speculations and along particular paths of my own, it is necessary that I see what the others have done too.'[23] As Cesi himself specifies, it was crucial to compare different synoptic methods. We can take it for granted that, in his process of 'sublimation', he carefully examined Clusius' approach. Cesi and his fellow Linceans possessed copies of the majority of Clusius' works, as is shown by the inventories of their libraries, which included the *Rariorum plantarum historia*, the *Exoticorum libri decem* and the *Curae posteriores*.[24]

As these aspects of Cesi's plan were already evident, perhaps even more so, at the time when the Academy was taking its first steps, I believe that Clusius was likely to remain aloof from the Roman circle. It may seem a paradox, but Clusius' contribution to the history of botany appears much more practical and empirical when compared to some of the utopian characteristics of Cesi's experience. Clusius collected and cultivated seeds and bulbs in his own botanical garden, pursuing a research whose goals were absolutely clear. His constant practice of botany and his meticulous search for illustrations painted 'ad vivum' finds its theoretical counterpart in his critical approach to ancient sources. In considering Fabio Colonna's merits, for instance, he underlined 'the wisdom (*giudicio*) demonstrated in considering the notes on plants after the Antique', a consideration which obviously implies praise of caution in referring to earlier literary sources.

By contrast, Cesi never considered gardening a crucial aspect of his activity, but just as a part of it. In his letters to Faber he admits avoiding expensive bulbs and flowers; the practical aspect of cultivation was mainly considered by the Prince of the Linceans as part of the Lipsian approach to Stoicism, for which the garden was a place of meditation, wisdom and friendship.[25] While Faber's private archive and correspondence include dozens of flower lists and notes on the practical issues of gardening,[26] Cesi seems plunged

---

[23] 'Mi trovo […] immesso alla distinzione e divisione di tutte le cose e sebene vado con certe mie speculationi e vie particulari tuttavia è necessario veda anco quello che hanno fatto li altri' (F. Cesi to J. Faber, 20 January 1620; Baldriga, *L'occhio della lince*, 130).

[24] Cesi's copies are often enriched with notes in the margin, including schemes and morphological observations.

[25] M. Morford, *Stoics and neostoics. Rubens and the circle of Lipsius* (Princeton, 1991). On Cesi and Stoicism, see also Baldriga, *L'occhio della lince*, 209-227 ('Il giardino stoico').

[26] Faber became director of the Vatican Garden in 1600 and later professor of medical herbs ('semplici') at Rome's University 'La Sapienza'. He created a small anatomical museum in his private house in Rome; cf. I. Baldriga, 'Il museo anatomico di Giovanni Faber Linceo', in S. Rossi (ed.), *Scienza e miracoli nell'arte del Seicento. Alle origini della medicina moderna* (Milan, 1998) [Exhibition catalogue, Palazzo Venezia, Rome], 82-87. Faber's private documents are held at the library of the Accademia dei Lincei (Rome, Biblioteca Corsiniana, Archivio of Santa Maria in Aquiro parish). Cf. Baldriga, *L'occhio della lince*, 196-220.

into a completely different conception of cultivation. In 1621 he writes to Faber: '[...] with our friend Stelluti, I enjoy the pleasures of staying at home, and when I feel like gardening, I enjoy my two little gardens for which I like collecting and amassing bulbs and seeds where I can [...].'[27] Gardening was for Cesi more of a hobby than a profession or a scientific activity. His invitation to favour direct experiences in the countryside instead of passive readings is not comparable with the constant dedication required by the practice of botany.[28]

These considerations might shed new light on the reasons for Clusius' rejection of Eckius' invitation. On the other hand, they should also clarify the attraction exerted by Clusius himself on the young Linceans. His many travels, his investigations of living specimens, his deep experience of scientific illustration, not to mention his huge knowledge of the subject, made him appear as a glittering star in the firmament of wisdom. I think that evidence of his influence on the Lincean activity can be found not only in the literary field but also in the elaboration of some specific projects of the Academy. I shall here focus on two cases: the *Mexican treasure* and the *Syntaxis plantaria*.

Publishing at the speed of a sloth: The Mexican treasure

Beyond any doubt, the greatest success achieved by the Lincean Academy was the publication of the so-called *Mexican treasure*, an ambitious treatise on the animals and plants of the New World. The book, published in its first edition in 1628, derived from a precious manuscript, written and illustrated by Nardo

---

[27] '[...] con il Stelluti nostro mi godo li gusti di casa e quando mi lece quelli della agricoltura in doi giardinetti per li quali vado colligendo e cumulando bulbi e semi dove posso, e piantando di mano in mano [...]' (Rome, Biblioteca Corsiniana, Archivio di Santa Maria in Aquiro, Fondo Faber, vol. 423, f. 56). Cf. Baldriga. *L'occhio della lince*, 209.

[28] '[...] the science of plants can be better acquired through personal speculation – better in the countryside – than through the reading of books by others' ('[...] la scienza dei vegetativi, quale più si acquista da se stesso speculando (massime essendo in campagna), che leggendo libri altrui'), Gabrieli, *Il carteggio linceo*, 39. It is also possible that Cesi was likely to delegate the practice of pure cultivation to his fellows more skilled in the field: this might be the case of Fabio Colonna, an acclaimed botanist whose letters always include very detailed descriptions. In a letter to Cesi, sent from Naples in May 1626, Colonna reports that he has collected a number of bulbs, seeds, roots and flowers which he has classified in order to describe them in his appendix to the *Mexican treasure* ('Ho raccolto alcune diversità di semi atti alla loro phisonomia plantaria, come ho fatto delle radici, foglie, fiori e frutti; ma non son molte, perché è più difficile la varietà nelli semi a ritrovarsi, et applicarsi con qualche buona occasione; quando che a V.E. piacerà stampar le mie Annotazioni, ve le giungerei per non parer manco in tutto e per tutto in questa parte, se ben non sia così copioso, almeno accennar il modo et aprir la strada a chi più di me farà e potrà farvi diligenza di osservazione'; Gabrieli, *Il carteggio linceo*, 1118).

Antonio Recchi, on the basis of the famous manuscript prepared in Mexico by Francisco Hernández. The genesis of the Lincean publication is well-known, and so is the story of Hernández's manuscript.[29] What has been underestimated so far is the huge success obtained in 1610 by Cesi in purchasing the manuscript brought by Recchi to Naples. In acquiring that rich document for the Lincean library, he had succeeded where other famous botanists and scientists had failed. In this respect, it is very interesting to focus on the attempts made by Clusius to obtain Recchi's treatise for himself. A letter from Ferrante Imperato, in Naples, testifies to Clusius' strong interest in acquiring the book (1598).[30] Imperato's letter conveys Clusius' anxiety to obtain information regarding the possible publication of the manuscript, probably motivated by his concern to protect the originality of his project for the *Exoticorum libri decem* (1605) and, most of all, to acquire some brand-new material to include in his text.[31]

It is a matter of fact that, beyond its actual value (it was looked down upon by some contemporaries both for the poor value of the images and for the scientific comments it contained), the significance of Recchi's work was hugely enhanced by the attention paid to it by scholars like Clusius; the status of its admirers was the best demonstration of the value of the book.[32]

The use of the manuscript by Cesi and the Linceans had, in my opinion, mainly a propagandistic goal. At the moment of the purchase, none of the Academicians was particularly engaged in the study of *exotica*, and the publication of the *Treasure* was probably conceived as a strategy to promote the Academy. A clear demonstration of this is given by the fact that, in his own microscope observations – testified by the *Syntaxis plantaria* – Cesi only included specimens from Central Italy. His research did not include the investigation of rare plants from far, unknown, lands. A further, particularly interesting proof of

---

[29] F. Guerra, 'La leyenda del Tesoro Messicano', in *Federico Cesi, Convegno celebrativo del IV centenario della nascita, Acquasparta, October 7-9 1985* (Rome, 1986), 307-314.

[30] De Toni, 'Il carteggio degli italiani col botanico Carlo Clusio nella Biblioteca Leidense', 174.

[31] In answering his requests, Imperato refers to the information he had snatched from Recchi: '[...] procurai de intendere se aveva in dissegno ponerlj in luce, o quelli, o altri, o in Napoli o altrove. Insomma li cavai di bocca che in corte di Sua Maestà, non so chi medico spagnolo haveva incarico di far un libro di quei semplici dell'India, e che poi a questo li fu fatto un lavoro di Mal officio, da quei altri medici tal che quei del Mal Consiglio, sbarrorno il negozio, né se ne parlò maj, questo è quanto n'ho inteso'. A month later, Giovanni Vincenzo Pinelli writes to Clusius on the same matter, relating about the frictions possibly created between Recchi and Imperato because of their interest in the publishing of the manuscript: 'io mi dubito che tra l'Imperato et il Recco non fusse troppo buona intelligenza sospettando facilmente l'uno dell'altro per la cosa della stampa' (De Toni, 'Il carteggio degli italiani col botanico Carlo Clusio nella Biblioteca Leidense', 174-175 and 226).

[32] On the literature concerning *exotica* from America, see P. Mason, 'From presentation to representation: Americana in Europe', *Journal of the history of collections*, 6/1 (1994), 1-20.

the lack of interest of the Linceans in *exotica* is contained in a letter by Galileo Galilei who, after having seen Hernández's manuscript in Cesi's library, wrote: '[...] I saw the paintings of 500 Indian plants [...] neither I nor anybody there had any idea of their qualities, virtues and effects [...] (Rome, 6 May 1611).'[33]

The purchase of Recchi's book was a boost to Lincean botanical research. The admission of the German botanist Johannes Faber to the select circle should be regarded in the light of the need for a specialist who could deal properly with this new material. The someway inappropriate and extemporaneous nature of the halo which has characterised Cesi's undertaking on the *Treasure* from the first is further confirmed by the many complaints often raised by Faber regarding the impossibility of checking his assertions on the basis of 'real' specimens: 'I send here what I recently wrote on the plants which lacked descriptions; I managed with the rest as I could, because once more I was not able to see all the plants in reality [...].'[34]

It should also be added that Roman publishing activity did not even have a local tradition in the publication of books on *exotica*. That is why, once he had been informed about Cesi's *Treasure* project, Marcus Welser suggested to Faber that he should look for a North European publisher to avoid exhausting delays. The words were prophetic: the project did turn out to be exhausting and took decades to go into print.[35]

Elsewhere, I have investigated circumstances and vagaries of the publication of the *Treasure*;[36] nevertheless, I would like here to add something about the obvious isolation suffered by Cesi and his fellows in fulfilling their project. Their correspondence testifies to ignorance of printing issues and practical matters in the field, which is also evidence of their status as scientific hermits. Once again, a comparison with Clusius' circle appears pertinent. In March 1606, Giovanni Pona ventured to ask to borrow the wooden matrix which had been used for the illustrations of Clusius' *Historia plantarum*, while he was taking

---

[33] '[...] veddi le pitture di 500 piante indiane [...] né io né alcuno de i circostanti conosceva le loro qualità, virtù et effetti [...]'; Gabrieli, *Il carteggio linceo*, 162.

[34] '[...] mando qui a V. Ecc.za quello che ho composto di nuovo per conto delle piante che non haveano descrittione; l'altro ho aggiutato quanto ho potuto, perché di novo non ho potuto vedere tutte le piante nell'originale [...]'; Johannes Faber to Federico Cesi; Rome, 21 September 1624; Gabrieli, *Il carteggio linceo*, 941. On Faber's unease in dealing with illustrations instead of real specimens, see Baldriga, 'La fatiga di pigliar i disegni delle piante'.

[35] The slow pace of the publication invited comparison with the movement of a sloth. Quoting from Clusius, Johannes Schreck wrote to Faber 'Io non so come cammini il libro: io per certo anderei intorno al mondo in quattro anni, et il libro non esce mai fuora la porta del Popolo. *Erit forte frater istius animalis americani de quo Clusius, quod una die vix conscendet unam arborem*' (Gabrieli, *Il carteggio linceo*, 614).

[36] Baldriga, *L'occhio della lince*.

note – on Clusius' suggestion – of the convenience of commissioning etchings in Italy.[37] This last comment contrasts completely with Cesi's understanding of the matter; because of his failure to explore the local printing market, he had a low opinion of the skill of the Italian engravers.[38]

By the way, despite these many difficulties, the Linceans' determination to bring the Mexican project to completion was constant and obstinate. They must have been deeply aware of the book's success potential, and in fact the *Treasure* soon raised expectations among those who where interested in *exotica*, as it appeared to be the natural development (or completion) of Clusius' work on the natural history of the New World. This is clearly explained in a letter sent in 1636 to Lucas Holstenius, in Rome, by the Dutch geographer Johannes de Laet:

For several years we have been awaiting the compilation of that great work put in order by the most learned Nardo Antonio Recchi (as I learnt from Fabio Colonna). Its printing was begun in Rome some time ago, and I saw its printed title page a few years ago. Last year I asked my cousin Elzevier to inquire in Rome about the prospects of that book: I understand that the work has been interrupted or dropped, because a similar book had been published by a certain Nieremberg in Belgium. But he hardly excelled in judgement, and failed to provide it with any images, apart from the ones that Cl. Clusius and others had already provided before. I am therefore surprised that a work which is such a desideratum should be dropped for that reason.[39]

---

[37] De Toni, 'Il carteggio degli italiani col botanico Carlo Clusio nella Biblioteca Leidense', 160 ('Ho inteso anco quanto sia meglio far fare i taglij in Italia et seguire anco il rimanente de miei desiderij in questo genere, consiglio buono al quale m'appigliarò. Haverei bene a singolar favore, che la S.V. mi facesse sapere se in Anversa io potrei havere le tavole de i foglij ch'anno servito nella descrittione di Monte Baldo, quando non habbino patito et che se ne possia valere in questo nuovo bisogno, et se si, con quanto dispendio, perché ne darei qualche particolar ordine').

[38] 'With regards to the images of the book which is in print for our Linceans, I have never thought that Italian engravers could achieve the quality of the Germans [...] They know no more than what they do' ('Quanto alle figure del libro che per i nostri Lincei si stampa, non avendo io mai preteso che l'artifici italiani possano arrivare ad una minina parte dell'ingegno e diligenza Germana: et vedendo venir da questa così bel lavoro, non mi sono atterrito altrimenti, parendomi che possiamo assai restar scusati mentre ci serviamo di quelli artefici ch'abbiamo, non sanno più di quello che fanno'); cf. Baldriga, *L'occhio della lince*, 248.

[39] 'Plurimi sunt anni quibus expectavimus compendium magni illius operis concinnatum a doctissimo viro Nardi Antonio Reccho (uti a Fabio Colonna didici) et Romae excudi iamdudum coeptum, cuius et titulum excusum ante aliquot annos hic vidimus: dederam superiori anno cognato Elzevirorum nostrorum in mandatis ut Romae inquireret quid porro spei esset de illo libro: intelligo operam intermissam aut etiam omissam, quia similis liber a quodam Nierenbergio in Belgio erat editus; verum ille parum modo iudicio praestitit, neque ullas icones dedit, *praeter eas quas iam ante Cl. Clusius et alii dederant*; quare miror illius causa tam desideratum opus omitti.' Gabrieli, *Il carteggio linceo*, 1243.

The letter not only clearly expresses the anxiety which surrounded the making of the *Mexican treasure*, but it also testifies to the incontrovertible role played by Clusius in the history of illustrated scientific books; like a watershed, he marks a 'before' and a 'after'. No doubt the Linceans grasped the exceptional opportunity represented by the making of the *Treasure*: a hazardous adventure, expensive and risky, but too attractive to be dropped. Even more, the book would put them in the footsteps of Clusius' scientific research on the exotic world, so that, despite his deliberate refusal to enter Cesi's circle, the great botanist would end up posthumously lending his aegis to the Lincean fellowship.

Mushrooms and microscopes

A second case of a possible influence of Clusius' activity on the Academy concerns his attention to the visual aspect of botanical descriptions.[40] The extraordinary quality of the *Libri Picturati*, in whose making Clusius was directly involved, could be related to another, equally exceptional collection of botanical illustrations: the *Syntaxis plantaria* sponsored by Federico Cesi, now held in the library of the Institut de France in Paris.[41] There is no doubt that Cesi was very well acquainted with Clusius' scientific production. His interest in mushrooms certainly led him to examine Clusius' work on the matter (but we could argue that it might even have been inspired by that work). His *Fungorum in Pannoniis observatorum brevis historia*, published in the last part of the *Rariorum plantarum historia*, must have been familiar to the Prince of Linceans. It is well known that Clusius paid great attention to the quality of images; among the many possible passages we could choose, I quote here his considerations on a drawing made after a Narcissus: 'It does not meet my taste; maybe the flowers are well done, but I cannot rely on the way they are disposed on the stem; nobody will ever let me believe such a thing, as it would be contrary to the nature of all Narcissi […].'[42]

The visual aspect was crucial to Cesi and his followers and, despite the poor quality of the wood engravings made for their books, the *Syntaxis plantaria* is

---

[40] On the importance of visual aspect and scientific illustration in the Lincean context, see Baldriga, *L'occhio della lince*; eadem, 'Le virtù della scienza e la scienza dei virtuosi: i primi lincei e la diffusione del naturalismo in pittura', in M. Calvesi and C. Volpi (eds.), *Caravaggio nel IV centenario della Cappella Contarelli, Atti del Convegno Internazionale di Studi, Roma, May 24-26 2001*, Accademia dei Lincei-Università degli Studi di Roma 'La Sapienza' (Rome, 2002), 197-208.
[41] A. Ubrizsy, 'Il codice micologico di Federico Cesi', *Rendiconti dell'Accademia Nazionale dei Lincei, Classe Sc. Fis. Mat. e Nat.*, VIII, 68/2 (1980), 129-138.
[42] De Toni, 'Il carteggio degli italiani col botanico Carlo Clusio nella Biblioteca Leidense'.

clear evidence of the importance given by the Linceans to illustrations *ad vivum*. I have elsewhere suggested that Cesi could have been informed about the existence of the *Libri Picturati*: one of his many contacts with Flanders (including Rubens, who had worked for the Arenberg family before his trip to Italy) could have seen them in Brussels and described them to the Linceans.[43] Cesi's observations on mushrooms have rarely been connected with Clusius' earlier contribution to the field, but there is evidence for a clear relationship between the two. Cesi was certainly working on them in 1615, if not before: a letter from Theophilo Müller (Molitor) asks for advice on 'two fungi that he found in a wood, of a species that I have never seen before, and now send you'.[44] As Clusius is considered to be one of the first to have described mushrooms in the history of botany, it is not hazardous to suggest that Cesi based his first observations on the work of his predecessor.[45]

Once again, the research undertaken by Cesi could encourage the hope of achieving pioneering goals, not only because of the objective originality of the subject, but also because, in unfolding the many secrets of mushrooms (at the time commonly considered as mysterious creatures), Cesi could make use of a new and futuristic instrument, the microscope. The amazing collection of precious botanical watercolours in Paris was created through an extensive use of the magnifying lens. As was recently observed, some sheets represent details magnified up to 100 times.[46] The use of magnifying lenses bring us, once more, to suggest some contacts with the Netherlands. Although the primacy of the invention of the microscope remains very controversial, it was probably the Dutch spectacle-maker Zacharias Jansen who produced the first compound microscope, around 1590, while Cornelis Drebbel is credited with having contributed to the development of the instrument before 1620. Cesi received his

---

[43] I heartily thank Florike Egmond for her help and advice in this matter. Cf. F. Egmond, 'Clusius, Cluyt, Saint Omer. The origins of the sixteenth-century botanical and zoological watercolours in the Libri Picturati A. 16-30', *Nuncius. Journal of the history of science*, 20/1 (2005), 11-67, with previous literature. See also Baldriga, 'La fatiga di pigliar i disegni delle piante'.

[44] 'Duos fungos invenit in sylva, quorum species numquam vidi, et iam E.V. mitto [...].' Gabrieli, *Il carteggio linceo*, 492.

[45] In a letter sent from Rome on 19 March 1606, Eckius asks an unknown correspondent for advice on the matter of mushrooms. The main subject of the letter and the clear reference to their visual representation suggested to Chiovenda that it might be addressed to Clusius. Nevertheless, it is quite difficult to confirm such a hypothesis. By the way, the letter informs us about the precocity of the mycological research undertaken by Eckius, probably accompanied by Cesi himself. See Gabrieli, *Il carteggio linceo*, 97-98.

[46] A. Graniti, 'Federico Cesi, Fungorum genera et species; Plantae et flore', (catalogue entries) in A. Cadei (ed.), *Il Trionfo sul Tempo. Manoscritti illustrati dell'Accademia Nazionale dei Lincei* (Rome, 2002) [Exhibition catalogue, Palazzo Fontana di Trevi, Rome, November-January 2003], 80-84.

first 'occhialino' (microscope) only in 1624, as a gift from Galileo,[47] but his fellows probably started to employ it for their botanical/anatomical observations many months afterwards: this can be deduced from two letters from Faber (Rome, 13 April 1625) and Colonna (Naples, 6 June 1625).[48]

The enthusiasm created by the microscope among scholars and scientists was properly described by the Dutch humanist Constantijn Huygens who, in his *Autobiography* (1629), regretted the missed opportunity – for the artist Jacques de Gheyn the Younger – to practise the use of the microscope. According to Huygens, the artist's skills in natural illustration would have allowed him to represent the smallest things and insects and to collect all his drawings in a book to be entitled 'The new world'.[49] It is quite fascinating, at this stage, to remember the close friendship which bound De Gheyn II to Carolus Clusius and the intimacy they achieved through their passion for the natural sciences.[50] In expressing his heartily felt regret, Huygens was certainly thinking of that fortunate alliance that De Gheyn had established with the botanist from Arras. It might be a simple coincidence, but the chronological and contextual references suggest some connection between the lost opportunity of the Dutchman and the daring enterprise undertaken by Cesi and his fellows. The Lincean publication of the celebrated *Apiarium*, an illustrated entomological essay on the bee printed in Rome in 1626 by the publisher Mascardi, associated once and for all the scientific use of the microscope with Cesi's Academy.

---

[47] Gabrieli, *Il carteggio linceo*, 942. For further information on the use of the microscope among the Linceans, see G. Gabrieli, 'Pratica e tecnica del telescopio e del microscopio presso I primi Lincei', in idem, *Contributi alla storia dell'Accademia dei Lincei*, vol. II, 347-371.

[48] Gabrieli, *Il carteggio linceo*, 1038-1039 and 1047-1048. In both cases the instrument was purchased from foreign merchants. Despite Galileo's claim to have built the microscope on his own, the Linceans must finally have been informed about the Northern invention of the tool. Faber writes to Cesi: 'I would like you to have a look at my descriptions of Galileo's new inventions and to check if I included everything, or if I should eliminate something. I gave the name of *Microscopio* to the new instrument used to see little things [...].' ('Ho voluto avertire quest'ancora a V. Ecc.za che lei dia una vista solamente a quello che io ho scritto delle nove inventioni del Sig.r Galileo, se ho messo ogni cosa, o se ha da levare, che faccia a modo suo. Et perché io fo anche mentione di questo novo ochiale di vedere le cose minute et lo chiamo *Microscopio* [...]').

[49] The passage by Huygens is quoted in S. Alpers, *The art of describing. Dutch art in the seventeenth century* (Chicago, 1983), 1-25, esp. 6-7.

[50] Besides, Clusius was certainly also in contact with the highly esteemed engraver Hendrick Goltzius (De Gheyn's master); see De Toni, 'Il carteggio degli italiani col botanico Carlo Clusio nella Biblioteca Leidense', 250.

Gaps and silences

Instead of the easy assumption which would assign to Clusius the role of a mere forerunner of the Lincean Academy, I am more inclined to put emphasis on the several openings that his experiences represented for the Roman fellowship. Even more than the huge amount of information he had collected, it was his example as a scholar which turned out to be magnetic for Cesi's group. Despite their pledge to live 'in the universal exercise of contemplation and practice',[51] the Linceans were in some respects limited by their ambition to synthesise the complexity of the universe in a scheme. Such a position probably affected the credibility of their project in the eyes of many Northern scholars. On the other hand, Cesi's desire to achieve their official recognition led him to embark on a number of ventures which the Academy was ill equipped to handle.

The story of the missed relationship between Clusius and Cesi can certainly contribute to a better understanding of some of the mechanisms of European scientific communication at the turn of the seventeenth century. Despite the limited amount of time available for the two scholars to have established direct relations (Clusius died in 1609), it is still possible to consider the effects produced by Clusius' experience on the Lincean activity. Most of all, the cases here examined seem to suggest Cesi's desire to emulate Clusius' experience and to appropriate the role of his follower. The historical analysis of Cesi's botanical research shows us an untiring chase which I have tried to unfold through the examples of the purchase of Recchi's manuscript (missed by Clusius and then obtained by Cesi), the mycological investigation (opened by Clusius and then continued by Cesi), and the application of the microscope to scientific description (started by the Linceans and missed by Clusius).

This story should serve to remind all scholars that gaps and silences may deserve the same historical attention as pieces of material evidence.

---

[51] 'in essercitio universale di contemplatione e prattica' (F. Cesi, 'Del natural desiderio di sapere et institutione de' Lincei per adempimento di esso', in S. Ricci [ed.], *Federico Cesi e la Fondazione dell'Accademia dei Lincei. Mostra bibliografica e documentaria, catalogo della mostra (Venezia, Biblioteca Nazionale Marciana, 27 agosto – 15 ottobre 1988)* [Naples, 1988], 125).

Some aspects of Clusius' Hungarian and Italian relations

Andrea Ubrizsy Savoia

In memoriam Istvàn (Stephan) A. Aumüller

The importance of Clusius' oeuvre, the network created by him to exchange knowledge via friendships throughout Europe, and the influence of his work on the development of the natural sciences can be studied from different points of view and via many examples. Here, the focus is on several aspects which belong to different phases of Clusius' life and work, but actually are connected directly or indirectly with Hungary. They are also linked by the correspondence between Clusius and his friends, which Clusius collected and preserved throughout his life, up till the last stage of his long journey through Europe in Leiden. Thanks to Clusius himself and to Bonaventura Vulcanius, his colleague at the University of Leiden, these letters are still today conserved at its university library.

Clusius' work and its relevance for botanical and mycological knowledge in Hungary

The period between 1573 and 1587 which Clusius spent (with interruptions) in 'Pannonia', has been examined and described in a considerable number of publications by Austrian and Hungarian authors.[1] Clusius was the first scholar to be interested in describing the flora of the eastern margins of the Alps. He pursued his investigations at sites which are located today in Austria, Hungary, Slovenia, Croatia and Slovakia.

Clusius' most important contribution to the knowledge of Hungarian plants consists of his description of circa 300 species in *Rariorum aliquot stirpium per Pannoniam, Austriam & vicinas quasdam provincias observatarum historia*

---

[1] See for example A. Barb, 'Die römischen Inschriften des südlichen Burgenlandes', *Burgenländische Heimatblätter*, 1 (1932), 75-80; S.A. Aumüller, 'Carolus Clusius, der Begründer der botanischen Forschung im Raume des heutigen Burgenlandes', *Burgenländische Heimatblätter*, 29/3 (1967), 98-107; G. Traxler, 'Die burgenländischen Pflanzenstandorte bei Carolus Clusius', *Burgenländische Heimatblätter*, 35 (1973), 49-59; E. Horvàth, 'Clusius Lithoxylonja és lelöhelye a kèsöbbi szakirodalomban', *Vasi Szemle*, 27/4 (1973), 585-595; S.A. Aumüller and J. Jeanplong (eds), *Carolus Clusius' Fungorum in Pannoniis observatorum brevis historia et Codex Clusii* (Budapest/Graz, 1983), and their references.

(Antwerp, 1583).² It was published in one volume together with *Stirpium nomenclator Pannonicus,* which Clusius wrote together with the Hungarian local expert István Beythe. A further source on Hungarian species is Clusius' *Rariorum plantarum historia* (Antwerp, 1601), which describes 25 plant species and also comprises a chapter on mushrooms. Clusius was also interested in plants that were cultivated in gardens, such as those of his friend and patron Boldizsár Batthyány. Yet, the importance of Clusius for Hungary goes beyond his contributions to the description of the Hungarian flora and mushrooms. He also published observations on the existence and interpretation of the 'Lithoxylon'³, thereby setting some steps on the road to palaeobotany. By collecting and quoting the vernacular names of many organisms, he made an important contribution to the history of the Hungarian language and to what is now called ethno-botany. In his writings Clusius also bore witness to the existence of Roman remains in Hungary, thus demonstrating an early archaeological interest. His work has become a precious source of information about historical sites, events and important persons in late sixteenth-century Hungary. Clusius' own activities and descriptions demonstrate, moreover, that cultural centres at a European level existed in Hungary at the time, in spite of wars, the occupation of the country by the Turks and by foreign troops opposing the Turks, and in spite of the religious clashes amongst the Hungarians themselves. When Clusius finally, in 1587, left the court of Batthyány, his interest in Hungary and his Hungarian friends did not cease, as is clear from the letters he received. One of the Hungarian friends who continued to write to him long after his departure from Hungary was Nicolas (Miklòs) Istvànffy, a nobleman and historian.⁴

Hungarian botany hardly profited from Clusius' research or insights. The difficult political and economic situation of the country – first because of the Turkish occupation, then during the rule of the Habsburgs – left few resources for scientific research, including that of the natural sciences. The first complete *Flora of Hungary,* which was published in 1807 by Diòszegi and Fazekas, entirely ignored Clusius' contribution. The importance of his work was apparent only in studies of the local flora in locations which Clusius had visited. The

---

² His 'Spanish flora' (printed in 1576) already contains some plants from Hungary, including the name of the places where the species grew.
³ In 1580 Clusius observed at Mount Vashegy (now Eisenberg, Austria) fossil wood, identifying it as oak, instead of attributing unnatural origins to it. In modern times this fossil has been identified as *Quercoxylon cerris* L. (E. Hofmann, 'Verkieselte Hölzer der Vashegy- (Eisenberg-) Gruppe', *Vasvàrmegyei Mùzeumok Evkönyve,* 3 [1927-29], 81-87; cf. Horvàth 'Clusius Lithoxylonja ès lelöhelye a kèsöbbi szakirodalomban'.
⁴ Eight letters from Istvànffy to Clusius are held in the collection of Leiden University Library and cover the period 1588-91.

physician K.Fl. Loew from Sopron (a Hungarian town on the border with Austria), for instance, grasped the extent of the gap that existed in the study of plants in Hungary for the period after Clusius. He took Clusius' work as a starting point and in 1739 published his *Flora Pannonica*. Another example is the essay published by Sàndor Sebeok in 1779 about the species *Crambe tataria* with a beautiful illustration of this plant, which had previously been described by Clusius as a typically Hungarian one. Sebeok knew and quoted Clusius' work. An article written in 1791 by Istvàn Lumnitzer about the local flora of Pozsony[5] was inspired by the fact that he had found two species of plants in the outskirts of that city which Clusius had described in his *Rariorum aliquot stirpium [...] historia*. Finally, the nineteenth-century botanist Vince Borbàs used Clusius' work in his study of the flora of the regions which had also been visited by Clusius. His paper on the subject was published in 1887 and won a prize of 100 gold coins which was – appropriately – donated by a prince from the Batthyány family.

Hungarian connections: Clusius, court life, and the various networks and associations of intellectuals linking Hungary, Austria and Italy

Although Boldizsár Batthyány had greatly stimulated Clusius' research about mushrooms, the latter did not dedicate his *Fungorum in Pannoniis brevis historia* (dated 1598), which was published as an attachment to his *Rariorum plantarum historia* (1601), to his Maecenas.[6] Batthyány had died in 1590, and Clusius dedicated this work instead to his Italian friend Giovanni Vincenzo Pinelli. Although Pinelli himself was not interested in mushrooms, he had informed Clusius that the famous botanist Ulisse Aldrovandi in Bologna was dealing with mushrooms and also owned a collection of drawings.

Pinelli was interested in what was happening in Hungary, its culture and its battle against the Turks. In fact, the letter of dedication in Clusius' *Fungorum historia* indicates that Pinelli knew very well that the Batthyány court in Nèmetùjvàr (present day Güssing, Austria) constituted the most important Hungarian cultural centre in Eastern Central Europe. It was a typical Renaissance cultural centre inspired by humanism.[7] Pinelli received information about Hungary and Batthyány's court not only from Clusius, but also from Nicasius Ellebodius, a Flemish physician and humanist, and one of the most

---

[5] Pozsony is the present-day Bratislava, the capital of Slovakia; already at the time an important town, it became the jurisdictional centre of Hungary during the Turkish occupation.
[6] Clusius had dedicated another work to his generous Hungarian patron Batthyàny, during the latter's lifetime: it was his *Aliquot notae in Garciae aromatum historiam*, published in 1582.
[7] See the essay by Dóra Bobory in this volume.

learned experts of Aristotle's works. Ellebodius lived in Hungary for a long time, mainly in Pozsony where his patron and host was Istvàn Radèczy. In his letters to Clusius (in Vienna) of 15 March 1575 and 20 July 1576, Ellebodius writes about the political situation and the war in Hungary and Austria, also providing news about mutual friends and fellow countrymen such as Hugo Blotius from the Southern Netherlands, and 'Philippus noster', that is the composer Filippo di Monte.[8] In his many (still unpublished) letters to Pinelli[9] Ellebodius wrote about the political situation in Hungary and about common friends: Hungarians, such as Màrton Berzeviczy, Johannes Sambucus (János Zsámboky), Antal Verancsics, and Georg (György) Purkircher, many of whom he had met while studying at the university of Padua; Italians, such as Paolo Manuzio, Antonio Riccoboni, Domenico Francesi, Girolamo Mercuriale, Ferrante Imperato, Paolo Aicardo; foreigners who were connected with Italy, such as Hugo Blotius (who had brought Ellebodius in contact with Pinelli) or Melchior Wieland (Melchiorre Guilandino), the prefect of the botanical garden in Padua. As these links by means of correspondence show, Clusius was able to remain in touch with Western European culture via hosts, friends and correspondents during his stay in 'Pannonia'.

Pinelli's own house in Padua likewise was a meeting point for intellectuals, humanists, and scholars. It resembled the first academies of the Renaissance, which were created in Florence (such as Marsilio Ficino's Academy in the Medici villa in Careggi). However, the gatherings held at Pinelli's house remained informal. The society of intellectuals meeting there never turned into a real Academy, regulated by a constitution.

Courtiers are usually required to conform to the expectations, tastes, philosophy (which often also included the religious orientation) and interests of their 'Prince' and 'Maecenas'. This was also the case in the relations between Batthyány and Clusius. Much of Clusius' research in Batthyány's territory had been commissioned by Batthyány, who often followed it personally. Clusius explored Batthyány's domain from every point of view: from archaeology to botany, zoology, medicine, linguistics, and numismatics. As Clusius himself explains in his *Fungorum in Pannoniis brevis historia* (1601), he started studying mushrooms (and then published the results of this research) because of the frequent presence of edible mushrooms at the table of his maecenas and

---

[8] Both these letters from Ellebodius in Pozsony to Clusius (who was then in Vienna) are part of the collection of Leiden University Library. On Di Monte see P. Bergmans, 'Quatorze letters inédites du compositeur Philippe de Monte', *Mémoires. Académie Royale de Belgique. Classe des Beaux-Arts*, 2e sér., 1 (1921), 1-30; and T. Hindrichs, *Philipp de Monte (1521-1603). Komponist, Kapellmeister, Korrespondent* (Göttingen, 2002).

[9] They are kept in the Pinelli archive in the Biblioteca Ambrosiana in Milan.

because of the wide variety (today called biodiversity) of mushrooms growing on the lands of Batthyány. Clearly, the orientation of the scientist's interests was directly influenced by those of his host.

In comparison with the cultural centre of a court, the circles of literary men and scientists who met at home – they could be called 'academies' in the wider sense of the term – were more independent. The host could simply provide the 'academics' with an assembly room, his private library, and his natural objects to examine. He generally did not support them financially, nor did he pay a regular 'salary' to the people who frequented his circle. Such circles did not specify their fields of interests, but attracted people with similar interests. Clusius had the opportunity to make contact with circles of Hungarian literati and scientists in both Vienna and Pozsony.

The 'academy of the court' itself in Vienna comprised a large variety of experts and aristocrats, including foreigners, such as several Italians who were employed by the court, and Clusius' own patrons.[10] Among its 'members' we find, for instance, the physician born in Breslau (Wroclaw) Johannes Crato von Kraftheim, Italian physicians such as Giulio Alessandrino, compatriots of Clusius like the physician Nicolas Biesius and his successor the physician and botanist Rembert Dodoens, and Hungarians such as the humanist and physician Johannes Sambucus and Miklòs Istvànffy, the court historiographer. In a certain sense aristocrats too (like Batthyány) can be considered members. Elias Corvinus, humanist and poet in this Viennese circle, is a good example. Encounters with these men helped Clusius to enlarge his network of acquaintances, also among Italians, Hungarians and Dutchmen.

Ellebodius (Ill. 48) was one of the 'Flemish' connections between Clusius, Hungary and Italy. Originally from the Southern Netherlands, he had studied medicine in Padua from 1561 to 1563, where he had become friends with Hungarian students such as Thomas Jordanus (Tamàs Jordàn), Johannes Sambucus, Andrea Dudicius (Andràs Dudith), and Georg Purkircher. All of these names also occur in Clusius' publications and biography. Both Clusius and Ellebodius participated in life at the academy of Pozsony. The scholars gathered under a linden tree in the garden of the palace of Steven (István) Radéczy, archbishop of Várad and later of Eger, and royal governor in Pozsony. His garden was a real *Hortus Musarum*, and Radéczy's palace formed a meeting point for scholars, poets and humanists. Among the most eminent members of this circle were the physician Johannes Sambucus, the historian Nicholas

---

[10] Gy. Istvànffi, *A Clusius-codex mykologiai méltatása adatokkal Clusius életrajzához* (Budapest, 1900) / *Études et commentaires sur le Code de l'Escluse augmentés de quelques notices biographiques* (Budapest, 1900), 171.

(Miklós) Istvánffy, and Ellebodius himself, who also belonged to the academic society of Pinelli.[11] Sambucus had met Clusius at gatherings in the house of the bibliophile Jean Grolier, royal counselor in Paris. During Clusius' stay in Vienna and in Hungary (1573-87) this acquaintance turned into friendship. The Sambucus manuscripts archive (now in the National Library of Vienna together with the entire library of Sambucus) testifies to this friendship: it contains the following note: 'Ex dono Caroli Clusii habebat Sambucus, Lutetiae 1561.'[12]

Ill. 48. The funerary monument of Nicasius Ellebodius, Pozsony. The epitaph was written by Istvànffy in 1577.

---

[11] T. Klaniczay, 'Le mouvement académique à la renaissance et le cas de la Hongrie', *Hungarian studies*, 2/1 (1986), 13-34. The four letters from Ellebodius to Pinelli, written in Italian between 22 February 1572 and 22 April 1573 and sent from Pozsony and Vienna (now in Milan, see note 9) also contain news about mutual friends, such as Sambucus, Purkircher, Wieland, Blotius, Imperato, and Radèczy.

[12] I. Bàlint-Nagy, 'Purkircher György (1530-1578) pozsonyi orvos èlete' [The life of György Purkircher (1530-1578), physician of Pozsony], *Orvosi Hetilap* [Medical weekly], 22-23 (1930), 553-584; P. Gulyàs, *Samboky Jànos konyvtàra. Facsimile edition: A Zsàmboky-konyvtàr katalògusa. Adattàr a XVI-XVIII. szàzadi szellemi mozgalmaink tortènetèhez* [The Library of Jànos Samboky. Facsimile edition: The catalogue of the Zsàmboky library. Database to the history of our ideological movements during the 16th-18th centuries] (Szeged, 1992²).

The major circles of the Viennese court itself and the 'academy' of Radéczy in Pozsony were not the only relevant ones, even in these two towns. In his work about the flora of Pannonia Clusius mentions the name of Andreas (Andràs) Heind(e)l (denoted as Hemal), a pharmacist in Pozsony, and remarks upon the plants that were growing in the garden of this city: 'I hear also that around that city the mistletoe occurs in both chestnut trees and on the fruits of cornel and roses, as I was advised by Andrea Hemal, apothecary of that city.'[13] Overlaps between the various circles were, moreover, numerous. Sambucus belonged to the Radéczy circle in Pozsony and was well known in Italy, but his own house in Vienna with its very rich library also formed a kind of academy. It was open to Hungarian (Verancsics, Listhy, Dudith, Purkircher, Istvànffy, Ujlaki, Olàh etc.), Italian, and many other leading intellectual figures in Vienna. Among them we find Crato von Kraftheim, Giulio Alessandrino, Pier Andrea Mattioli, Rembert Dodoens, Clusius himself, Paulus Fabricius, Ogier Ghislain de Busbecq, Hugo Blotius, Justus Lipsius and many others. Lipsius was very impressed by the Hungarian academic circle set up in Vienna, and established important contacts with András Dudith, Mihály Forgách, Péter Révay, János Rimay, and Johannes Sambucus.

In Pozsony the house of Purkircher formed the center of yet another circle. In fact, Sambucus accompanied Clusius on his excursions to collect plants in 1573, at a time when they were both guests of Purkircher in Pozsony.[14] Georg Purkircher was a Hungarian physician who had studied in Wittemberg and Padua, where he obtained his degree in medicine in 1563. He settled down in Pozsony in 1566. Purkircher is mentioned in Clusius' work on the plants of Pannonia because he tried to acclimatise plants in Hungary which he had seen in Italy or which had just arrived in Europe from other parts of the world, such as the bean (*Phaseolus vulgaris* L.). He had brought the seeds with him from Naples and grown it in his garden in Pozsony, sending seeds to Clusius as well. Clusius called it 'Phaseolus I sive Purkircherianus'.[15]

Purkircher, like Clusius, was in contact with the librarian and jurist Hugo Blotius (de Bloot, Blotz), who was born in the Dutch town of Delft in 1533. Blotius had studied in Leuven, Toledo, Orléans and Strasbourg, and was subsequently employed by János Liszthy, bishop of the Hungarian city Veszprèm,

---

[13] 'Audio etiam circa eandem urbem & in Castaneis arboribus & Coryli rosarumque fructicibus viscum nasci, referente ornatiss. viro Andrea Hemal eius urbis pharmacopaeo'; *Rariorum aliquot stirpium per Pannoniam [...] historia* (Antwerp, 1583), 100-102. Since the chestnut is not a native tree in Pozsony, the trees Clusius refers to must have been specially planted in gardens.
[14] Bàlint-Nagy, 'Purkircher György (1530-1578) pozsonyi orvos èlete'.
[15] *Rariorum aliquot stirpium per Pannoniam [...] historia*, 722.

and Làzàr Schwendi, captain of Upper-Hungary as a tutor of their young relatives during Blotius' studies in Italy. Blotius and his pupils spent the years 1570-72 in Padua. In 1571 Blotius sent a drawing with the plan of the botanical garden in Padua to Ulisse Aldrovandi in Bologna, who was not only a professor at the university of Bologna, but also founded its botanical garden.[16] The garden plan sent by Blotius is special, because it mentions the names of the plants which were grown in each section of the garden, with references for each to the page number of the relevant illustration in Mattioli's work *Commentarii in Dioscoridem* (1567).[17]

Like Blotius, Thomas Jordanus – a native of Transylvania whose name has already emerged as a companion of Ellebodius, Purkircher and other Hungarians in Padua – got in touch with Aldrovandi. Jordanus' name occurs in manuscript notes by Aldrovandi which date back to 1568 and concern a plant species (*Kochia*) from the Carpatian mountains which was used by Hungarians and Turks.[18] Further information and the dried plant itself can still be found in the Aldrovandi archive, iconographic collection and herbarium in Bologna. Jordanus' name is also mentioned in the visitor's book of Aldrovandi's museum: 'Septemcastris Hungarus Thomas Jordanus.'[19] In this register other Hungarians figure as well, among them Georg Purkircher, who is also mentioned in one of the manuscript volumes kept as a 'diary' by Aldrovandi, covering the years 1562 to 1565.[20] The visit of Jordanus to Aldrovandi is furthermore documented by a letter of presentation which Jordanus carried in his pocket when he travelled from Padua to Bologna. It was signed by Giacomo Antonio Cortuso and dated 9 September 1563.[21] Cortuso would in later years (1590-1603) become prefect of the botanical garden of Padua, but at the time could still dedicate himself exclusively to botanical excursions and his private garden, which was famous both in Italy and abroad. Upon leaving Bologna for Florence, Jordanus was carrying another letter of presentation, this time by Aldrovandi.[22] Jordanus kept

---

[16] A. Ubrizsy Savoia, 'The Botanical Garden of Padua in Guilandino's day', in A. Minelli (ed.), *The Botanical Garden of Padua 1545-1995* (Venice, 1995), 172-195.

[17] Ibid.

[18] 'Netata cuius copia in Carpato monte qua Ungari et etiam Turcae pro Catartico utuntur valde.' Bologna University Library (henceforth BUB), Ms. 1024, fondo Aldrovandi, vol. 136, tomo III, c. 158 and tomo XV, c. 29v; 'Netata ex Carpato monte qua Ungari et Turcae pro cathartico utuntur valde' (Aldrovandi's Herbarium, University of Bologna, vol. IX, f. 140).

[19] BUB, Ms. 1024, fondo Aldrovandi, vol. 110, n. 2.

[20] BUB, Ms. 1024, fondo Aldrovandi, vol. 136, tomo I, c. 140.

[21] This letter is cited in G.B. De Toni, *Spigolature Aldrovandiane. XIII. Un altro corrispondente di Ulisse Aldrovandi, il medico Giovanni Battista Balestri* (Leipzig, 1912).

[22] This fact is recorded in a letter exchanged between Aldrovandi and Gregorio Cantarini. See BUB, Ms. 1024, fondo Aldrovandi, vol. 38, tomo II, c. 100.

up his connection with Aldrovandi even after returning to his homeland.[23] The two letters from Jordanus to Clusius held in Leiden University Library may still provide us with further information concerning this Hungarian scholar about whom very little is known as yet. Clusius' connections with the Hungarians Jordàn and Hertel (to whom we will shortly return), as well as the role of Ellebodius as a link between Clusius and Hungary, were previously unknown.

The Clusius Codex of mushrooms: History and copies

From a botanical point of view Clusius' above mentioned *Fungorum historia* – which was published as an attachment to Clusius' *Rariorum plantarum historia* (1601) – is one of the most valuable results of his stay in Hungary. It includes 41 xylograph plates, of which 9 were taken from Lobelius' publications, and refers to circa 105 'species' distributed among 47 genera by Clusius (identified in 68 species).[24]

Clusius compiled this work in Leiden in 1597 on the basis of mycological research carried out on Batthyány's estates in the counties of Vas and Zala, in Burgenland (nowadays in Austria), and in Croatia. The role of Clusius' patron Boldizsár Batthyány went beyond stimulating Clusius' interest in mushrooms.[25] In order to raise the level of research concerning mushrooms collected in Hungary, Batthyány also invited a certain 'French' painter from Vienna.[26] This painter was probably the person who made the coloured drawings of mushrooms which together with some notes constitute the so called *Clusius Codex* of mushrooms, which remained unpublished during Clusius' lifetime.[27] In fact, Clusius believed this album lost after Batthyány's death in 1590. As I have argued in more detail elsewhere, on the basis of the information provided by the letters written by Esaya le Gillon (Clusius' nephew on his sister's side) to Clusius himself, the artist who made the coloured drawings in the *Clusius Codex*

---

[23] See A. Ubrizsy Savoia, *Rapporti italo-ungheresi nella nascita della botanica in Ungheria* (Pécs, 2002).
[24] G. Bohus, 'Interprétation des bolets de Clusius', *Acta mycologica Hungarica* 2 (1945), 20-27, 69-76; idem, 'Mikològiai èrdekessègek a Clusius Codexbol' [Mycological curiosities in the Clusius codex], *Vasi Szemle* [Newspaper of Vas] 27, 4 (1973), 582-585; idem, 'A Clusius-Codex gombafajainak revizioja' [Identification of the fungi species in the Clusius codex], *Mikològiai kozlemenyek* [Micological comunications] 3 (1975), 121-136; and idem, 'Revision der Pilzarten des Clusius-Codex', in S.A. Aumüller and J. Jeanplong (eds), *Carolus Clusius Fungorum in Pannoniis observatorum Brevis historia et Codex Clusii mit Beiträgen von einer internationalen Autorengemeinschaft* (Budapest/Graz, 1983), 60-69.
[25] Istvànffi, *A Clusius-codex*, 172.
[26] See the letters by Clusius to Batthyàny from 2 June 1578 (Istvànffi, *A Clusius-codex*, 206-207).
[27] It forms part of the collection of Leiden University Library: *Icones fungorum in Pannoniis observatorum*, BPL 303, 87 fols.

could be Le Gillon himself. Le Gillon was invited in 1574 by Clusius to come to Vienna (and later to Prague), where he continued to live for 25 years.[28]

The unpublished mycological iconographic collection which has become known as the *Clusius Codex* was (re)discovered in 1874 in the library of the Leiden University.[29] The Hungarian botanist Gyula Istvànffi published it in 1900 at his own expense in a facsimile edition with 90 illustrations in order to celebrate the tercentenary of the publication of Clusius' *Rariorum plantarum historia*.[30] Istvànffi augmented his facsimile edition with additional information, explanations, and the identification of species. He also discussed the similarities between the *Codex* and a considerably later mycological publication by Franciscus van Sterbeeck, *Theatrum fungorum* (1675) which has often been regarded as one of the earliest European works on mushrooms. Furthermore, Istvànffi published a number of as yet unpublished letters to Clusius. His work did much to (re)establish Clusius as a prime figure in the history of mycology, designating Van Sterbeeck as a later follower.

It is now known that Leiden University Library had bought the bound volume known as the *Clusius Codex* in 1679 from the private library of Arnoldus Seijen, a Leiden professor of botany since 1670, who had died in 1678.[31] Van Sterbeeck testifies in his *Theatrum fungorum* (1675, vol. II, 5-12) that he had access to the volume already in 1672 thanks to the mediation by Adrianus David, a pharmacist in Antwerp. Van Sterbeeck based the copper plates for his own publication (1675) on the *Clusius Codex* and also made a coloured copy of it for himself. This copy I have been able to find in the

---

[28] A. Ubrizsy Savoia, 'Les acquarelles mycologiques de Charles de l'Escluse', *Histoire et nature*, 7 (1975), 89-95; 'Il Codice di Clusius', *Rassegna di micologia ed ecologia Romana dell'AMER (Associazione Micologica ed Ecologica Romana)*, III (7-8) (1976), 6-11; *Die Beziehungen des Lebenswerkes von Carolus Clusius zu Italien und Ungarn. Clusius' pilzkundliche Aquarelle* (Güssing/Vienna, 1977); 'Wissenschaftliche Beziehungen zu Italien. Der Maler der Pilzaquarelle im Clusius-Codex', in S.A. Aumüller and J. Jeanplong (eds.), *Carolus Clusius Fungorum in Pannoniïs observatorum brevis historia et Codex Clusii* (Budapest/Graz, 1983), 54-56. Six letters in Italian from Le Gillon to Clusius from the years 1590-1606 are preserved in Leiden University Library (VUL 101). The examination of these letters has been suggested to me by the late Stephan (Istvàn) Aumüller from Güssing/Nèmetùjvàr, an untiring and not always acknowledged Austro-Hungarian promoter of research concerning Clusius' work.

[29] See E. Morren, 'Charles de l'Escluse, sa vie et ses oeuvres', *Bulletin des Sociétés d'Horticulture de Belgique*, 184 (1875), 1-45, here 41-42. In fact, thirty years before Morren, in 1841-45, the Hungarian botanist Jòzsef Sadler already stated in his study about the history of botany in Hungary during the sixteenth century, that 'The original drawings for the illustrations of the mycological study by Clusius seem to be at Leyden library'; J. Sadler, A növénytan törtènetei honunkban a 16-ik szàzadban', *Magyar Termèszettudomànyi Tàrsulat Evkönyvei*, 1 (1841-45), 78-118, here 106.

[30] Istvànffi, *A Clusius-codex*.

[31] He is called A. Syen in F.W.T. Hunger, *Charles de l'Escluse (Carolus Clusius), Nederlandsch kruidkundige 1526-1609*, 2 vols. (The Hague, 1927-43). The volume in Seijen's library is, in fact, mentioned by Van Sterbeeck.

Koninklijke Bibliotheek of Belgium in Brussels thanks to the help of Michiel Verweij.[32] It lacks a title page and comprises 149 folio's, preceded by a brief biography of Van Sterbeeck in French written by Charles Van Hulthem. It turns out that this album contains not only pictures of mushrooms copied by Van Sterbeeck after the *Clusius Codex* (up to f. 60r) but also original drawings of fungi by Van Sterbeeck. A short note on its first leaf by Van Hulthem dated at Ghent on 7 October 1831 reads:

Franciscus Van Sterbeeck de Fungis, ou Recueil de Champignos (sic!) trouvés par François Van Sterbeeck, pretre d'Anvers, dans les excursions botaniques et peints par lui-meme avec leurs couleurs naturelles. Il y a joint les Champignons que Clusius avoit peints d'après nature dans un volume que le Docteur Syen, professeur de Botanique à l'universitè de Leyde avoit dans sa Bibliothèque, et dont Van Sterbeeck fit l'acquisition en 1672.

Some newly discovered information about Van Sterbeeck's links with the *Clusius Codex* throws further light on the interest in and influence of Clusius' work on mushrooms. There is one more copy of the *Clusius Codex*: an album which is not mentioned in any studies about Clusius. The library of the Department of Plant Sciences at the University of Oxford owns an interesting manuscript, catalogued as 'Watercolours of fungi, Caroli Clusii'.[33] It bears the following title: *Liber fungorum depictorum Caroli Clusii quem ab Arnoldo Syen per Adrianum David communicatum habuit Franciscus Sterbeeck. Cui ipse, tanquam basi, Theatrum suum fungorum superstruxit; ut apparet ex Theatri dicti pag 27 & p. 168. Conferantur utrinque figurae, praecipue Sterb. p. 269, hujus vero p. 69 & ultim. vid. & Clusii Hist.* The top of the title page shows the following note in French: 'Peint et dessigné par un stable Peintre de Vienna avec freis et depars de Balthazar de Batthyany […].' (Ill. 49) The 'stable Peintre de Vienna' must refer to Esaya le Gillon.

A comparison between this newly discovered manuscript in Oxford and the *Clusius Codex* in Leiden shows that the Oxford manuscript is a copy of the

---

[32] The copy bears the shelfmark KBR Ms. 15475. The pictures are attributed to François Van Sterbeeck in the card index, but it has not been included in the catalogue. See also J. Kickx, 'Esquisses sur les ouvrages de quelches anciens naturalistes Belges. II. François Van Sterbeeck', *Bulletin de l'Académie Royale des Sciences* I, sér. 9, 2 (1842), 393-426, here 395-396. In his facsimile edition (1900, 125-126) of the *Clusius Codex* Istvànffi referred to a note where Kickx states that 'Sterbeeck himself had copied the pictures writing on them his own notes'. Van Sterbeeck had copied from the *Codex* and published the pictures of 70 *Hymenomycetes* (he copied seven figures also from the *Fungorum historia* by Clusius, among others).

[33] Sherard Collection Ms. 43. According to curator Stephen A. Harris, this manuscript arrived in the collection in 1719 as part of a collection of books and manuscripts bequeathed by Jacob Bobart the Younger (1640-1719). He had succeeded his father Jacob Bobart the Elder as prefect of the Oxford Botanical Gardens in 1680 and may have acquired the manuscript from his father, as he had inherited half of his father's library.

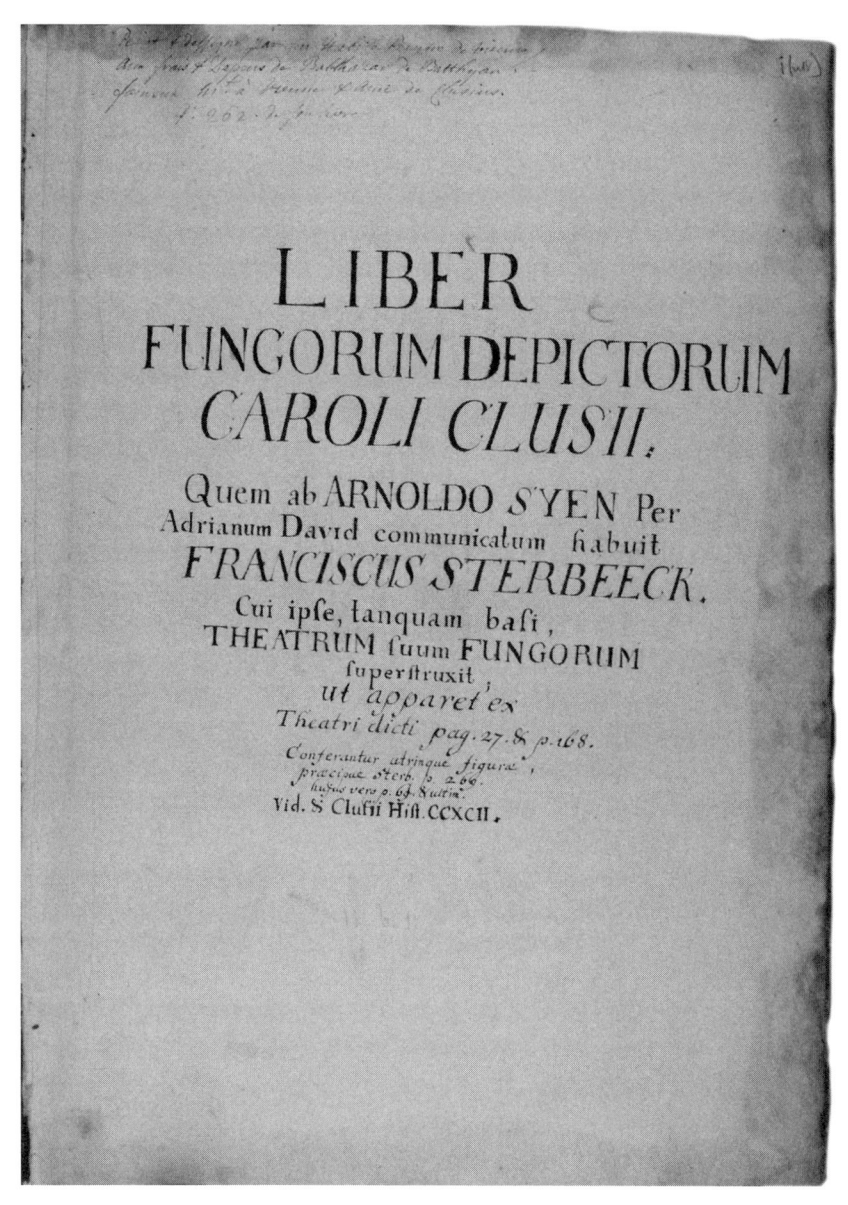

Ill. 49. Frontispiece of the Oxford album *Watercolours of fungi, Caroli Clusii*.

Leiden codex. The illustrations are in watercolour, and the names of the mushrooms mentioned in the *Clusius Codex* have all been copied, even those in Hungarian, which contain many mistakes since the person who copied them clearly did not know the language. The order of the pages is, moreover, practically identical with that of the Leiden codex, which demonstrates that the *Clusius Codex* had been bound in an order which does not follow the numbering of the bound pages before the 'Oxford' copy was made. The presence of one illustration in oil, which is completely different from the watercolors and bears the date 1616 – and therefore must have been attached to the *Clusius Codex* after Clusius' death – indicates that the Oxford copy was certainly made after 1616 and more probably after 1675, the publication year of Van Sterbeeck's *Theatrum*.[34]

A preliminary comparison of these three albums renders important information. While the Oxford copy is identical with the Leiden original, the Oxford and Brussels albums differ from each other. The Brussels album copies the figures of the *Clusius Codex,* but the lay-out is modified: the figures are concentrated to fill up the empty space on the sheets. For example, while table (i.e. group of illustrations) no. 6 is identical in the Leiden and Oxford albums, the drawings of this table appear in the Brussels copy on folio 7, combined with the figures of another table from the Leiden *Codex* (Ills. 50a-c). In fact, the original figure was recopied (the colours are not identically reproduced) and moved toward the bottom of the sheet (while there is also a small difference in the note in Latin) to the space which was left empty in the original version; and the upper part of the sheet was filled up with a figure of table no. 10 of the Leiden codex (even in this case with differences in the notes). Another example is table no. 5, which is again identical in the Leiden and Oxford albums, whereas it appears in the Brussels copy on folio 6: there, the figures of this table no. 5 are mixed and concentrated to fill up the upper part of the sheet. On the lower part of the sheet two figures of fungi have been added which together constitute table no. 9 in the Leiden and the Oxford albums (Ills. 51a-c). Here again, we find differences and mistakes in the transcribed notes.

The Oxford album is thus an exact copy – there are some differences only in the use of colours of the drawings – of the original Leiden *Clusius Codex*. The Brussels copy, made by Van Sterbeeck as the basis for his printed work on mushrooms, differs from both the Oxford and Leiden albums – the single figures of fungi on at least two different pages in the *Clusius Codex* of Leiden have been 'concentrated' on only one page, perhaps in order to save space –,

---

[34] In so far as the title page is concerned, it might even date to not long before.

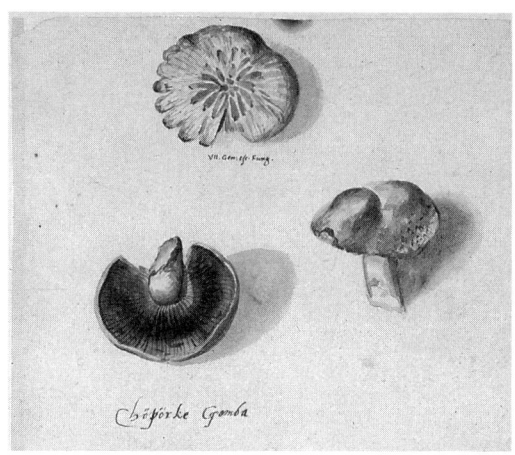

Ill. 50a. Table 6 in the Leiden Clusius Codex. (See also colour plate 10).

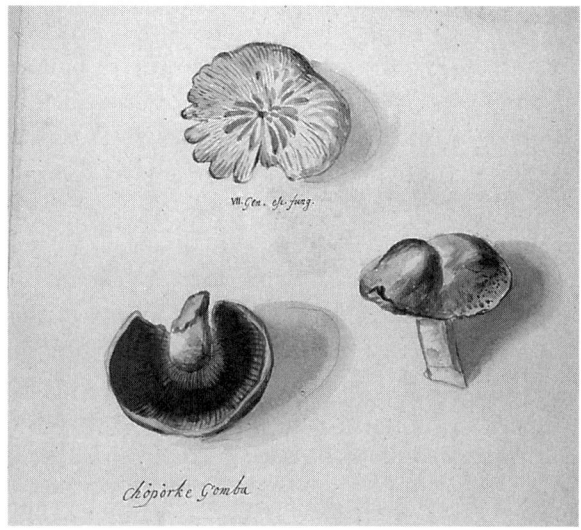

Ill. 50b. A page of the Oxford album, a copy after table 6 in the Leiden Clusius Codex. (See also colour plate 11).

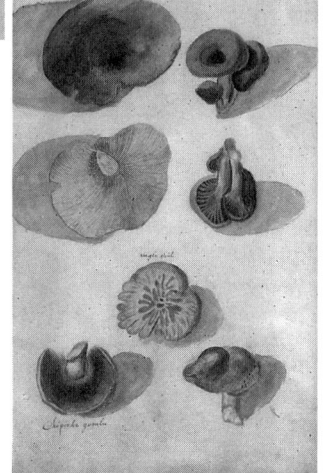

Ill. 50c. Page n. 7 in the Brussels album, which combines table 6 of the Clusius Codex in Leiden with the figures on table 10 of that same codex. (See also colour plate 12).

and shows many similarities with the illustrations of the *Theatrum fungorum* by Van Sterbeeck. Yet, there are also differences between the Brussels album and the printed version of Van Sterbeeck's book. In some cases the differences occur when the same species were depicted on different sheets in the original Leiden codex while Van Sterbeeck put them together on one page. In other cases, completely different species were combined by Van Sterbeeck on the same page.

Both the *Clusius Codex*, and the slightly later mycological codex of Federico Cesi, founder of the Roman Accademia dei Lincei, which I had the chance to find and identify after it had been believed lost for 200 years,[35] have remained hidden for a long time from historians of botany and of mycology who regarded Van Sterbeeck as the founder of mycology. As a matter of fact Van Sterbeeck mainly copied, interpreted and explained Clusius' monograph and *Codex*. Cesi himself knew only the printed version of Clusius' works, and he owned a copy of the *Rariorum plantarum historia* (1601).[36]

Ill. 51a. Table 5 in the Leiden Clusius Codex. (See also colour plate 13).

---

[35] It can be found at present at the library of the Institute de France in Paris. See A. Ubrizsy Savoia, 'Il Codice micologico di Federico Cesi', *Rendiconti dell'Accademia Nazionale dei Lincei, Classe Sc. Fis. Mat. e Nat.*, ser. VIII, 68/2 (1980), 129-134.

[36] This copy (today in the Biblioteca Corsiniana, Rome) bears the seal of the Accademia dei Lincei and has annotations by Cesi, mainly regarding the colours and shapes of flowers and seeds.

Ill. 51b. A page of the Oxford album, copied after table 5 in the Leiden Clusius Codex. (See also colour plate 14).

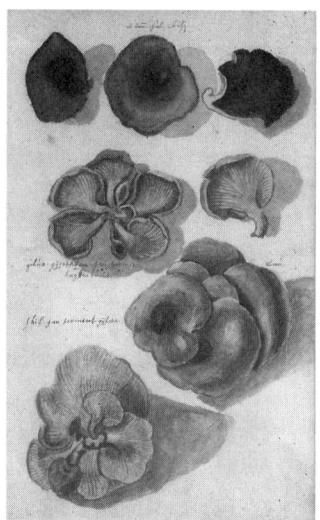

Ill. 51c. Page n. 6 in the Brussels album, which combines table 5 in the Clusius Codex in Leiden with figures on table 9 of that same codex. (See also colour plate 15).

Some further 'Hungarian' connections

As we have seen, the connections between Clusius and Hungary neither stopped after his period in Austria and Hungary was over, nor were they limited to his contacts and correspondence with Hungarians. For instance, the Frenchman Claude de Roussel, captain of the castle of Tokaj (Upper Hungary) and in the pay of the court at Vienna, informed Clusius (then at Frankfurt) in his letters about the battles against Turks up to 1590.[37] Strictly speaking, the correspondence between Clusius and the English diplomat Henry Wotton should perhaps belong to his 'English relations', but Wotton's letters, some of which written in Italian from Vienna, contain almost no information concerning England but mainly deal with Hungary and Italy.[38] Henry Wotton is another example. He is known as a poet, diplomat and connoisseur of arts, since he was first described as such by Izaak Walton in his *Reliquiae Wottonianae* (1651). Wotton traveled on the Continent for the first time in 1588, spending most of his time in Austria and subsequently in Italy. His correspondence with Clusius seems to have begun in October 1590, when Clusius was in Frankfurt. Wotton, who was travelling in Italy, acted as an intermediary for Clusius, sending him plants, especially from Casabona in Pisa (a person to whom we will shortly come back). In return, Wotton asked Clusius to buy certain books for him at the book fair in Frankfurt.[39] In August 1593 Wotton was in Geneva, writing to Clusius to suggest that he visit Leiden. Briefly mentioning Josephus Scaliger, Wotton further wrote about the plants that Clusius was going to receive and remarked that Prospero Alpini – who would later become prefect of the Padua hortus botanicus (1603-1616) – was describing plants from Egypt.[40]

In his letter to Clusius (in Frankfurt) of 23 August 1593 Wotton mentions the name of Johannes Hertel. He was a native of Kolozsvàr in Transylvania (presently Cluj in Romania) and the son of the well known Protestant preacher and Bishop Ferenc Dàvid.[41] Hertel first studied medicine in Padua in 1586 and returned in 1589 to Padua as an already graduated physician in order to

---

[37] See his letters published in Istvànffi, *A Clusius-codex*, 279-282.
[38] In Leiden University Library there are seven letters by Henry Wotton to Clusius; according to Aumüller (personal communication) there are two in the National Library in Vienna.
[39] See for instance his letter of 29 August 1592 (Leiden University Library, VUL 101). Similar exchanges, also involving Hungary, can be deduced from the correspondence between Clusius (from Frankfurt) and the Dutchman Theobald van Hoghelande. Clusius asked him to collect seeds during his travels in Austria, Hungary and Transylvania. See L.A. Tjon Sie Fat, 'Clusius's garden: A reconstruction', in idem and E. de Jong (eds.), *The authentic garden. A symposium on gardens* (Leiden, 1991), 3-12.
[40] Letter of 23 August 1593 (Leiden University Library, VUL 101).
[41] Ubrizsy Savoia, *Rapporti italo-ungheresi nella nascita della botanica in Ungheria.*

complete his medical studies.[42] He did not leave the university and in 1590 was appointed to assist Girolamo Fabrizi d'Acquapendente in teaching anatomy. The following year he was offered a position to teach botany and asked to take the duties of the prefect of the botanical garden of Padua upon himself. These had been neglected (on account of illness) by Giacomo Antonio Cortuso. Since Cortuso was still in function, however, the two men would have had to share his single university salary. Hertel did not accept this condition and left Padua. Clusius' connection with Hertel was previously unknown and the information available about the activities of Hertel in Hungary after his return from Italy is scarce and uncertain. The documents of the Padua university archive indicate that the responsibilities given to Hertel were revoked formally as of 11 February 1593, but a letter written by Hertel to Clusius is dated 8 February 1593 from Vienna, which means that he had left Padua earlier.[43] Justus Lipsius seems to have been the intermediary between Clusius and Hertel, and Hertel's letter to Clusius contains the name of Giovanni Vincenzo Pinelli and a request for seeds from the 'admirable and fertile garden' set up by Clusius in Frankfurt. Hertel also suggested that the Southern Netherlandish Philippus Caretto, tailor of Archduke Matthias of Austria, later the successor of his brother Emperor Rudolf II, might act as an intermediary with respect to Transylvania. It thus seems that Hertel was setting up a garden, possibly a botanical or medical one, probably in Transylvania – which throws new light on the history of Hungarian botany.

'New' Hungarian contacts also appear from the examination of the letters by Cluisus' nephew Esaya le Gillon – the probable painter of the *Clusius Codex*.[44] In his letter of 24 June 1606, for example, Le Gillon told his uncle that a book sent by Clusius and addressed to 'Signor Barvitio', counsellor and personal secretary of Emperor Rudolf II, had arrived and that 'Sig. Barvitio' would answer soon. From an earlier letter, dated 7 September 1601, by Le Gillon it appears that Ioannis Barvitio was a garden owner, possibly in Vienna or Prague, and that the seeds sent by Clusius for his garden had arrived. In the post scriptum of this letter Le Gillon assured Clusius that he would forward any letters from Signor Barvitio to Clusius as soon as he received them.

The letters written by Le Gillon to Clusius between 1590 and 1606 contain much detailed and historically interesting news about eminent Hungarians as

---

[42] See E. Veress, *Olasz egyetemeken járt magyarországi tanulók anyakönyve és iratai (1222-1864)* [Documents and matriculations of students at Italian universities from Hungary (1222-1864)] (Budapest, 1941); and L. Rossetti, *Matricula nationis Germanicae artistarum in gymnasio Patavino (1553-1721)* (Padua, 1986).
[43] See R. Trevisan, R., 'Luigi Anguillara', in A. Minelli (ed.), *The botanical garden of Padua 1545-1995* (Venice, 1995), 57-59. Hertel's letter can be found in Leiden University Library, VUL 101.
[44] These letters are in Leiden University Library, VUL 101.

well as a chronicle of the battles against the Turks – incidentally showing how important contacts with and news about Hungary still were for Clusius in his old age. Attached to the Le Gillon's letter of July 1606 is a strange sketch representing three monsters, born in April of that year in different places of Hungary. The drawing was copied by one of le Gillon's children after an original with comments written partly in Hungarian and partly in Latin and German.[45] The three monsters are a two-headed sheep and calf, and a pig with a human head and three legs. As is well known, most Renaissance scholars were fascinated by 'monstrous' creatures. For instance, Ulisse Aldrovandi, with whom Clusius corresponded just before his journey to Pannonia, collected monstrous samples for his museum. Monsters were believed by many to be messengers of disasters or premonitions of great calamities. However, Clusius, like Aldrovandi, had doubts about regarding them as fabulous organisms. For them and for some other sixteenth-century scholars they became the opposite: a means of distinction from between the 'normal' and natural form of an organism and the 'abnormal'. These extraordinary specimens came to represent the exception to the natural normalcy of the majority of organisms.

Italy and Clusius

Various aspects of the relationship between Clusius and Italy have been the subject of monographs by De Toni in 1911, Battistini in 1927, Ginori Conti in 1939, Ubrizsy Savoia in 1977, and of shorter contributions, for example by Tongiorgi Tomasi and Garbari in 1991. Many references to Clusius and his Italian connections can, moreover, be found in studies dealing with the history of botany (and mycology) in Italy.[46] Since his connections with Hungary and Italy were so strongly intertwined, it is worthwhile to have a closer look at Clusius' bonds with Italy here.

Already during his studies in Montpellier in the course of the 1550s Clusius had been attracted by the idea of visiting Italy and its famous universities and professors, and of finishing his studies in medicine there, as so many scholars did during that period. Above all, he wanted to visit Pisa with the first (1543) botanical garden of Europe. Although Clusius actually never managed to visit Italy, his works are full of information concerning species from the Italian peninsula. His *Rariorum plantarum historia* describes more than ninety Italian plant species, and in some cases his descriptions are the first ever, preceding

---

[45] A. Ubrizsy Savoia, 'Clusius levelezotársai', *Vasi Szemle*, 32/78-1 (1978), 124-127.
[46] See A. Ubrizsy Savoia, 'I rapporti tra Carolus Clusius ed i naturalisti italiani del suo tempo', *Physis*, 20 (1978), 49-69.

descriptions by Italian scholars.[47] Often, Clusius quotes the name of the person who sent him the information about a particular Italian plant (or specimens or seeds), or that of the author who had published about a particular species in Italy (for example, Luigi Anguillara). The names most frequently mentioned in this context are Alfonso Pancio, Giacomo Antonio Cortuso, Ferrante Imperato, Ulisse Aldrovandi, Fabio Colonna, Giovanni Vincenzo Pinelli, and several foreigners living in Italy, above all Giuseppe Casabona, who is also known as Giuseppe Benincasa. The latter was of Southern Netherlandish background (his original Flemish name was Joseph Goedenhuize), and fulfilled the function of prefect of the Pisa botanical garden from 1592 until his death in 1595.[48] Some names mentioned by Clusius in his printed works are not among the list of correspondents whose letters to Clusius are known to have been preserved: for example N. Raffio (physician in Reggio), Giovanni de Mera (physician in Naples), G. Barbaro (ambassador of Venice), Ippolito Salviani (physician in Rome). It was the Italian friends – with whom Clusius kept up cordial relations by means of correspondence – who gave him the opportunity to describe so many Italian plants.

In return, one might say, Clusius' work was very quickly received and highly estimated in Italy, while his influence on the evolution of botanical studies in Italy was both immediate and important. Clusius was evidently respected, and regarded as an authority by Italian scholars and colleagues. This is all the more significant if we take the fierce controversies into account which existed at the time among Italian naturalists as well as among Italian and foreign colleagues: many openly showed contempt for each other's publications. The conflicts of the famous botanist Pietro Andrea Mattioli with Amatus Lusitanus (a Portuguese naturalist at Ferrara) and the two prefects of the botanical garden in Padua, Luigi Anguillara and Melchiorre Guilandino, are a case in point. Among Guilandino's enemies we find both Joseph Scaliger and the prefect of the Pisa botanical garden, Giuseppe Casabona.

The relationship between Casabona and Clusius was fruitful for both of them: Casabona sent Clusius more than 100 different kinds of plant seeds from Crete, where he had gone in order to collect rare species. Crete was at the time believed to be the 'garden' of Ancient Greece and some of the species collected by Casabona were depicted in Crete by the young Flemish or German soldier Georgius Dyckman whom he had met there.[49] Copies of these

---

[47] A. Ubrizsy Savoia, 'Piante italiane in un'opera olandese del 1601', *Annali di botanica (Roma)*, 35-36 (1978 [= 1976-77]), 144-154.
[48] See F. Garbari, L. Tongiorgi Tomasi and A. Tosi, *Giardino dei semplici/Garden of simples* (Pisa, 2002).
[49] See also Garbari et al., *Giardino dei semplici / Garden of simples*, 53.

illustrations were offered to Clusius. From one of the four extant letters from Casabona to Clusius, dated 25 August 1590 and sent from Florence to Clusius in Frankfurt, we know that he sent him 'these two lines of mine and together with these a small box with three or four types of spring Crocus with quite beautiful colours and moreover two bulbs of the plant 'bulbous Leucoium', which he had found in northern Italy.[50] Casabona describes these plants with such accuracy that we can even now provide a fairly accurate identification and deduce the approximate location where he collected them.

Casabona's pupil and successor, Francesco Malocchi from Pisa, continued the tradition of corresponding with Clusius, sending him letters with attached watercolours, lists of plants in the garden, and plant specimens from the outskirts of Pisa. In a letter from 1606 Malocchi also listed the plants he had received from the hortus botanicus in Leiden (of which Clusius was the prefect during these years), which gives us a good idea of the plant species which were at that time growing in this Leiden garden.[51] Just like the botanical gardens in Pisa, Padua and other towns in Italy, the Leiden one served didactic purposes and contained the most common, widely-used medicinal plants. Clusius, however, also wanted it to include rare and exotic species, which were generally hard to obtain and expensive. By engaging in exchanges with other gardens, colleagues and friends he could enlarge the Leiden collection. The Italian colleague from whom he received most plants and seeds for the Leiden garden was Giacomo Antonio Cortuso (prefect of the Padua garden from 1590-1603), but the latter also received many gifts in exchange from Clusius, including the seeds of the American sun-flower – the first in Italy. A letter from Cortuso to Clusius contains one of the earliest references to the horse-chestnut, a plant which was called *castagna equina* by some at the time, and *Lebanese cedar* by others.[52]

Under the name 'Clusius' the index of the catalogue of Aldrovandi's manuscripts at the Bologna University Library, made by Frati in 1907, indicates three lists of plants or seeds: one sent to Clusius to put in the imperial garden in Vienna, where Clusius had an appointment[53]; one sent by Clusius from Vienna

---

[50] '[...] questi miei doi rigi (sic!) et insieme mandarvi una Schatolina con III o IIII sorte de Croco Vernio de assai belli colori et de più missevi dentro II Cipollini di una pianta di Leucoio bulboso.' Leiden University Library, VUL 101.

[51] G.B. De Toni, 'Il carteggio degli italiani col botanico Carlo Clusio nella Biblioteca Leidense', *Memorie Regia Accademia, scienze, lettere e arti, Modena*, 10 (1911), 1-147; L. Tongiorgi Tomasi and F. Garbari, 'Carolus Clusius and the botanical garden of Pisa', in L. Tjon Sie Fat and E. de Jong (eds.), *The authentic garden*, 61-74.

[52] H.W. Lack, 'Lilac and horse-chestnut: Discovery and rediscovery', *Curtis's botanical magazine*, 6/17 (2000), 109-114.

[53] 'Catalogus seminum missorum ad Excell. D. Carolum Clusium pro horto Imp.is' (BUB, fondo Aldrovandi, Ms. 136, tomo V, c. 371-374).

to Aldrovandi in Bologna[54]; and a third which was delivered to Aldrovandi by a certain Nicolaus Gaddus[55] The arrival of 'thirty seeds arrived two days ago, part of them received from Hungary from Carolus Clusius', was mentioned by Aldrovandi to the Grand duke of Tuscany.[56] And Aldrovandi also sent the catalogue of seeds of his botanical garden at the University of Bologna to Clusius.[57] A more detailed investigation shows that more plant lists from Clusius can be found in this library: one of them is indicated as '*Atrebate catalogus*'.[58] It also comprises lists of plants exchanged with Vienna which were probably likewise received from Clusius: one of these lists of plants from Austria bears the name of 'Aicholz' and is undoubtedly connected with Clusius. From the letters sent by Aldrovandi to Clusius it emerges that Clusius sent plant specimens to Bologna: these were included in volume XV of Aldrovandi's Herbarium.[59] Both the answers and the plants sent by Clusius are, however, missing from the Aldrovandi archives. In his library Aldrovandi had, among other works, 'the translation by Clusius of the book by Monardes and of the work about aromatic plants by Garcia, recently published and enlarged (by Clusius), which were sent by means of the merchants of Gualteri in Venice.'[60] Aldrovandi's herbarium (preserved at the University of Bologna) contains two specimens connected with his relationship with Clusius. One is the 'Garyophyllus palustris odoratissimo a Clusio' (vol. XV, f. n. 1 (*Dianthus superbus* L.). It was most probably grown in the Botanical Garden in Bologna (that had been founded by Aldrovandi) from the seeds ('Garyophyllus palustris odoratissimo') he received in April 1578 from Clusius (BUB, fondo Aldrovandi, Ms. 136, vol. VI, c. 171v). In the same volume XV, folio 15 there is the specimen 'Cariophillus montanus odoratus flore albo' (*Dianthus plumarius* L.) which can be connected with the 'Cariophillus montanus flore albo odorato' listed among

---

[54] 'Semina mihi missa Vienna a Carolo Clusio' (BUB, fondo Aldrovandi, Ms. 136, tomo VI, c. 138-153r).
[55] 'Catalogus seminum quae missi D. Nicolao Gaddo et habui a Carolo Clusio' (BUB, fondo Aldrovandi, Ms. 136, tomo VI, c. 171v-180r).
[56] '[...] trenta semi, che due giorni sono ho riceuti, [...] parte mi vengono d'Ungaria da Carolo Clusio', quoted in both O. Mattirolo, *Le lettere di Ulisse Aldrovandi ai granduchi della Toscana Francesco I e Ferdinando I* (Turin, 1904) and A. Tosi, *Ulisse Aldrovandi e la Toscana. Carteggio e testimonianze documentarie* (Florence, 1989).
[57] BUB, fondo Aldrovandi, Ms. 124, vol. 136, tomo III, c. 371-374.
[58] BUB, fondo Aldrovandi, Ms. 124, vol. 143, tomo III, c. 399r.
[59] Leiden University Library, VUL 101, has seven letters from Aldrovandi to Clusius, written between 1569 and 1596.
[60] 'la traddotione del Clusio fatta del lib. de Monardis, et de aromatici del Garzia del Clusio di nuovo riconosciuto, et amplificato qual si manda p(er) mezzo dei mercanti de Gualteri in Venetia.' Aldrovandi's library catalogue (BUB, fondo Aldrovandi, Ms. 147) includes the books by Monardes, Cr. Acosta and Garcia ab Orto translated by Clusius.

the seeds received in 1578. The same may be assumed for the 'Ptarmica austriaca' (vol. XV, f. 18; *Xeranthemum inapertum* (L.) Moench) present in the list of seeds as 'Ptarmica austriaca'.[61] Undoubtedly, many parts of Aldrovandi's collection were lost or seriously damaged when the whole collection was moved from Bologna to Paris as booty during the Napoleonic invasion and only returned to Italy after 1818. However, a detailed inspection of Aldrovandi's manuscripts reveals new references to Clusius and to his relationships.[62]

Clusius provided other services as well to his Italian friends. For instance, he sent a description of the exact composition of a certain drug to the chemist Giovanni Pona, one of his friends from Verona, who had asked for Clusius' expert opinion concerning its genuineness. Clusius' appreciation of the contributions and information sent him by Italian scholars was great, as is demonstrated, for example, by his inclusion of the same Giovanni Pona's description of a botanical excursion on the Monte Baldo near Verona as an attachment to the *Rariorum plantarum historia* (1601). In a different context we have already seen above, that Clusius showed his gratitude to Gian Vincenzo Pinelli, who often acted as intermediary between Clusius and various Italian as well as non-Italian colleagues, by dedicating the *Fungorum Historia* to him. Mutual respect, gratitude, and the exchange of services and information are also evident from the contacts between Clusius and two Italian friars, who got in touch with him towards the end of his life. In 1597 Evangelista Quattrami invited Clusius to settle down in Italy as prefect of the botanical garden of the duke of Ferrara. He also sent Clusius a short description of the potato, 'Papas Peruvanorum radix', which the latter published in *Rariorum plantarum historia* (1601), while the capuchin friar Gregorio da Reggio sent Clusius information about Italian as well as exotic plants, such as a drawing and short description of the sweet pepper ('Capsici').[63] Clusius included it in the posthumously published *Curae*

---

[61] Adriano Soldano, 2005, personal communication. In the study by Aldrovandi entitled *Piante odorate* (BUB, fondo Aldrovandi, Ms. 90, cc. 183v-184) it is said that the 'Ptarmica austriaca' was raised up from seeds at the Bologna Botanical Garden. The picture of this plant is present in the ten-volume collection of drawings in Aldrovandi's Musaeum (BUB, fondo Aldrovandi, Tavole di piante, fiori e frutti, vol. IV, tav. 364).

[62] Some of the manuscript volumes are written like a diary. Aldrovandi took notes about everything concerning scholars and colleagues, news about publications and editions, collections, list of specimens sent etc., often including summaries of his letters.

[63] The fact that Clusius was interested in and obtained exotic specimens undoubtedly goes back to his stay on the Iberian Peninsula. The importance of American plants for Clusius can be seen from the *Rariorum plantarum historia*, the *Exoticorum libri decem* (Antwerp, 1605) and the *Curae posteriores* (Leiden, 1611). A summary of the American species mentioned in these works can be found in A. Ubrizsy Savoia and J. Heniger, 'Carolus Clusius and American plants', *Taxon*, 32/3 (1983), 424-435, listing 162 quotations of American plants.

posteriores (1611), which moreover contains a dedication to Matteo Caccini, who had shared so much information and so many of his plants with Clusius. The latter's interest in exotic *naturalia* obviously strengthened similar interests in Italy, since Aldrovandi's manuscripts contain many references to information provided by Clusius.

In one of the twenty-two letters (in Italian) to the specialist cultivator of flowers Matteo Caccini in Florence, Clusius apologises for his Italian, since he had never been able to visit Italy and had learned the language by reading books in Italian.[64] In fact, many of the letters exchanged between Clusius and his Italian friends[65] and colleagues were written in Italian. Clusius clearly knew the language very well, and had learnt it quite early on, as is evident also from his Latin translation (in 1561) of the *El ricettario di Firenze*, one of the early Italian pharmacopoeias. During the Renaissance the ability to speak and write Italian imparted status to non-Italians: it implied that they had been highly educated, generally at an Italian university, and were familiar with humanist culture. It is therefore not surprising to see that the Clusius correspondence also contains letters written in Italian by non-Italians, such as Balthazar ab Herden, Franciscus Broyardus, Alexander Fugger, Paulus Schedius Melissus, Laurentius Gryllus, and Ferdinand Weidner de Bilterburg, beside the ones already mentioned above by Esaya Le Gillon, Giacomo and Filippo di Monte, Casabona and Henry Wotton.

Clusius' botanical contributions have left many traces in the works of Italian authors, such as *Ecphrasis* (1592) by Fabio Colonna, *Herbario novo* (1584) by Castore Durante, *Exactissima descriptio rariorum quorundam plantarum* (1625) by Tobia Aldino (Pietro Castelli), *Historia naturalis* (1672) by Ferrante Imperato. Clusius' name also appears frequently in unedited manuscript works, such as the 16 volumes of Aldrovandi's Herbarium and the 10 volumes of his iconographic collection (including his *Dendrologia*, printed posthumously in 1668). In Italy Clusius' mycological contribution has mainly been referred to by G. Turre in his *Historia plantarum* (1685), by P.A. Micheli in *Nova genera plantarum* (1729), by L.F. Marsigli in his unpublished *Collectio fungorum vegetantium in regnis Croatiae Hungariae* (1669-1670; BUB, fondo Marsigli) and his *Dissertatio de generatione*

---

[64] The letters to Caccini have been published by P. Ginori Conti, *Lettere inedite di Charles de l'Escluse (Carolus Clusius) a Matteo Caccini, floricultore fiorentino. Contributo alla storia della botanica* (Florence, 1939).
[65] Letters to Clusius by the Italians Orazio Bembo, Baldassarre Peverello and Octavia Peverello de Bruti, Giovanni Calandrini, Francesco Calzolari, Arnoldo Paradiso di Sette Monti, Antonio Cappa, Antonio Abbondio, Giacomo Scutellari, Bartolomeo Guarinoni, Giovanni Viviani, and Francesco Malocchi were first published and discussed by De Toni, 'Il carteggio degli italiani col botanico Carlo Clusio'.

fungorum (1714) which he wrote together with L.M. Lancisi. J.B. Morandi describes some of Clusius' fungi in his book *Historia botanica* (1744). One of the most important comments can found in a letter written by Ferrante Imperato to G.B. Faber, a member of the Roman Accademia dei Lincei founded in 1603: 'it is not by seed that the fungi are reproduced, they grow from decaying materials as discussed at length by our Carlo Clusius in the *Libri Exoticorum*, to which I add that the cold steep of fungi when spread over the ground results in the rise of many fine edible mushrooms'.[66]

Conclusion

Clusius' *Fungorum historia* (1598), which was published as an attachment to his *Rariorum plantarum historia* (1601), and his iconographic collection which was begun during his stay in Austria and Hungary, represent pioneer contributions to the field of mycology. Clusius is particularly important for Hungary. At the court of Batthyány – which formed a typical Renaissance cultural centre – and thanks to the support of his maecenas and the information of his local friends, Clusius could observe and describe both mushrooms and plants, note their Hungarian names, and collect and describe historical and ethno-botanical knowledge concerning. As a witness to the high cultural and scientific level in a country heavily hit by the Turkish occupation and the equally devastating presence of foreign solders, Clusius remains a fundamental point of reference in Hungarian scientific and cultural life, a fact which has been recognized and appreciated in Hungary only since the end of the nineteenth century.

Clusius' connections with Italian and Hungarian colleagues, among others, enabled him to obtain information about plants (and animals) in places where he himself could not go. Through his correspondence Clusius was able to obtain many rare, often exotic plant species. By setting up a network of exchanges Clusius contributed to the introduction of many of these species in different parts of Europe, sometimes via the newly created university botanical gardens, but often also via the private gardens of rich 'amateurs'. The most important introduction – or more precisely, propagation and acclimatization – of plant species connected with Clusius concerns bulbs, such as hyacinths, saffron and others, and species coming from South-Eastern Europe such as horse chestnut, plane, *Platanus orientalis*, *Paeonia* species, the use of primrose species in formal gardens for early flowering, and the American exotic species such as potato, tobacco plants and beans. Many of these are connected with his Austrian-Hungarian period (and with

---

[66] G. Gabrieli, *Il carteggio Linceo* (Rome, 1996; 1st edn. 1838-42)

persons such as Purkircher). The introductions in Italy of the sunflower and many rare (endemic) species from Crete were made possible by Clusius' extensive contacts with Italian scholars such as Fabio Colonna. And the most emblematic plant connected with the name of Clusius, the tulip, was linked with both his Hungarian and Italian relations.

The international transfer of medicinal drugs by the Society of Jesus (sixteenth to eighteenth centuries) and connections with the work of Carolus Clusius

Sabine Anagnostou

From the sixteenth to the eighteenth century the Society of Jesus was one of the most successful Catholic orders. Within about 230 years after its foundation in 1540, the small community of Ignatius of Loyola and his first companions developed into a worldwide, powerful order. Besides their principal activity of preaching the gospel and spreading it all over the world, Jesuits were engaged in a wide range of different activities in many mission countries: for example, as scientists and researchers, pharmacists, physicians, sculptors, painters, carpenters and bricklayers. These activities were, of course, instrumental to the propagation of Christianity and thereby, ultimately, to the pursuit of the highest aim of the Society of Jesus: 'omnia ad maiorem Dei gloriam', everything to the greater glory of God.[1]

As researchers, pharmacists and physicians Jesuits also worked in the field of botany. They explored their new surroundings, searching the mission areas around the globe for unknown plants and investigating their medicinal properties.[2] These investigations were mainly motivated by two reasons. On the one hand, many Jesuits took the duties of physicians and apothecaries upon themselves since they considered taking care of the sick one of the important Christian duties. Medical provisions in the missions were poor, however, and to provide the local population with affordable remedies the Jesuits had to make use of the indigenous flora and traditional local remedies. On the other hand, according to Jesuit philosophy and spirituality, nature reflected God's

---

[1] Concerning the history of the Society of Jesus see, for example, J.W. O'Malley, *The first Jesuits* (Cambridge [Mass.], 1998); idem (ed.), *The Jesuits: Cultures, sciences, and the arts 1540-1773* (rpt. Toronto, 2000) [Papers of the international conference *The Jesuits: Culture, learning, and the arts 1540-1773*, held May 1997 in Boston], C.E. O'Neill and J.M. Domínguez (gen. eds.), *Diccionario histórico de la Compañía de Jesús: Biográfico-temático*, 4 vols. (Rome/Madrid, 2001); and J. Meier (ed.), *Sendung – Eroberung – Begegnung. Franz Xaver, die Gesellschaft Jesu und die katholische Weltkirche im Zeitalter des Barock. Studien zur Außereuropäischen Christentumsgeschichte* (Wiesbaden, 2005) [Asien, Afrika, Lateinamerika. Studies in the history of Christianity in the non-Western World, 8].

[2] See S. Anagnostou, 'Jesuits in Spanish America and their contribution to the exploration of the American materia medica', *Pharmacy in History*, 47 (2005), 3-17.

omnipotence and divine providence. To describe and explore nature was, therefore, one way of worshipping God.³ The results of the wide-ranging investigations by the Jesuits were published in books and manuals, and spread via the order's own communication system, the so-called *Litterae annuae*. The drugs themselves (together with information about their efficacy and preparation) were transmitted and distributed as well, mainly by Jesuit pharmacists in the context of their worldwide drug transfer. This global network was based on the order's own, well-connected pharmacies in both Europe and the mission countries.⁴ Many of the indigenous drugs which Jesuits used in non-European countries were integrated in the *materia medica* of the Old World, and enriched medical therapies there. Some of these medicinal plants – such as Fever bark, Passionflower, Ipecacuanha and Jaborandi – still today are important elements of medicine and pharmacy. The Jesuits, moreover, significantly contributed to the preservation of ethnomedical and ethnopharmaceutical traditions in many parts of the world by investigating the use of traditional healing plants and including such information in their publications. The latter nowadays constitute a precious source for the study of ethnopharmacy and ethnomedicine.

Jesuit perception and interpretation of their new surroundings were, of course, based on and influenced by contemporary European scientific concepts, as represented by the ideas and knowledge of well established and famous scholars.⁵ It would be interesting, therefore, to investigate the influence of the great Netherlandish botanist Carolus Clusius on the botanical activities of the Jesuits, and, vice versa, to discover whether Clusius himself made use of contemporary publications about foreign regions written by Jesuits. Another question deserves attention as well and touches upon both the worldwide botanical research of the Jesuits – whether scholars, medical laymen or professional pharmacists – in the context of pharmacy and medicine, and the incorporation of foreign drugs in the European *materia medica* via Jesuit apothecaries. On which sources did they rely for their botanical exploration of foreign floras, apart from empirical research and intercultural exchange?

---

³ See M. Sievernich, 'Vision und Mission in der Neuen Welt Amerika bei José de Acosta', in M. Sievernich and G. Switek (eds.), *Ignatianisch. Eigenart und Methode der Gesellschaft Jesu* (Freiburg/Basel/Vienna, 1990²), 239-313; and S. Anagnostou, *Jesuiten in Spanisch-Amerika als Übermittler von heilkundlichem Wissen* (Stuttgart, 2000) [Quellen und Studien zur Geschichte der Pharmazie, 78], 110-117. See also S.J. Harris, 'Jesuit scientific activity in the overseas missions, 1540-1773', *Isis*, 96 (2005), 71-79.
⁴ S. Anagnostou, 'Mission und Heilkunde. Das Heilmittelversorgungssystem der Jesuiten in den Missionen Spanisch-Amerikas', *Neue Zeitschrift für Missionswissenschaft. Nouvelle Revue de science missionnaire*, 4 (2001), 241-259.
⁵ Anagnostou, *Jesuiten in Spanisch-Amerika als Übermittler von heilkundlichen Wissen*, 110-117.

Carolus Clusius (1526-1609) explored the floras of various European regions which had not been investigated before, and devoted a large part of his life to the collection and exploration of foreign plants and exotic plant and animal products. He received these naturalia via a wide range of people: his European network of informants not only comprised scientists, botanists, collectors, pharmacists and 'amateurs' of gardening and botanical studies, but also included travellers and sailors.[6] Several of Clusius' publications were extremely valuable to both Jesuit scientists and pharmacists in the overseas missions and Jesuits who worked in the field of European botany.[7] His publications concerning exotica were especially relevant, such as *Rariorum plantarum historia* (1601) (Ill. 52), *Exoticorum libri decem* (1605) and the posthumous *Curae posteriores* (1611), and the same can be said of his translations and revisions of widely known publications about foreign drugs and plant remedies, such as the *Coloquios dos simples e drogas e cousas mediçinais da India* (*Aromatum et simplicium aliquot medicamentorum apud Indios nascentium historia* [1567]) by Garcia da Orta, the *Tractado de las drogas y medicinas de las Indias Orientales* (*Aromatum et medicamentorum in Orientali India nascentium liber* [1582]) by Christóbal Acosta, and the *Historia medicinal de las cosas que se traen de nuestras Indias Occidentales que sirven en medicina* (*De simplicibus medicamentis ex Occidentali India delatis quorum in medicina usus est* plus supplement by Clusius [1574]) by the Spanish physician Nicolás Monardes.

Most of the Jesuit botanists and pharmacists who were involved in the international drug transfer (which was still in its early stages during Clusius' lifetime, but would expand and become a worldwide network between the mid-seventeenth century and the extinction of the Jesuit order in 1773) had studied the contemporary European literature about botany and natural history that included the corpus of Clusius' works. The fact that copies of Clusius' books could, in fact, be found in many libraries of the Jesuit colleges in Europe and the mission countries reflects the high esteem in which the Netherlandish scholar was held. Such libraries provided the researchers and apothecaries of the order with both traditional knowledge and the latest information. Moreover, Jesuit authors of botanical books and medical-pharmaceutical handbooks regularly refer to Clusius' work as a reliable source. Especially those Jesuits whose native language was not Spanish or Portuguese relied on Clusius' Latin translations and revisions of the *Coloquios dos simples*, the *Tractado de las drogas y medicinas de las Indias Orientales* and the *Historia medicinal de las cosas que se*

---

[6] See F. Egmond, 'Correspondence and natural history in the sixteenth century: cultures of exchange in the circle of Carolus Clusius', in F. Bethencourt and F. Egmond (eds.), *Correspondence and cultural exchange in early modern Europe* (in press Cambridge, 2007).

[7] J.C. Gillispie (ed.), *Dictionary of scientific biography*, 14 vols. and 4 suppls. (New York, 1970-90), vol. VIII, 120f.; A. Fetzner, *Carolus Clusius und seine* Libri exoticorum (dissertation, Marburg, 2004).

Ill. 52. Engraved title-page of C. Clusius, *Rariorum plantarum historia* (Antwerp, 1601). According to the inscription on the title-page, this copy was donated by Clusius to Leiden University Library.

traen de nuestras Indias Occidentales que sirven en medicina for information about exotica. Latin was, after all, the universal language of erudition at the time. It is also likely that copies of his *Antidotarium* (1561) – the Latin translation of the Italian pharmacopoeia *El Ricettario di Firence* – could be found in the collections of pharmaceutical literature of the Jesuit libraries in the missions. These collections generally did not focus on a single geographical region, but comprised a wide range of works from both Europe and the missions around the globe.[8]

The evident intellectual connection between the Jesuits and Clusius raises various questions. In how far did Clusius' investigations influence the exploration and interpretation by Jesuits of the non-European flora? To what extent did the Jesuits make use of Clusius' information in their horticultural practices? What effect did Jesuit interest in and high opinion of Clusius' work have on its distribution and reception? What was the connection between the European scientific network established by Clusius in the course of the sixteenth century and the later, worldwide network of the Jesuit order. What consequences did all of this have for the 'global' transmission of plants and drugs, and, ultimately, for the development of pharmacy? It is well known, moreover, that Jesuits corresponded with European scholars about botanical questions and sent comprehensive collections of plant samples and exotica, botanical descriptions and drawings to the Old World.[9] Could there have been such a scientific exchange between Jesuit missionaries and Clusius himself? Or were the religious barriers too high? While the Jesuits devoted themselves to the re-establishment of Catholicism after the Reformation and its dissemination over the world, Clusius had embraced the Reformation. Of course, these questions cannot be fully answered at the present stage of research. Yet, even preliminary investigations already promise interesting new information about the influence of Jesuits on the worldwide reception of Clusius, the impact of Clusius' work on Jesuit botanical and pharmaceutical research, and Clusius' studies as one of the fundamental sources of the Jesuit network of worldwide drug transfer from the sixteenth to the eighteenth century.[10]

In the following, the development of worldwide drug transfers by the Jesuits will be sketched from the very first beginning until the extinction of the order in 1773. We will outline the motives for the Jesuit exploration of foreign floras

---

[8] The collection of the Jesuit library in Santiago de Chile, for example, consisted of approximately 130 books in several different languages including dictionaries. See S. Anagnostou and M. Müller, 'Joseph Zeitler – Auf den Spuren eines bayerischen Apothekers in Chile', *Geschichte der Pharmazie*, 56 (2004), 16-23.

[9] J. Gicklhorn and R. Gicklhorn, *Georg Joseph Kamel S.J. (1661-1706), Apotheker, Botaniker, Arzt und Naturforscher der Philippineninseln* (Eutin, 1954) [Veröffentlichungen der Internationalen Gesellschaft für Geschichte der Pharmazie, 4].

[10] In this context a thorough evaluation of the wide-ranging Clusius correspondence in the University Library in Leiden will be essential.

and show how the results of Jesuit investigations promoted international drug transfer by the Society of Jesus and stimulated the transmission of scientific (botanical) knowledge and the distribution of plants around the world. We will investigate to what extent Clusius' work could have influenced Jesuit botanical studies, thus becoming part of the intellectual background of the order's worldwide drug transfer. It will be shown, moreover, that Clusius' research was one of the topics of scientific exchange between European scholars and Jesuits, and that it was of great importance to those Jesuits who studied the cultivation of European and non-European plants.

Clusius' works in Jesuit libraries

Library catalogues of Jesuit colleges in both the mission countries and Europe reveal Jesuit fields of interest, while also showing which types of publications they preferred for work and study. According to the catalogues of various Jesuit libraries in Europe, several of Clusius' works were available in their colleges in Central Europe. Jesuits in Europe with an interest in the exploration of foreign floras appreciated Clusius' knowledge of exotic plants and obviously regarded him as an authority in the field of botany. Clusius' annotated Latin translation *Aromatum et simplicium aliquot medicamentorum apud Indios nascentium historia* of Garcia da Orta's *Coloquios dos simples* could, for instance, be found in both the libraries of the Jesuit *Domus probationis* in Mainz and the Jesuit college in Trier.[11] The latter also owned his Latin translation of Monardes *Historia medicinal*.[12]

Clusius' works were also sent to countries far away from Europe, where Jesuits consulted them for their own botanical and pharmaco-botanical research. As we will see in more detail below, Clusius' works were studied in the missions of Spanish America and the Philippines. They also found their way to China. The famous library of Pét'ang in Beijing consists of various collections, parts of which were brought together at the time of the Portuguese and French missions. The latter was controlled by the Jesuits until the late eighteenth century. During the early decades of the seventeenth century the Jesuits Nicolas Trigault and Johann Schreck (also known as Terrentius), about whom more will be said later, created a library in China that was intended to be worthy of an ecclesiastic dignitary and form an eternal monument for the Catholic church in China.[13] Before travelling to China, they had

---

[11] Stadtbibliothek Mainz Ms. III 70: *Catalogus II Librorum Bibliothecae Domus Probationis Societatis Iesu. Moguntiae secundum Cognomina Auctorum una cum forma librorum, columnae et serie* (1743), fol. 5r.
[12] Stadtbibliothek Trier, Ms. 2412/2358: *Catalogus librorum bibliothecae Trevierensis collegii S.J. in classes distributus* (Trier, 1764), fol. 405. Today Clusius' translations carry the signature D 1278.
[13] H. Verhaeren, 'Aperçu historique de la Bibliotèque du Pét'ang', in idem, *Catalogue de la Bibliothèque du Pé-T'ang* (Beijing, 1949; rpt. Paris, 1969), viii-xi.

brought together the most important publications of their time from the Netherlands, France and Germany, for all scientific disciplines they considered relevant. This impressive collection included several works by Clusius: *Curae posteriores* (1611), *Exoticorum libri decem* (1605) including the Latin translations of Acosta's, da Orta's and Monardes' works as well as the Latin translation of the accounts of Pierre Belon's *Observationum libri III* (1553), the *Libellus de rosa et dissertatiuncula de citriis* (1565) together with the *Altera Appendix ad Rariorum plantarum historiam,* and finally the *Rariorum plantarum historia* (1601), which comprised the *Rariorum plantarum historia libri sex, Commentariolus de fungis, Honori Belli ad Carolum Clusium aliquot epistolae de rarioribus quibusdam plantis agentes,* Tobias Roelsius' *De certis quibusdam plantis epistola*, and Giovanni Pona's *Plantae seu simplicia ut vocant, quae in Baldo monte et in via ab Verona ad Baldum reperiuntur.*[14] In this well documented case we thus know precisely who transferred the works of Clusius to the Far East, where they served as a basis for botanical studies.

Research concerning medicinal plants in the missions

Jesuit botanical exploration of the non-European flora was mainly stimulated by the relatively poor state of medicine in the missions, where professional pharmacists and physicians were hard to find. Taking care of the sick was a fundamental aspect of Christianity, and the Jesuits considered the healing of the body as important as the saving of the soul. From an early stage onwards in the history of their order, they regarded alleviating and curing the physical illnesses of the suffering as one of their important tasks. Naturally, most of the missionaries did not have an adequate medical training, but they constantly endeavoured to improve their medical-pharmaceutical knowledge by studying the contemporary European scientific literature, learning from traditional local expertise, and searching their surroundings for new remedies. Besides those Jesuits who gained their medical expertise in practice, the order also comprised professional apothecaries, who founded pharmacies in the urban centres of the mission countries. Quite often, these pharmacists received indigenous drugs from rural regions in the mission countries, included them in their medication and therapies, and thereby paved the way for the introduction of these drugs in the European *materia medica*.[15]

---

[14] Verhaeren, *Catalogue de la Bibliothèque du Pé-T'ang*, 586; and I. Iannaccone, *Johann Schreck Terrentius. Le scienze rinascimentali e lo spirito dell'Accademia dei Lincei nella Cina dei Ming* (Naples, 1998) [Istituto Universitario Orientale Dipartimento di Studi Asiatici, Series minor, 54], 53.

[15] S. Anagnostou, 'Missionsmedizin und Missionspharmazie im kolonialen Amerika', in Meier (ed.), *Sendung – Eroberung – Begegnung*, 260–291.

Jesuits in the missions primarily relied on European medical and pharmaceutical traditions. European drugs were scarce, however, and in most cases proved to be too expensive for the poor, especially in those missions which were far removed from urban centres. Moreover, many drugs had lost their potency after months of transportation overseas. In order to become less dependent on this problematic source of supply, Jesuits soon decided to explore the local resources and search for easily available and affordable remedies which could satisfy their own and their patients' medical requirements. Sharing their day to day existence with the natives, the missionaries also came to share their knowledge of the healing properties of indigenous plants, animals and minerals. At the same time, they investigated the local flora in order to discover useful medicinal properties of indigenous plants. In the gardens of their missions they grew medicinal herbs from Europe as well as local healing plants. In a report about his life in Paraguay at the end of the seventeenth century the German Father Anton Sepp describes the large and wonderful garden next to his house which had a separate section of medicinal herbs for his patients.[16] Gradually, many Jesuits collected drugs (some of which also came from the Jesuit pharmacies in the urban centres), created small pharmaceutical stocks, and thus established modest pharmacies which could satisfy the immediate needs of their patients. Father Ignaz Pfefferkorn, a missionary in Mexico, describes the situation in Sonora in the eighteenth century:[17]

For that purpose I had a small pharmacy, which comprised various indigenous plants and some remedies that I had ordered from Mexico-City. According to the illness and the circumstances, I applied them as well as possible, for the recovery of my suffering Indians: and I was lucky enough to be able to restore the health of many patients.

Missionaries were not only concerned with the patients in their immediate surroundings but also with their fellow-Jesuits – especially those who worked in regions far removed from urban centres and had to deal with a problematic supply of drugs and the scarcity of medical knowledge. Several Jesuits therefore composed medical-pharmaceutical handbooks or manuals which were

---

[16] A. Sepp, *RR. PP. Antonii Sepp und Antonii Böhm, der Societät Priestern Teutscher Nation [...] Reißbeschreibung* (Nuremberg, 1698), 61.
[17] I. Pfefferkorn, *Beschreibung der Landschaft Sonora samt andern merkwürdigen Nachrichten von den inneren Theilen Neu-Spaniens und Reise aus Amerika bis in Deutschland nebst einer Landcharte von Sonora*, 2 vols. (Cologne, 1794-95), vol. II, 403f.: 'Ich hatte zu dem Ende eine kleine Apotheke, theils von verschiedenen einheimischen Pflanzen, theils auch von einigen Arzneyen, die ich aus Mexico [Stadt] kommen ließ; welche ich, nach Befinden der Krankheit und der Umstände, so gut es nur möglich war, zur Genesung meiner leidenden Inder gebrauchte: und ich hatte wirklich das Glück, vielen zur Gesundheit zu verhelfen.' See also Anagnostou, *Jesuiten in Spanisch-Amerika als Übermittler von heilkundlichen Wissen*, 58-69, 223-226.

written specifically for fellow-Jesuits and any other person taking care of the ill. Via such manuals (or *vademecums*) Jesuit authors sought to transmit their knowledge of indigenous medicinal plants and provide advice concerning the treatment of patients. These handbooks contain descriptions and drawings of many indigenous plants, information about the best period to collect them and optimal storage conditions, explanations about their medicinal effects, and advice for the preparation of different medications. Such manuals were often copied; some were even printed, and through their wide distribution became very well known in the mission countries.

During the early eighteenth century Pedro Montenegro (1663–1728) described the medicinal plants of the Guaraní missions of Paraguay in his *Materia médica misionera*.[18] Montenegro had medical experience. He had worked in the Hospital General in Madrid, probably as a nurse, before he arrived in Paraguay, where he devoted his life to the assistance of the suffering and the exploration of the medicinal plants of the Guaraní. Montenegro collected his vast knowledge of (mainly) American plants in his impressive herbal, also including some European and Asian plants which he evidently regarded as important drugs that were available for pharmaceutical supplies. In circa 150 monographs, which are often accompanied by drawings, Montenegro carefully describes each individual plant, explains its healing properties, and provides detailed instructions for its use in medication and the doses that should be administered to a patient. Finally, Montenegro lists the names of the plants in Guaraní, Tupí and – in sofar as they existed – Castilian, in order to characterize each plant as clearly as possible and make sure that the reader of his manual could identify it in nature.[19]

A considerable part of Montenegro's knowledge concerning American medicinal plants was based on the ethnomedical tradition of the Guaraní and his own personal research and experience. Yet, he also relied on traditional European authorities in the fields of medicine, botany and pharmacy, such as Theophrastus of Eresos, Dioscorides and his sixteenth-century commentators (such as Pier Andrea Mattioli and Andrés de Laguna), Pliny and Galen. He had, moreover, consulted the literature of the sixteenth and seventeenth centuries about the natural history and *materia medica* of the New World, such as Agustín Farfan's *Tratado brebe de medicina* (Mexico, 1579). And he copied information about medicinal plants together with some drawings from the

---

[18] P. Montenegro, *Materia médica misionera* (Buenos Aires, 1945). This is the first complete printed version of the eighteenth-century manuscript.

[19] Concerning plants and remedies in the *Materia médica misionera* see C. Martín Martín and J. L. Valverde, *La farmacia en la América colonial: el arte de preparar medicamentos* (Granada, 1995); and Anagnostou, *Jesuiten in Spanisch-Amerika als Übermittler von heilkundlichen Wissen*, 265-325.

Historia naturalis Brasiliae (Amsterdam, 1648) by Willem Piso. This publication included the unfinished work *Historiae naturalis et medicae Indiae Orientalis libri sex* by Jacobus Bontius, which dealt with medical treatments in India. For healing plants from Asia, such as China-root (*Smilax spec.*), Clove tree (*Syzygium aromaticum [L.] Merr. et L. M. Perry*) and Pepper (*Piper nigrum* L.), Montenegro specifically refers to Christobál Acosta's *Tractado de las drogas, y medicinas de las Indias Orientales* as well as to Garcia da Orta's famous *Coloquios dos simples e drogas e cousas medicinais da India*.[20] Since Montenegro's native language was Spanish, he probably did not consult the original Portuguese texts, but used their respective Latin translations by Carolus Clusius.

While Montenegro concentrated on the flora of Paraguay, the Jesuit apothecary Johann Steinhöfer (1664-1716), who worked in the missions of Northern Mexico, included several Mexican healing plants in the therapies which he describes in his medical-pharmaceutical handbook *Florilegio medicinal de todas las enfermedades* (1712). His *Florilegio* acquired a great reputation; it was widely distributed in the Spanish colonies. Since Steinhöfer first and foremost relied on his European pharmaceutical education, he mostly used elements of the European *materia medica*. Yet, he had also learned to appreciate genuine American (and especially Mexican) remedies, such as agaves (*Agave americana* L.), contrayerba (*Dorstenia contrayerva* L.), jojobas (*Simmondsia chinensis* Link. Nutt.) and marigold (*Tagetes spec.*).[21] Steinhöfer does not explicitly mention Clusius' works, but as an experienced apothecary he had probably consulted the *Antidotarium* (1561) – Clusius' Latin translation of an Italian pharmacopoeia – with respect to the composition of various remedies. In several instances these follow recipes contained in European pharmacopoeias. Tracing back the origins of some preparations in the *Florilegio medicinal* might, therefore, provide information about a possible connection with the *Antidotarium*.

The worldwide network of drug transfer

While many Jesuits explored their new environment far away from the centres of civilization, Jesuit apothecaries founded professional pharmacies in the major cities of the mission countries, such as Santiago de Chile, Lima, Buenos Aires, Mexico City, Pernambuco, Rio de Janeiro, Manila, Goa, and Macao.

---

[20] Montenegro, *Materia médica misionera*, 325, 440, 435.
[21] For details see J. Esteyneffer, *Florilegio medicinal de todas las enfermedades [México 1712], Edición, estudio preliminar, notas, glosario e indice analítico por C. Anzures Bolaños*, 2 vols. (Mexico, 1978); I. Schuler, *Das Florilegio medicinal von 1712 des Johann Steinhöfer, Jesuitenmissionar in Mexiko* (Munich, 1973); S. Anagnostou, 'Ethnomedizinische Aspekte jesuitischer Missionstätigkeit in Spanisch-Amerika', *Zeitschrift für Phytotherapie*, 5 (2001), 229-235.

These pharmacies generally originated as dispensaries that could provide the residents of the colleges themselves with the necessary drugs. Since they often were the only pharmacies in the region, however, they soon started to extend their domain. Eventually, they supplied both the local population and those living in a large territory around the cities with skilfully prepared medication. Inventories show that the stocks of these pharmacies contained traditional European drugs as well as remedies that were genuinely indigenous to the mission countries. A certain quantity of European drugs was generally ordered from abroad, and many traditional healing plants from Europe were grown in the Jesuit gardens.[22] Indigenous drugs were investigated by the Jesuit apothecaries themselves and also sent to the urban centres by missions far away in the provinces, where Jesuits took care of the sick.

The Jesuit apothecaries Joseph Zeitler in Santiago de Chile and José Rojo in Lima, corresponded about pharmaceutical topics, such as the analysis of different salts, and exchanged drugs for their pharmacies. Soon, these two men also included Jesuit apothecaries in the other cities of Spanish America in their scientific and pharmaceutical exchanges, such as Andreas Lechner in Quito and Georg Schultz in Mexico-City.[23] Ultimately, many Jesuit pharmacies became international centres of scientific exchange and medical supply. The pharmacies of San Pablo in Lima, San Ignacio in Manila, and San Pablo in Macao are good examples. By the second half of the seventeenth century San Pablo in Lima had become the international centre for the distribution of the famous fever bark or Jesuit bark (*Cinchona spec.*), the first effective remedy against malaria.[24] Jesuits from Peru carried Fever bark to Europe and distributed it all over the world. During a devastating fever epidemic in Rome cardinal Juan de Lugo (1583–1660), himself a Jesuit, imported huge amounts of fever bark from Lima at his own expense to be handed out to the poor and ill. Jesuits also took fever bark from Rome to China, where they cured the emperor K'ang-hsi of an intermittent fever. This American drug was called Jesuit bark or Jesuit powder for many years because Jesuits had a monopoly on its international distribution. The bezoar stone, a calcareous concretion in the intestines of ruminants, which was regarded as an antidote and a panacea at the time, and Jesuit tea (*Chenopodium ambrosioides* L.) also arrived in Europe from San Pablo in Lima. The Saint Ignatius bean (*Strychnos ignatii* Berg.) came

---

[22] See Anagnostou and Müller, 'Joseph Zeitler – Auf den Spuren eines bayerischen Apothekers in Chile', 16-23.

[23] See Anagnostou, 'Missionsmedizin und Missionspharmazie im kolonialen Amerika', in Meier, *Sendung – Eroberung – Begegnung*, 281-283.

[24] Concerning the history of the pharmacy of San Pablo in Lima see L. Martín, *The intellectual conquest of Peru: The Jesuit College of San Pablo 1568-1767* (New York, 1968), 97-118.

from Manila, while the mysterious *Lapis de Goa* (which was prepared according to a secret recipe developed by the Jesuit apothecary Gaspar António) arrived in Europe from Goa.[25]

In Europe, meanwhile, Jesuits had established an efficient network of pharmacies, linking cities such as Munich, Vienna, Cologne, Munster, Madrid, Rome, Milan and Sopron (Hungary). All of these pharmacies supplied both the local population and those who lived in the environs of these cities; they often supplied each other as well.[26] At the core of this network was the pharmacy of the Collegio Romano in Rome. From there, both foreign drugs, such as fever bark, and traditional European medication, such as *Theriaca Andromachi* or the Roman theriac, were sent to Jesuit pharmacies all over Europe and the rest of the world.[27] Thus, the Jesuits gradually established a worldwide network of drug transfer which unquestionably influenced the development of many different *materiae medicae* in various parts of the world. In the mission countries this elaborate system of trade and distribution perished after the expulsion of the Jesuits from the colonies. In Europe it disappeared after the extinction of the Society of Jesus.

The phenomenon of worldwide drug transfer by the Jesuits is all the more intriguing since it was neither the duty nor the intention of a religious order. Of course, medicine and religion were closely connected at the time. Yet, the strong links between pharmacy (including worldwide drug transfer), botanical research, and religion in the form of the Christian missions, are aspects of the history of pharmacy that deserve further investigation.

Scientific exchanges between Jesuits in the missions and scholars in Europe

The exploration of foreign floras by Jesuit apothecaries was not only inspired by the need and desire to discover useful new remedies. Some of these apothecaries engaged in much more wide-ranging botanical studies, which were presented in Europe by respected scholars.

---

[25] See A.M. Amaro, 'A famosa Pedra Cordial de Goa ou de Gaspar António', *Revista de cultura*, 7/8 (1988-89), 87-108.
[26] S. Anagnostou, 'Jesuitenapotheken vom 16. bis 18. Jahrhundert und ihr Publikum', in C. Friedrich and W.-D. Müller-Jahncke (eds.), *Apotheke und Publikum. Die Vorträge der Pharmaziehistorischen Biennale in Karlsruhe vom 26. bis 28. April 2002* (Stuttgart, 2003), 53-70.
[27] S. Anagnostou, 'Vom Römischen und Brasilianischen Theriak', in C. Friedrich and S. Bernschneider-Reif (eds.), *Rosarium litterarum. Beiträge zur Pharmazie- und Wissenschaftsgeschichte. Festschrift für Peter Dilg zum 65. Geburtstag* (Eschborn, 2003), 17-32; and S. Anagnostou, 'Pharmazie auf internationaler Ebene – die Apotheke des Collegio Romano vom 16. bis 18. Jahrhundert', *Geschichte der Pharmazie*, 57 (2005), 57-63.

Georg Joseph Kamel (1661-1706)

The Bohemian Jesuit Georg Joseph Kamel (1661-1706), apothecary of San Ignacio in Manila, devoted himself to the exploration of the local flora of the Philippines. In order to become less dependent on expensive European drugs, he studied indigenous remedies. By caring for the sick regardless of their origin he gained the trust and appreciation of the indigenous population, and by learning local languages he was able to communicate without intermediaries with the local inhabitants, coming to share their knowledge of the medicinal properties of many local plants. Kamel soon extended his studies from medicinal herbs to all of the flora and even fauna of the Philippines, and eventually collected and described large numbers of exotic plants and animals.

By the end of the seventeenth century Kamel got in touch with two of the most famous naturalists in England, both of them members of the celebrated Royal Society: the naturalist John Ray (1627-1705) and the apothecary James Petivier (1663-1718). Kamel's huge collection of natural objects (including numerous drawings and descriptions) was sent to John Ray. This formed the beginning of an intense scientific exchange and friendship, which lasted until Kamel's death. Several of his descriptions and pictures were published by James Petivier in the *Philosophical Transactions;* such as the description of the Saint Ignatius bean, which had been described first by Kamel.[28] John Ray was so deeply impressed by Kamel's botanical research that he decided to express his high esteem for the Jesuit's wide-ranging knowledge and investigations by publishing the latter's descriptions of the plants of the Philippines in the appendix to the third volume of his (Ray's) famous work *Historia plantarum* (1701) under the title *Herbarum aliarumque stirpium in insula Luzone Philippinarum primaria nascentium, a Reverendo Patre Georgio Josepho Camello, S. J. observatarum et descriptarum syllabus.*[29] Although Ray originally intended to include the drawings from Kamel's collection as well, this turned out to be impossible, probably for financial reasons. Nevertheless, he explicitly mentioned in the title that he possessed many drawings made by Kamel himself of the plants described by this Jesuit.[30]

While collecting and exploring the flora of the Philippines, Kamel also studied the contemporary European literature about exotic plants, animals and

---

[28] 'De Igasur, seu Nux Vomica legitima Serapionis. A further and more exact account of the same, sent in a letter from Father Camelli to Mr. John Ray and Mr. James Petivier', *Philosophical transactions*, vol. XXI, no. 250 (1699), 88.

[29] G.J. Kamel, 'Herbarum aliarumque stirpium in insula Luzone Philippinarum primaria nascentium', in J. Ray, *Historia plantarum*, vol. III (London, 1704), Appendix.

[30] For detailed information concerning life and work of Joseph Kamel see: Gicklhorn and Gicklhorn, *Georg Joseph Kamel S. J. (1661-1706)*.

minerals in order to check whether they had been described before, and investigated what kind of information had been transmitted about them. Unfortunately, there are no catalogues extant of the library of the Jesuit college in Manila to give us some idea of the sources that Kamel could have used. In his letters Kamel does, however, mention a few publications which he had consulted, such as *Historia plantarum* by John Ray, *Rerum medicarum Novae Hispaniae thesaurus* by Francisco Hernández, and *Historia naturalis Brasiliae* by Willem Piso.[31] In his descriptions of the Philippine plants in the appendix to the *Historia plantarum* he also refers explicitly to famous authors of works about foreign or non-European plants and exotica, among them Carolus Clusius – thus demonstrating his excellent knowledge of the contemporary literature concerning botany and pharmacy.

A further analysis of Kamel's Philippine herbal demonstrates clearly that Clusius' works were of great importance to this Jesuit botanist. Kamel not only refers in general terms to Clusius,[32] but also repeatedly quotes specific pages of his *Exoticorum libri decem*.[33] Moreover, he mentions publications by Christóbal Acosta and Nicolás Monardes. Since Kamel communicated in Latin and studied Latin publications, we may assume that he knew their works via the Latin translations by Clusius.[34] Clusius' own publications – or at least the *Exoticorum libri decem* – and his translations of Acosta and Monardes must, therefore, have been available in Manila, probably in the library of the Jesuit college of Manila. We do not know, however, whether these books had been sent to the Philippines before Kamel's arrival, had been transported to Manila by Kamel himself, or had been received by him from Europe some time after his arrival. Whatever the case may be, Kamel clearly contributed in a significant way to the reception of Clusius' publications. Either he himself or Jesuits closely connected with him introduced Clusius' work in the mission countries.

Johann Schreck (Terrentius) and Clusius

The Jesuit Johann Schreck (1576-1630), who is also known as the botanist and astronomer Terrentius, was a member of the exclusive and celebrated *Accademia dei Lincei* in Rome and one of the most famous scientists of the early seventeenth century. In 1604 this same *Accademia* had invited Clusius – without

---

[31] Gicklhorn and Gicklhorn, *Georg Joseph Kamel (1661-1706)*, 72-75.
[32] Kamel, 'Herbarum aliarumque stirpium', 2, 15, 17, 37, 39, 61, 65.
[33] Kamel, 'Herbarum aliarumque stirpium', 9, 19, 25, 28, 44, 45, 46, 52, 55, 58, 74, 77.
[34] Kamel, 'Herbarum aliarumque stirpium', 13, 14, 25, 27, 28.

success – to become a member.[35] After Schreck's admission to the Society of Jesus in 1611 he decided to join the mission in China. Travelling all over Europe together with the Jesuit Nicolas Trigault, he collected mathematical instruments, scientific books and money for the missions in China. He arrived in Goa in 1619 and stayed there for two years before he could continue his journey to China. Once there, he lived first in Hangzhou and then in Beijing, devoting himself to various scientific tasks, such as the reform of the Chinese calendar and the translation of European scientific publications into Chinese.[36]

Johann Schreck was also a keen naturalist, who participated in the edition of the sixteenth-century manuscripts by Francisco Hernández about the flora, fauna and minerals of the Spanish colonies in America. This *Rerum medicarum Novae Hispaniae thesaurus* was finally published in 1649. Schreck's comments, annotations and additional *Aliarum Novae Hispaniae plantarum*, which form part of the Hernández edition, reflect his comprehensive knowledge while also revealing some of his sources. He clearly regarded Clusius' works as precious and reliable sources for his own studies, and relatively often refers to *Exoticorum libri decem*, *Rariorum plantarum historia* and Clusius' Latin translations of the works by Garcia, Monardes and Acosta.[37] It has even been suggested that Schreck's mode of commenting on the *Rerum medicarum Novae Hispaniae thesaurus* was inspired by Clusius' comments on Garcia's work in *Aromatum et simplicium aliquot medicamentorum apud Indios nascentium historia*.[38] Above all, Schreck's high regard for Clusius is also evident from the fact that he (as mentioned above) included several of the latter's works in the collection that was to form the core of the library of Pé-T'ang in Beijing.

Gardens of the Jesuit colleges

Throughout Europe and the mission countries Jesuits created their own gardens in the Jesuit colleges where they grew European as well as exotic plants. The Jesuit college in Rome, the Collegio Romano, had a large garden with many plants from non-European regions, such as the American wormseed (*Chenopodium ambrosioides* L.), tobacco (*Nicotiana tabacum* L.), passionflower

---

[35] Iannaccone, *Johann Schreck Terrentius*, 35. For a further discussion of this episode see Irene Baldriga's contribution in this volume.
[36] C. von Collani, 'Schreck, Johann', in *Biographisch-bibliographisches Kirchenlexikon* (Herzberg, 1995), vol. IX, 919-922.
[37] F. Hernández, *Rerum medicarum Novae Hispaniae thesaurus* (Rome, 1649; facsimile edn. Rome, 1992), for example 93, 119, 124, 125, 127, 138, 162, 163, 165, 170, 172, 212, 218, 271. See also Iannaccone, *Johann Schreck Terrentius*, 34f., 47-58.
[38] Iannaccone, *Johann Schreck Terrentius*, 35.

(*Passiflora spec.*), hibiscus (*Hibiscus rosa-sinensis* L.) and yucca (*Yucca spec.*). This garden specifically belonged to the pharmacy of the Collegio Romano, and its main function was to supply the apothecaries of the college with medicinal plants for the preparation of drugs. Yet, there is also evidence for a wider ranging, botanical interest in the different species that could be observed in this garden. Jesuits from all over the world came to Rome, and some of them brought seeds from these foreign countries for the garden of the Collegio Romano.[39]

The Jesuit Giovanni Battista Ferrari (1584-1655) specialized in the cultivation of flowers, including non-European species. According to his own description he was the first to sow *Rosa chinense* (*Hibiscus rosa-sinensis* L.) in Rome and see it grow from seed there.[40] His impressive work about the cultivation of flowers (*De florum cultura* [Rome, 1633]) reflects the cultural atmosphere of the pontificate of pope Urban VIII and the scientific spirit of the Accademia dei Lincei. It was widely distributed and much appreciated at the time; several editions appeared of its Italian translation (*Flora overo cultura di fiori* [1638]) (Ill. 53).

Ferrari describes different ways to design a garden, provides descriptions of a variety of flowers, such as the bulbiferous tulip, hyacinth, anemone, iris, narcissus, crocus, roses and orchids, and gives advice about their cultivation. He concludes with a reflection on plants as a miracle of nature. Ferrari derived information about flowers and their cultivation from both classical authors like Theophrastus of Eresos (*Historia plantarum*), Pliny (*Naturalis historia*) and Columella

Ill. 53. Title-page of Giovan Battista Ferrari, *Flora overo cultura di fiori* (Rome, 1638).

---

[39] Anagnostou, 'Pharmazie auf internationaler Ebene – die Apotheke des Collegio Romano vom 16. bis 18. Jahrhundert', 60f.

[40] G.B. Ferrari, *Flora overo cultura di fiori*, facsimile edn., introd. L.T. Tongiorgi Tomasi, texts by A. Campitelli and M. Zalum (Florence, 2001), 477.

(*De re rustica*), and the contemporary scientific literature, for instance *Stirpium historia* by Rembert Dodoens and *De natura stirpium* by Jean Ruel.[41] Ferrari's explanations and descriptions reveal his excellent knowledge of Clusius' works, which he clearly regarded as an important and reliable source. He mentions them rather frequently, especially where exotic, non-European plants are concerned, quoting for example *Rariorum plantarum historia* concerning *Spartium junceum* L. (Ill. 54), *Exoticorum libri decem* while discussing *Hibiscus rosa-sinensis* L. (Ill. 55), the *Appendix ad plantarum historiam* when describing *Muscari botryoides* L. (Ill. 56), and *Curae posteriores* in the section about *Jasminum spec*.[42] Ferrari also explicitly refers to Clusius' annotated Latin translations of Garcia da Orta and Christobál Acosta.[43] Ferrari's integration of Clusius' information in his own work and his references to Clusius as an important scientist and source of information definitely underpinned the reception of Clusius' work.

Ill. 54. *Spartium junceum* L. (See also colour plate 7).

---

[41] For further information concerning the *Flora overo cultura di fiori* see L.T. Tongiorgi Tomasi, 'L'arte ingenua e ingegnosa di coltivare i fiori. Note su *Flora overo cultura di fiori* di Giovan Battista Ferrari', in Ferrari, *Flora overo cultura di fiori*, ix-xxv.
[42] Ferrari, *Flora overo cultura di fiori*, 167, 197, 207, 378.
[43] Ferrari, *Flora overo cultura di fiori*, 378, 385.

Ill. 55. *Hibiscus rosa-sinensis* L. (See also colour plate 8).

Ill. 56. *Muscari botryoides* L. (See also colour plate 9).

Botanical research as a part of the Jesuits' scientific programme

While Jesuits thus played an important part in the reception and worldwide distribution of Clusius' work, we should not overlook the possibility that Clusius himself may have studied the reports of the early Jesuits about the history of non-European regions. These works often contained detailed information about the flora, fauna and minerals of overseas territories.

Jesuit scientific exploration of foreign regions was not only sustained or inspired by practical considerations, such as discovering new medicinal drugs. It was also stimulated by the religious, spiritual and philosophical concepts of the order. Through scientific research Jesuits did not intend merely to describe and explain the phenomena of nature: they attempted to uncover the laws of a world constructed according to the divine Creator's plan. The excellence of the world could only confirm the excellence of its Maker. A Jesuit scientist was therefore praising God, the Creator, by his scientific investigations. Especially for the Jesuits, scientific research thus meant pursuing their most noble aim: 'everything for the greater honour of God', 'omnia ad maiorem Dei gloriam'. These ideas are represented by the work of José de Acosta (1540-1600).[44]

Acosta was one of the most famous Jesuit scientists writing about the natural history of the New World. He was ordained a priest and devoted himself to rigorous studies of theology, philosophy and other scientific disciplines. At his own request Acosta was sent to the overseas missions in Peru, where he worked as a missionary and scientist. Various expeditions and journeys gave him the opportunity to explore the viceroyalty of Peru and study the nature of the New World. Before he returned to Spain in 1587, he spent a year in Mexico teaching, preaching, and studying Indian culture. Among all of his works the *Historia natural y moral de las Indias* (Seville, 1590) about the natural phenomena and cultural history of Peru and Mexico became his best known and most successful book. The *Historia natural y moral de las Indias* was published several times and translated into seven European languages. It includes descriptions of many American plants, animals, and animal products, such as the miraculous bezoar stone. Among the nearly 150 plants that are mentioned, Acosta describes many native American plants that could be used for medical treatment or general health care. While information about plants and plant products can be found throughout this work, chapters 28 and 29 of book four are dedicated specifically to the plants that (according to divine providence) were meant as remedies for human beings: maize (*Zea mays* L.), Chile peppers (*Capsicum spec.*), manihot (*Manihot utilissima* Pohl), cacao-tree (*Theobroma cacao* L.), balm of Peru (fluid exuded from *Myroxylon spec.*), tobacco

---

[44] See Sievernich, 'Vision und Mission in der Neuen Welt Amerika bei José de Acosta', 239-313.

(*Nicotiana spec.*), contrayerba (*Dorstenia spec.*), copal (resin of *Hymenea spec.*, *Bursera spec.* and *Myrocarpus frondosus* Fr. All.), sarsaparilla (*Smilax spec.*) and guayacan (*Guaiacum officinale* L., *Guaiacum sanctum* L.), to name only a few.[45]

Acosta's work was famous all over Spanish America and Europe, and it was rapidly incorporated in the body of European scientific literature. It would be surprising, therefore, if Clusius, who was obviously very well informed about the literature concerning foreign plants and plant products, had not consulted the *Historia natural y moral de las Indias*. We may assume that this book was one of the sources which provided Clusius with information for his *Exoticorum libri decem* (1605). José de Acosta is, moreover, mentioned in the list of authors named in the correspondence between Carolus Clusius and various Spanish scientists.[46] The connections between Clusius and the Jesuits were therefore probably two-directional – an issue on which further examination of the Clusius correspondence in the University Library in Leiden may throw more light.

Conclusion

The clear connection between Clusius' work on the one hand, and Jesuit botanical research and their global network of drug transfer, on the other hand, may well have been stronger and more complex than is known as yet. Clusius' own publications and his translations of the works by others appear to have influenced Jesuit botanical research to an extent that needs to be explored and analysed further. Jesuits undoubtedly contributed in a significant way to the reception and worldwide distribution of Clusius' knowledge. Finally, the early scientific network of Clusius and the global network of the Society of Jesus seem to be related and linked in many aspects. This multilateral connection between Clusius and Jesuit scientists, botanists and pharmacists definitely influenced the development of European as well as worldwide *materia medica* and deserves to be the focus of new research which takes into account the Clusius correspondence in Leiden.

---

[45] José de Acosta, *Historia natural y moral de las Indias, en que se tratan las cosas notables del cielo, y elementos, metales, plantas y animales dellas: y los ritos, y ceremonias, leyes, y govierno, y guerras de los Indios*. (Seville, 1590), facsimile edn., introd. B.G. Beddal, transl. J.M. López Piñero and F. Bujosa (Valencia, 1977), 236-239, 346f., 239f., 250f., 263f., 265-267.
[46] J.L. Barona and X. Gómez Font, *La correspondencia de Carolus Clusius con los científicos españoles* (Valencia, 1998) [Clasicos y documentos, 2], 127f.

Bibliography

Primary sources

anon., *Histoire naturelle des Indes: The Drake manuscript in the Pierpont Morgan Library*, introd. Verlyn Klinkenborg (New York, 1996).

Acosta, Christovál [Christophorus a Costa], *Aromatum et medicamentorum in Orientali India nascentium Liber: Plurimum lucis adferens iis quae a Doctore Garcia da Orta in hoc genere scripta sunt. Caroli Clusii Atrebatis opera ex Hispanico sermone Latinus factus in Epitomen contractus, et quibusdam notis illustratus* (Antwerp, 1582).

Acosta, José de, *Historia natural y moral de las Indias, en que se tratan las cosas notables del cielo, y elementos, metales, plantas y animales dellas: y los ritos, y ceremonias, leyes, y govierno, y guerras de los Indios* (Seville, 1590; facsimile edn., with an introduction, appendix and anthology by Barbara G. Beddal, translated into Catalan by José M. López Piñero and Francesc Bujosa, Valencia, 1977).

—, *Tratado de las drogas medicinas y plantas de las Indias orientales* (Burgos, 1578).

Arias Montano, B., *Naturae historia, prima in magni operis pars* (Antwerp, 1601).

Bauhin, Caspar, *Pinax theatri botanici [...] siue index in Theophrasti Dioscoridis Plinii et botanicorum qui a seculo scripserunt opera plantarum* (Basle, 1623).

Belon, Pierre, *De aquatilibus, libri duo cum eiconibus ad vivam ipsorum effigiem, quoad eius fieri potuit, expressis* (Paris, 1553).

—, *Plurimarum singulari & memorabilium rerum in Graecia, Asia, Aegypto, Iudae, Arabia, aliisq. exteris provinciis ab ipso conspectarum observationes: Tribus libris expressae. Carolus Clusius Atrebas à gallicis latinas faciebat* (Antwerp, 1589).

—, *Observations*, vol. I, ed. A. Merle (Paris, 2001).

Bontius, Jacobus, *Historiae naturalis et medicae Indiae orientalis libri sex. Commentarii, quos auctor, morte in Indiis praeventus, indigestos reliquit a Gulielmo Pisone in ordinem redacti et illustrati, atque annotationibus et additionibus rerum et iconum necessariis adaucti* (Amsterdam, 1658).

Brahe, Tycho, *Opera omnia*, ed. I.L.E. Dreyer, vol. IX (Copenhagen, 1913-29).

Bright, Timothy, *A treatise, wherein is declared the sufficiencie of English medicines for the cure of all diseases cured with medicine* (London, 1580).

Castellan, Honoré, *Oratio Lutetiae habita, qua futuro medico necessaria explicantur* (Paris, 1555).

Cieza de León, P., *Primera parte de la chrónica del Perú* (Sevilla, 1553).

Clusius, Carolus, *Antidotarium, sive exacta componendorum miscendorumque medicamentorum ratione libri tres* (Antwerp, 1561).

—, *Rariorum aliquot stirpium per Hispanias observatarum historia, libris duobus expressa* (Antwerp, 1576).

—, *Aliquot notae in Garciae Aromatum historiam, ejusdem descriptiones nonnullarum stirpium et aliarum exoticarum rerum [...]* (Antwerp, 1582).

—, *Rariorum aliquot stirpium per Pannoniam, Austriam et vicinas quasdam provincias observatarum historia quatuor libri expressa* (Anwerp, 1583).

—, *Rariorum plantarum historia* (Antwerp, 1601).

—, *Exoticorum libri decem quibus animalium, plantarum, aromatum, aliorumque peregrinorum fructuum historiae describuntur. Item Petri Bellonii observationes, eodem Carolo Clusio interprete* (Leiden, 1605).

—, *Curae posteriores, seu plurimarum non ante cognitarum, aut descriptarum stirpium, peregrinorumque aliquot animalium novae descriptiones: quibus et omnia ipsius opera, aliaque ab eo versa augentur, aut illustrantur* (Leiden, 1611).

—, *Fungorum in Pannoniis observatorum brevis historia et Codex Clusii. Mit Beiträgen von einer internationalen Autorengemeinschaft*, facsmile edition, ed. S. A. Aumüller and J. Jeanplong (Budapest/Graz, 1983).

Columella, Lucius Junius Moderatus, *Res rustica. On agriculture*, ed. and transl. Harrison Boyd Ash, 3 vols. (London/Cambridge [Mass.], 1960-68).

Cordus, Valerius, *Dispensatorium sive antidotarium* (Nuremberg, 1592).

Covarrubias, S. de, *Tesoro de la lengua castellana* (Madrid, 1611).

Crato von Crafftheim, Johannes, *Consiliorum & epistolarum medicinalium liber* (Frankfurt a. M., 1594).

Dodoens, Rembert, *De frugum historia* (Antwerp, 1552).

—, *Histoire des plantes, en laquelle est contenue la description entiere des herbes, c'est à dire, leurs espece forme, noms, temperament, vertus & operations: non seulement de celles qui croissent en ce pays, mais aussi des autres estrangeres qui viennent en usage de medecine [...] nouvellement traduite de bas aleman en françois par Charles de l'Escluse* (Antwerp, 1557).

—, *Stirpium historiae pemptades sex sive libri XXX* (Antwerp, 1554).

—, *Frumentorum, leguminum, palustrium et aquatilium herbarum [...] historia* (Antwerp, 1566).

—, *Stirpium pemptades sex* (Antwerp, 1583).

Farfán, Agustín, *Tractado brebe de medicina, y de todas las enfermedades, que á cada passo se ofrecen* (México, 1592; facsimile edn. Madrid 1944) [Colección de incunables americanos: siglo XVI, 10].

Fernández de Oviedo y Valdez, G., *Sumario de la natural historia de las Indias* (Toledo, 1526).

—, *Primera parte de la historia natural y general de las Indias, yslas y tierra firme del mar oceano* (Sevilla, 1535).

Ferrari, Giovan Battista, *Flora overo cultura di fiori.* Facsimile edition, introd. Lucia Tongiorgi Tomasi, texts Alberta Campitelli and Margherita Zalum (Florence, 2001).

Fragoso, J., *Discursos de las cosas Aromática, árboles y frutales, y de otras muchas medicinas simples que se traen de la India Oriental, y sirven al uso de la medicina* (Madrid, 1572).

Fuchs, Leonhardt., *Primi de stirpium historia commentariorum tomi vivae imagines, in exiguam angustioremque formam contractae* (Basle, 1545).

Garcia ab Orta, D., *Coloquios dos simples e drogas he cousas medicinais da India* (Goa, 1563).

—, *Aromatum et simplicium aliquot medicamentorum apud Indos nascentium historia* (Antwerp, 1567; facsimile edn. with an introduction by M. de Jong and D.A. Wittop Koning, Nieuwkoop, 1963) [Dutch classics on history of science, 6].

—, *Aromatum et simplicium aliquot medicamentorum apud Indos nascentium historia [...] nunc vero Latino sermone in epitomem contracta, et iconibus ad vivum expressis, locupletioribusque annotatiunculis illustrata a Carolo Clusio* (Antwerp, 1574).

Gessner, Conrad, *De raris et admirandis herbis, quae sive quod noctu luceant, sive alias ob causas, Lunariae nominantur* (Zurich, 1555).

Hernandez, Francisco, *Rerum medicarum Novae Hispaniae thesaurus* (Rome, 1648; rpt. Rome, 1992).

Jordán de Asso, I. *Clariorum hispaniensium atque externorum epistolae* (Zaragoza, 1793).

Joubert, Laurent, *Erreurs populaires et propos vulgaires touchant la médecine et le régime de santé. Des erreurs populaires, et propos vulgaires, touchant la médecine et le régime de santé, réfutés ou expliqués* (Bordeaux, 1579).

—, 'Gulielmi Rondeletii vita, mors et epitaphiae', *Operum latinor[um] tomus primus* (Lyon, 1579)

Léry, J. de, *Histoire d'un voyage fait en la terre du Bresil, autrement dite Amerique* (La Rochelle, 1578).

Lobelius, M., *Stirpium adversaria nova* (London, 1570).

—, *Plantarum seu stirpium historia* (Antwerp, 1576).

López de Gómara, F., *Primera y segunda parte de la historia general de las Indias, con todo el descubrimiento y cosas notables que han acaecido desde que se ganaron asta el año de 1551. Con la conquista de México y de la nueva España* (Medina del Campo, 1553).

Matthioli, P.A. (ed.), *Commentarii in libros sex Pedacii Dioscoridis Anazarbei de materia medica* (Venice, 1554).

Monardes, N., *Dos libros. El uno trata de todas las cosas que traen de nuestras Indias Occidentales que sirven al uso de medicina [...] El otro libro, trata de las dos medicinas maravillosas que son contra todo veneno, la piedra bezaar y la yerva escuerçonera* (Sevilla, 1565).

—, *Segunda Parte del Libro de las cosas que se traen de nuestras Indias occidentales que sirven al uso de la medicina [...] Va añadido un libro de la nieve, do verán los que beven frío con ella, cosas dignas de saber y de grande admiración acerca del uso del enfriar con ella* (Sevilla, 1571).

—, *Primera y Segunda y tercera Partes de la Historia Medicinal... Tratado de la piedra bezaar y de la yerva escuerçonera. Diálogo de las grandezas del hierro y de sus virtudes medicinales. Tratado de la nieve y del bever frío [...] Van en esta impressión la tercera parte y el Diálogo del hierro nuevamente hechos [...]* (Sevilla, 1574).

—, *De simplicibus medicamentis ex Occidentali India delatis, quorum in medicina usus est [...] interprete Carolo Clusio* (Antwerp, 1574).

—, *Due altri libri parimente di quelle cose che si portano dall'Indie Occidentali* (Venice, 1575).

—, *Ioyfull newes out of the newe founde worlde, wherein in declared the rare singular vertues [...] with their applications, aswell of phisicke as chirurgerie* (London, 1577).

—, *Simplicium medicamentorum ex Novo Orbe delatorum, quorum in medicina usus est, historia. Hispanico sermone descripta a D. Nicolao Monardis [...] Latio deinde donata et annotationibus iconibusque affabre depictis illustrata a Carolo Clusio Atrebate* (Antwerp, 1579).

—, *Primera y segunda y tercera partes de la Historia medicinal de las cosas que se traen de nuestras Indias Occidentales, que sirven en medicina* (Sevilla, 1580).

—, *Simplicium medicamentorum ex Novo Orbe delatorum [...] historiae liber tertius*, transl. Carolus Clusius (Antwerp, 1582).

—, *Simplicium medicamentorum ex Novo Orbe India nascentium liber* (Antwerp, 1593).

—, *Histoire des simples medicaments apportés des Terres Neuves, desquels on se sert en la medecine* (Lyon, 1602).

Montaigne, M. de, *Oeuvres complètes*, ed. A. Thibaudet and M. Rat, introd. and notes M. Rat (Paris, 1962).

Noort, O. van, *Extract oft kort verhaal wt het groote journael* (Rotterdam, 1601).

—, *Beschryvinghe vande voyagie, om den geheelen werelt cloot [...]* (Amsterdam, 1602).

Paauw, P., *Hortus publicus Academiae Lugduno-Batavae, eius ichnographia, descriptio, usus* (Leiden, 1601).

Peiresc, N.F. de, *Lettres à Cassiano dal Pozzo (1626-1637)*, ed. and comm. J.-F. Lhote and D. Joyal (Clermont-Ferrand, 1989).

Pena, P., *Stirpium nova adversaria* (London, 1605).

Pfefferkorn, Ignaz, *Beschreibung der Landschaft Sonora samt andern merkwürdigen Nachrichten von den inneren Theilen Neu-Spaniens und Reise aus America bis in Deutschland, nebst einer Landcharte von Sonora*, 2 vols. (Cologne, 1794-95).

Pharmacopoeia seu medicamentarium pro Rep. Augustana (Augsburg, 1580).

Pharmacopoeia Amstelredamensis (1636), facsimile edn. by D.A. Wittop Koning (Nieuwkoop, 1961).

Piso, Willem, *Historia naturalis Brasiliae in qua non tantum plantae et animalia, sed et indigenarum morbi, ingenia et mores describuntur et iconibus supra quingentas illustrantur* (Leiden/Amsterdam, 1648).

Plinius Secundus, Gaius, *Naturalis historiae libri XXXVII*, 5 vols. (Leipzig, 1892-1906; rpt. Stuttgart, 1967-70).

—, *Naturkunde. Lateinisch-Deutsch*, transl. and ed. Roderich König and Gerhard Winkler, 31 vols. (Zurich/Munich/Düsseldorf, 1973-96).

—, *Histoire universelle [...]*, vol. X (Basle, 1742).

Potgieter, Barent Jansz, *Wijdtloopigh verhael van tgene de vijf schepen (die int jaer 1598 tot Rotterdam toegherust werden, om door de Straet Magellana haren handel te drijven) wedervaren is [...]* (Amsterdam, 1600).

Rabelais, Françpois, *Le tiers livre*, ed. M.A. Screech (Geneva, 1964) [Textes littéraires français].

Ray, John, *Historia plantarum species hactenus editas aliasque insuper multas noviter inventas et descriptas complectens*, 3 vols. (London, 1686-1704).

Recchius, A., *Rerum medicarum Novae Hispaniae thesaurus* (Rome, 1628).

Rondelet, Guillaume, *Libri de piscibus marinis, in quibus verae piscium effigies expressae sunt* (Lyon, 1554).

—, *Universae aquatilium historiae pars altera, cum veris ipsorum imaginibus* (Lyon, 1555).

Ruel, Jean, *De natura stirpium libri tres* (Paris, 1536).

Salviani, Ippolito, *Aquatilium animalium historiae, liber primus, cum eorundem formis, aere excusis* (Rome, 1554).

Sepp, Anton, and Böhm, *RR. PP. Antonii Sepp und Antonii Böhm, der Societät Jesu Priestern Teutscher Nation [...] Reißbeschreibung* (Nuremberg, 1698).

Smetius, Martinus, *Inscriptionum antiquarum [...] liber* (Leiden, 1588).

Steinhöfer, Johann [Juan de Esteyneffer], *Florilegio medicinal de todas las enfermedades* (México, 1712; rpt. 2 vols., México, 1978).

Sterbeeck, F., *Theatrum fungorum oft het toonel der campernoelien* (Antwerp, 1675).

Sterne, Lawrence, *The life and opinions of Tristram Shandy, gentleman*, vol. V (London, 1762).

Sweertius, E., *Florilegii, in qua agitur de praecipuis plantis et floribus fibrosas radices habentibus; nec non arboribus speciosis et odoriferis* (Frankfurt a. M, 1612).

Theophrastus of Eresos, *Opera, quae supersunt, omnia*, ed. Fridericus Wimmer (Paris, 1866; rpt. Frankfurt a. M, 1964).

Thevet, A., *Les singularitéz de la France antarctique, autrement nommée Amérique & de plusieurs terres & isles decouvertes de nostre temps* (Paris, 1557).

Thou, Jacques Auguste de, *Historiarum sui temporis libri XVIII* (Paris, 1604).

Tovar, S. *Examen y censura [...] del modo de averiguar las alturas de las tierras, por la altura de la Estrella del Norte, tomada con la Ballestilla* (Sevilla, 1595).

Zárate, A. de, *Historia del descubrimiento y conquista del Perú, con las cosas naturales que señaladamente allí se hallan, y los sucesos que ha avido* (Antwerp, 1555).

Secondary literature

Altona, 'Aus den Akten des Reichskammergerichts', *Zeitschrift für die gesamte Strafrechtswissenschaft*, 12 (1892), 898-913.

Amelang, J., 'Mourning laments becomes eclectic: Ritual, lament and the problem of continuity', *Past and present*, no. 187 (2005), 3-31.

Anagnostou, Sabine, *Jesuiten in Spanisch-Amerika als Übermittler von heilkundlichem Wissen* (Stuttgart, 2000) [Quellen und Studien zur Geschichte der Pharmazie, 78].

—, 'Mission und Heilkunde. Das Heilmittelversorgungssystem der Jesuiten in den Missionen Spanisch-Amerikas', *Neue Zeitschrift für Missionswissenschaft. Nouvelle revue de science missionaire*, 4 (2001), 241-259.

—, 'Ethnomedizinische Aspekte jesuitischer Missionstätigkeit in Spanisch-Amerika', *Zeitschrift für Phytotherapie*, 22 (2001), 229-235.

—, 'Jesuitenapotheken vom 16. bis 18. Jahrhundert und ihr Publikum', in Christoph Friedrich and Wolf-Dieter Müller-Jahncke (eds.), *Apotheke und Publikum. Die Vorträge der Pharmaziehistorischen Biennale in Karlsruhe vom 26. bis 28. April 2002* (Stuttgart, 2003), 53-70.

—, 'Vom Römischen und Brasilianischen Theriak', in Christoph Friedrich and Sabine Bernschneider-Reif (eds.), *Rosarium litterarum. Beiträge zur Pharmazie- und Wissenschaftsgeschichte. Festschrift für Peter Dilg zum 65. Geburtstag* (Eschborn, 2003), 17-32.

—, 'Jesuits in Spanish America and their contribution to the exploration of the American materia medica', *Pharmacy in history*, 47 (2005), 3-17.

—, 'Pharmazie auf internationaler Ebene – die Apotheke des Collegio Romano vom 16. bis 18. Jahrhundert', *Geschichte der Pharmazie*, 57 (2005), 57-63.

—, and Müller, Michael, 'Joseph Zeitler – Auf den Spuren eines bayerischen Apothekers in Chile', *Geschichte der Pharmazie*, 56 (2004), 16-23.

Arber, A., *Herbals, their origin and evolution; a chapter in the history of botany, 1470-1670* (Cambridge, 1912; 1938²).

—, 'The colouring of sixteenth-century herbals', in eadem, *Herbals: Their origin and evolution: A chapter in the history of botany 1470-1670*, ed. W.T. Stearn (Cambridge, 1990), 315-318.

Antonino, B. (ed.), *L'Erbario di Ulisse Aldrovandi* (Milan, 2003).

Antonioli, R., *Rabelais et la médecine*, Etudes rabelaisiennes, vol. XII: *Travaux d'humanisme et Renaissance*, no.143 (Geneva, 1976).

Ashworth Jr., W.B., 'The persistent beast: Recurring images in early zoological illustrations', in A. Ellenius (ed.), *The natural sciences and the arts* (Stockholm, 1985) [Acta universitatis Upsaliensis, Figura nova Series, 22], 46-66.

—, 'Natural history and the emblematic worldview', in D.C. Lindberg and R.S. Westman (eds.), *Reappraisals of the Scientific Revolution* (Cambridge, 1990), 303-332.

—, 'Remarkable humans and singular beasts', in J. Kenseth (ed.), *The age of the marvelous* (Hanover [N.H.], 1991) [Exhibition catalogue, Hood Museum of Art, Dartmouth College], 113-144.

—, 'Natural history and the emblematic world view', in D.C. Lindberg and R.S. Westman (eds.), *Reappraisals of the scientific revolution* (Cambridge, 1990), 303-332

—, 'Emblematic natural history of the Renaissance', in N. Jardine, J.A. Secord and E.C. Spary (eds.), *Cultures of natural history* (Cambridge, 1996), 17-37.

Atran, S., *Cognitive foundations of natural history: Towards an anthropology of science* (Paris/Cambridge, 1989).

Auhagen, Ulrike, and Schäfer, Eckart (eds.), *Lotichius und die römische Elegie* (Tübingen, 2001) [NeoLatina, 2].

Aumüller, S.A., 'Carolus Clusius, der Begründer der botanischen Forschung im Raume des heutigen Burgenlandes', *Burgenländische Heimatblätter*, 29/3 (1967), 98-107.

—, 'Einige Ergebnisse der neuen Clusius-Forschung', *Burgenländische Heimatblätter*, 34/2 (1972), 66-72.

—, 'Bibliographie und Ikonographie', in *Clusius-Festschrift. Burgenländische Forschungen*, Sonderheft 5 (1973), 9-92.

— and Jeanplong, J. (eds.), *Carolus Clusius' Fungorum in Pannoniis observatorum brevis historia et Codex Clusii* (Budapest/Graz, 1983).

—, 'Wissenschafliche Tätigkeit in Wien', in *Carolus Clusius' Fungorum in Pannoniis observatorum brevis historia et Codex Clusii. Mit Beiträgen von einer internationalen Autorengemeinschaft*, ed. S.A. Aumüller and J. Jeanplong. Facsimile edition (Budapest/Graz, 1983), 28-33.

Baas, P., 'De VOC in Flora's lusthoven', in L. Blussé and I. Ooms (eds.), *Kennis en Compagnie: De Verenigde Oost-Indische Compagnie en de moderne wetenschap* (Amsterdam, 2002), 124-137.

Baldriga, I., "La fatiga di pigliar i disegni delle piante": Federico Cesi, la pittura filosofica e la riproduzione del mondo vegetale', in *Federico Cesi un principe naturalista, Atti del Convegno Internazionale di Studi, Acquasparta, September 29-30, 2003, Accademia Nazionale dei Licei* (forthcoming), 505-525.

—, 'Il museo anatomico di Giovanni Faber Linceo', in *Scienza e miracoli nell'arte del Seicento. Alle origini della medicina moderna*, ed. S. Rossi (Milan, 1998) [Exhibition catalogue, Palazzo Venezia, Rome], 82-87.

—, 'Le virtù della scienza e la scienza dei virtuosi: i primi lincei e la diffusione dei naturalismo in pittura', in *Caravaggio nel IV centenario della Cappella Contarelli, Atti del Convegno Internazionale di Studi, Rome, May 24-26 2001, Accademia dei Lincei-Università degli Studi di Roma 'La Sapienza'*, ed. M. Calvesi and C. Volpi (Rome, 2002), 197-208.

—, *L'occhio della Lince. I primi lincei tra scienza, arte e collezionismo (1603-1630)* (Rome, 2002).

Bàlint-Nagy, I., 'Purkircher György (1530-1578) pozsonyi orvos èlete', *Orvosi Hetilap*, 22-23 (1930), 553-584.

Balis, A., 'Naar de natuur en naar model', in idem, M.-C. Maelis and R.H. Marijnissen (eds.), *De albums van Anselmus de Boodt (1550-1632). Geschilderde natuurobservatie aan het Hof van Rudolf II te Praag* (Tielt, 1989), 56-71.

Balis, J., *Hortus Belgicus* (Brussels, 1962) [Exhibition catalogue, Bibliothèque Albert I, Brussels].

—, *Van diverse pluimage. Tien eeuwen vogelboeken* (Antwerp, 1968) [Exhibition catalogue, Antwerp, The Hague and Brussels].

Balvoll, G., and G. Weisaeth, *Horticultura. Norsk hagebok frå 1694 av Christian Gartner* (Landbruksforlaget, 1994).

Bang, A.C., *Den norske kirkes historie i Reformations-aarhundredet* (Oslo, 1895).

Barb, A., 'Die römischen Inschriften des südlichen Burgenländes', *Burgenländische Heimatblätter*, 1 (1932), 75-80.

Barlay, Sz.Ö., 'Boldizsár Batthyány und sein Humanisten-Kreis. Die ersten Jahrzehnten der Güssinger Bibliothek', *Magyar Könyvszemle*, 95 (1979), 231-251.

Barona, J.L., 'Saber científico y lenguaje. Una reflexión histórica desde la botánica y la medicina', in *Historia de los lenguajes iberorrománicos de especialidad (S. XVII-XIX)* (Barcelona, 1997), 27-48.

— and X. Gómez Font, *La correspondencia de Carolus Clusius con los científicos españoles* (Valencia, 1998) [Clásicos y documentos, 2].

— and —, 'Carolus Clusius, los naturalistas hispanos y la naturaleza Americana', in *Acta Conventus Neo-Latini Abulensis. Proceedings of the Tenth International Congress of Neo-Latin Studies* (Tempe [Arizona], 2000), 105-111.

Batthyány-Strattmann, L., 'Güssing und die Batthyány zur Zeit des Clusius', in *Festschrift anläßlich der 400jährigen Wiederkehr der wissenschaftlichen Tätigkeit von Carolus Clusius (Charles de l'Escluse) im pannonischen Raum* (Eisenstadt, 1973) [Burgenländische Forschungen, Sonderheft 5], 104-121.

Battistini, M., 'Un botaniste flamand à la cour de Toscane (Joseph Goedenhuyse)', *Revue de l'Université de Bruxelles*, 3 (1926), 1-8.

—, 'Giuseppe Casabona botanico fiammingo a servizio dei Medici e le sue relazioni con Carlo Clusio', *Archivio botanico*, 3 (1927), 191-202.

Båtvik, J.I.I., 'Gamle bevarte herbarier, og Østfolds eldste herbariebelegg', *Natur i Østfold*, 19 (2000), 17-27.

Berendts, A., 'Carolus Clusius (1526-1609) and Bernardus Paludanus (1550-1633). Their contacts and correspondence', *Lias* 5 (1978), 49-64.

Bergmans, P., *Quatorze lettres inédites du compositeur Philippe de Monte* (Brussels, 1921).

Berkel, K. van, 'Een onwillige mecenas? De rol van de VOC bij het natuurwetenschappelijk onderzoek in de 17e eeuw', in J. Bethlehem and A.C. Meijer (eds.), *VOC en Cultuur: Wetenschappelijk onderzoek en culturele relaties tussen Europa en Azië ten tijde van de Verenigde Oostindische Compagnie* (Amsterdam, 1993), 39-58.

Berkvens-Stevelinck, C.M.G., H. Bots and J. Häseler (eds.), *Les grands intermédiaires culturels de la République des Lettres. Etudes de réseaux de correspondence du XVIe au XVIII siècles* (Paris, in press).

Bethencourt, F., and F. Egmond (eds.), *Correspondence and cultural exchange in early modern Europe* (Cambridge, 2007).

Biagioli, M., 'The anthropology of incommensurability', *Studies in the history and philosophy of science*, 21 (1990), 183-209.

—, 'Scientific revolution, social bricolage, and etiquette', in R. Porter and M. Teich (eds.), *The scientific revolution in national context* (Cambridge, 1992), 11-54.

—, 'Le Prince et les savants. La civilité scientifique au 17me-siècle', *Annales HSS* (1995), no. 6, 1417-1453.

—, 'Etiquette, interdependence, and sociability in seventeenth-century science', *Critical inquiry*, 22 (1996), 193-238.

Bibliotheca Belgica: Bibliographie générale des Pays-Bas, ed. F. van der Haeghen, re-ed. M.-T. Lenger, 6 vols. (Brussels, 1979).

Biographisch-bibliographisches Kirchenlexikon, eds. Friedrich-Wilhelm Bautz and Traugott Bautz (Hamm/Herzberg/Nordhausen, 1975-2005).

Bloom, H., *Agon. Towards a theory of revisionism* (New York, 1982).

Bohus, G., 'Interprétation des bolets de Clusius', *Acta mycologica Hungarica*, 2 (1945), 20-27, 69-76.

—, 'Mikològiai èrdekessègek a Clusius Codexbol', *Vasi szemle*, 27/4 (1973), 582-585.

—, 'A Clusius-Codex gombafajainak revìzìòja', *Mikològiai kozlemenyek*, 3 (1975), 121-136.

—, 'Revision der Pilzarten des Clusius-Codex', in S.A. Aumüller and J. Jeanplong (eds.), *Carolus Clusius Fungorum in Pannoniis observatorum Brevis historia et Codex Clusii mit Beiträgen von einer internationalen Autorengemeinschaft* (Budapest/Graz, 1983), 60-69.

Bots, H. and F. Waquet, *La République des Lettres* (Paris, 1997).

Bradford, C.A., *Hugh Morgan: Queen Elizabeth's apothecary* (London, 1939).

Bresson, A., *Nicolas-Claude Fabri de Peiresc* (Florence, 1992).

Brocas, Jean, *Contribution à l'étude de la vie et de l'oeuvre d'André Vésale* (Paris, 1958).

Browne, J., *Charles Darwin. Voyaging. Volume I of a biography* (London, 1995).

—, *Charles Darwin. The Power of Place. Volume II of a biography* (London, 2002).

Brunon, H., 'Il bell' ordine della natura: spazio e collezioni nel giardino di villa Medici', in M. Hochmann (ed.), *Villa Medici. Il sogno di un cardinale* (Rome, 1999), 67-73.

Bujok, E., *Neue Welten in europäischen Sammlungen. Africana und Americana in Kunstkammern bis 1670* (Berlin, 2004).

Caneva, G., *Il mondo di Cerere nella Loggia di Psiche* (Rome, 1992).

Cañizares-Esguerra, J., 'New world, new stars: Patriotic astrology and the invention of Indian and Creole bodies in colonial Spanish America, 1600-1650', *American historical review*, 104 (1999), 33-68.

Capanna, E., 'Zoologia Kircheriana', in E. Lo Sardo (ed.), *Athanasius Kircher. Il Museo del Mondo* (Rome, 2001) [Exhibition catalogue, Palazzo di Venezia], 167-177.

Carrillo, J., 'Taming the visible: Word and image in Oviedo's *Historia general y natural de las Indias*', *Viator*, 31 (2000), 399-431.

—, 'Naming difference: The politics of naming in Fernandez de Oviedo's *Historia General y Natural de las Indias*', *Science in context*, 16 (2003), 489-504.

Castillo Solórzano, Alonso de, *Jornadas alegres a D. Francisco de Erasso, Conde de Humanes, señor de las Villas de Mohernando y el Cañal* (Madrid, 1907).

Catelli, N., and M. Gargatagli, *El tabaco que fumaba Plinio. Escenas de la traducción en España y América: relatos, leyes y reflexiones sobre los otros* (Barcelona, 1998).

Cesi, F., 'Del naturale desiderio di sapere et Institutione de' Lincei per adempimento di esso', in *Federico Cesi e la Fondazione dell'Accademia dei Lincei. Mostra bibliografica e documentaria, catalogo della mostra (Venezia, Biblioteca Nazionale Marciana, 27 agosto-15 ottobre 1988)*, ed. S. Ricci (Naples, 1988).

Christianson, J.R., *On Tycho's island. Tycho Brahe, science, and culture in the sixteenth century* (Cambridge, 2003).

Cobo, B., *Historia del Nuevo Mundo. Con notas y otras ilustraciones por Marcos Jiménez de la Espada*, 4 vols. (Sevilla:, 1890-95).

Collins, M., *Medieval herbals: The illustrative tradition*, (London, 2000).

Collinson, John, *The life of Thuanus, with some account of his writings, and a translation of the preface to his history* (London, 1807).

Les Correspondances. Leur importance pour l'historien des sciences et de la philosophie. Problèmes de leur édition (Paris, 1976).

Dannenfeldt, Karl H., 'The University of Wittenberg during the period of transition from medieval herbalism to botany', *The social history of the Reformation: Essays in honor of Harold J. Grimm*, ed. Lawrence P. Buck and Jonathan W. Zophy (Columbus [Ohio], 1972), 223-248.

Dávila-Pérez, A., *La correspondencia de Arias Montano conservada en el Museo Plantin-Moretus, de Amberes* (Madrid/Alcañiz, 2002).

Davis, N.Z., *The gift in sixteenth-century France* (Oxford, 2000).

De Backer, Ch., and L.J. Vandewiele, 'Le botaniste flamand Carolus Clusius (1526-1609) et ses rélations avec l'Espagne', in *Medicamento, Historia y sociedad. Estudios en memoria del Profesor D. Rafael Folch Andreu* (Madrid, 1982), 183-186.

De Landtsheer, J., 'Justus Lipsius and Carolus Clusius: A flourishing friendship', in M. Laureys (ed.), *The world of Justus Lipsius: a contribution towards his intellectual biography. Proceedings of a colloquium held under the auspices of the Belgian Historical Institute in Rome (Rome, 22-24 May 1997)* [= *Bulletin van het Historisch Instituut te Rome*, 68 (1998)], 273-295.

Delaunay, P., *La zoologie au seizième siècle* (Paris, 1962).

Delavault, R., *André Vésale: Biographie* (Bruxelles, 1999).

Depauw, C., 'Peeter vander Borcht (1535/40-1608): The artist as inventor or creator of botanical illustrations?', in F. de Nave and D. Imhof (eds.), *Botany in the Low Countries* (Antwerp, 1993), 47-56.

Dietze, A., '1600-talls kjøkkenhagetradisjon på Baroniet Rosendal, Kvinnherad, Norge', in D. Moe, P.H. Salvesen and D.O. Øvstedal (eds.), *Historiske hager. En nordisk hagehistorisk artikkelsamling ved 100-årsfeiringen av Muséhagen i Bergen maj 1999* (Bergen, 2000), 40-45.

Dilg, P., 'Apotheker als Sammler', in A. Grote (ed.), *Macrocosmos in Microcosmo. Die Welt in der Stube. Zur Geschichte des Sammelns 1450 bis 1800* (Opladen, 1994) [Berliner Schriften zur Museumskunde, 10], 453-474.

Donattini, M., 'Orizzonti geografici dell'editoria italiana (1493-1560)', in A. Prosperi and W. Reinhard (eds.), *Il Nuovo Mondo nella coscienza italiana e tedesca del Cinquecento* (Bologna, 1992), 79-154.

Drees, J., 'Die 'Gottorfische Kunst-Kammer'. Anmerkungen zu ihrer Geschichte nach historischen Textzeugnissen', in idem and H. Spielmann (eds.), *Gottorf im Glanz des Barock. Kunst und Kultur am Schleswiger Hof 1544-1713*, 4 vols. (Schleswig, 1979), vol. II, 11-28.

Dübber, I., *Zur Geschichte des Medizinal-und Apothekenwesens in Hessen-Kassel und Hessen-Marburg von den Anfängen bis zum Dreissigjährigen Krieg* (Marburg, 1969).

Dulieu, Louis, *La médecine à Montpellier*, vol. II, *La Renaissance* (Avignon, 1979).

Durling, R.J., 'Konrad Gessner's Briefwechsel', in R. Schmitz and F. Krafft, *Humanismus und Naturwissenschaften* (Boppard, 1980) [Beiträge zur Humanismusforschung, 6], 101-112.

Eckblad, F.-E., 'Molter som skjørbuksmiddel i skriftlige kilder', *Blyttia*, 4 (1988), 177-181.

—, 'Henrik Høyer og de første tulipaner i Norge', *Blyttia*, 3 (1991), 145-150.

Egmond, F., and P. Mason, "These are people who eat raw fish': Contours of the ethnographic imagination in the sixteenth century', *Viator*, 31 (2000), 311-339.

—, 'Execution, dissection, pain and infamy – A morphological investigation', in F. Egmond & R. Zwijnenberg (eds.), *Bodily extremities. Preoccupations with the human body in early modern European culture* (Aldershot, 2003), 92-128.

—, 'Clusius, Cluyt, Saint Omer. The origins of the sixteenth-century botanical and zoological watercolours in the Libri Picturati A. 16-30', *Nuncius. Journal of the history of science*, 20/1 (2005), 11-67.

—, 'Correspondence and natural history in the sixteenth century: Cultures of exchange in the circle of Carolus Clusius', in Francisco Bethencourt and Florike Egmond (eds.), *Correspondence and cultural exchange in early modern Europe* (Cambridge, 2006), 104-142.

—, 'A European community of scholars: Exchange and friendship among early modern natural historians', in D. Curto and A. Molho (eds.), *Rethinking the history of Europe. Images, smbols, discourses* (Florence, forthcoming).

Eichberger, D., 'Naturalia et artefacta: Dürer's nature drawings and early collecting', in idem and C. Zika (eds.), *Dürer and his culture* (Cambridge, 1998), 13-37.

—, *Leben mit Kunst, Wirken durch Kunst. Sammelwesen und Hofkunst unter Margarete von Österreich, Regentin der Niederlande* (Turnhout, 2002).

Ellenius, A. (ed.), *The natural sciences and the arts. Aspects of interaction from the Renaissance to the twentieth century* (Uppsala, 1985).

Elliott, J.H., *The old world and the new, 1492-1650* (Cambridge, 1970).

Ernyei, J., 'Clusius ès Bàthory Istvàn', *Botanikai Közlemènyek*, 32 (1935), 1-7.

Fægri, K., 'Klostervesenets bidrag til Norges flora og vegetasjon', *Foreningen til norske fortidsminnesmerkers bevaring Årbok*, 1987, 225-238.

Fazekas, Á., 'A magyar nyelvű herbárium-irodalomról', *Orvostörténeti Közlemények*, 97-99 (1982), 43-64.

Feest, C.F., 'European collecting of American Indian artefacts and art', *Journal of the history of collections*, 5/1 (1993), 1-11.

Fernández de Oviedo, G., *Historia general y natural de las Indias, islas y tierra firme del mar Océano*, 4 vols. (Madrid, 1851-55).

Fetzner, Angela, *Carolus Clusius und seine Libri Exoticorum* (dissertation, Marburg, 2004).

Feyer, T.M.G. de, 'Biobibliographie de Vésale', *Janus*, 19 (1914).

Findlen, P., *Possessing nature: Museums, collecting, and scientific culture in early modern Italy* (Berkeley, 1994).

—, 'The formation of a scientific community: Natural history in sixteenth-century Italy', in A. Grafton and N. Siraisi (eds.) *Natural particulars: Natural philosophy and the disciplines in early modern Europe* (Cambridge [Mass.], 1999), 369-400.

—, 'Inventing nature: Commerce, art and science in the early modern cabinet of curiosities', in P.H. Smith and eadem (eds.), *Merchants and marvels: commerce, science and art in early modern Europe* (New York/London, 2002), 297-323.

—, 'Un incontro con Kircher a Roma', in E. Lo Sardo (ed.), *Athanasius Kircher. Il Museo del Mondo* (Rome, 2001), 39-47.

Flatt (Alföldi), K., 'Clusius hivatala a bècsi udvarnàl', *Pòtfüzetek Termèszet Tudomànyi Közlemènyek*, 27 (1895), 29-34.

Foucault, M., *Les mots et les choses* (Paris, 1966).

Franchini, D.A., et al., *La scienza a corte. Collezionismo eclettico natura e immagine a Mantova fra Rinascimento e Manierismo* (Rome, 1979).

Frati, L., *Catalogo dei manoscritti di Ulisse Aldrovandi* (Bologna, 1907).

Freedberg, D., *The eye of the lynx. Galileo, his friends, and the beginnings of modern natural history* (Chicago/London, 2002).

French, R.K., 'Natural philosophy and anatomy', in J. Céard, M.M. Fontaine and J.C. Margolin (eds.), *Le corps à la Renaissance. Actes du XXXe Colloque de Tours 1987* (Paris, 1990), 447-460.

Gabrieli, G., *Contributi alla storia della Accademia dei Lincei*, 2 vols. (Rome, 1989).

—, *Il carteggio Linceo* (Rome, 1996; 1st edn. 1838-42).

Garbari, F., L. Tongiorgi Tomasi and A. Tosi, *Giardino dei semplici / Garden of simples* (Pisa, 2002).

Germain, A.Ch., *La Renaissance à Montpellier* (Montpellier, 1871).

Gicklhorn, Josef, and Renée Gicklhorn, *Georg Joseph Kamel S.J. (1661–1706). Apotheker, Botaniker, Arzt und Naturforscher der Philippineninseln* (Eutin, 1954) [Veröffentlichungen der Internationalen Gesellschaft für Geschichte der Pharmazie, 4].

Gil, J., *Arias Montano en su entorno [bienes y herederos]* (Mérida, 1998).

Gillispie, Charles Coulston (gen. ed.), *Dictionary of scientific biography*, 14 vols. and 4 suppls. (New York, 1970-90).

Ginori Conti, P., *Lettere inedite di Charles de l'Escluse (Carolus Clusius) a Matteo Caccini, floricultore fiorentino. Contributo alla storia della botanica* (Florence, 1939).

Goldgar, A., 'Nature as art: The case of tulips', in P.H. Smith and P. Findlen (eds.), *Merchants and marvels: Commerce, science and art in early modern Europe* (New York/London, 2002), 324-346.

—, *Impolite learning. Conduct and community in the Republic of Letters, 1680-1750* (New Haven/London, 1995).

Gombocz, E., *A Magyar botanika története. A Magyar flóra kutatói* (Budapest, 1936).

Gömöry, G., 'Sir Philip Sidney magyarországi kapcsolatai és hírei Magyarországról', *Kortárs*, (1983), 428-437.

González Carvajal, T., 'Elogio histórico del Dr. Benito Arias Montano', in *Memorias de la Real Academia de la Historia*, vol. VII (Madrid, 1832).

Gouw, J.L. van der, 'Marie de Brimeu. Een Nederlandse prinses uit de eerste helft van de tachtigjarige oorlog', *De Nederlandsche leeuw*, 64 (1947), 5-49.

Grafton, A., *New World, ancient texts: The power of tradition and the shock of discovery* (Cambridge, MA, 1992).

Gravelle, S.S., 'The Latin-vernacular question and humanist theory of language and culture', *Journal of the history of ideas*, 49/3 (1988), 367-386.

Greenblatt, S.J., *Marvelous possessions: The wonder of the New World* (Chicago, 1991).

Greene, E.L., *Landmarks of botanical history*, parts I and II, ed. Frank N. Egerton (Stanford, 1983).

Grotefend, H., *Christian Egenolff, der erste ständige Buchdrucker zu Frankfurt. a. M.* (Frankfurt a. M., 1881).

Guerra, F., *Nicolás Bautista Monardes. Su vida y su obra (ca. 1493-1588)* (México, 1961).

Guicciardini, L., *Descrittione di tutti i Paesi Bassi*, ed. Dina Aristodemo (Amsterdam, 1994).

Guiraud, Louise, *Le procès de Guillaume Pellicier* (Paris, 1907).

—, 'Le premier jardin des plantes français: Creation et restauration du Jardin du Roy à Montpellier', *Archives de la ville de Montpellier*, vol. IV (Montpellier, 1920).

Gulden, G.; Høiland, K., 'Finn-Egil Eckblad 12.8.1923 – 14.7.2000', *Blyttia* 3-4 (2000), 147-152.

Gulyàs, P., *Samboky Jànos konyvtàra. Facsimile edition: A Zsàmboky-konyvtàr katalògusa. Adattàr a XVI-XVIII. szàzadi szellemi mozgalmaink tortènetèhez* (Szeged, 1992²).

Härting, U. (ed.), *Gärten und Höfe der Rubenszeit. Im Spiegel der Malerfamilie Brueghel und der Künstler um Peter Paul Rubens* (Munich, 2000) [Exhibition catalogue Gustav-Lübcke-Museum Hamm and Landesmuseum Mainz].

Harkness, D.E., "'Strange' ideas and 'English' knowledge. Natural science exchange in Elizabethan London', in P.H. Smith and P. Findlen (eds.), *Merchants and marvels: Commerce, science and art in early modern Europe* (New York/London, 2002), 137-160.

Harris, Steven J., 'Jesuit scientific activity in the overseas missions, 1540-1773', *Isis*, 96 (2005), 71-79.

Harrisse, H., *Les De Thou et leur célèbre bibliothèque* (Paris, 1905).

Heller, J.L., and F.G. Meyer, 'Conrad Gessner to Leonhart Fuchs, October 18, 1556', *Huntia*, 5/1 (1983), 61-75.

Hermann, Walter, 'Il commentario pliniano di Guillaume Pellicier (ca. 1490-1567) e la storia del codice Parigino latino 6808', *Studi umanistici Piceni*, 17 (1997), 179-194.

Heyden, D., 'Jardines botánicos prehispánicos', *Arqueología Mexicana* 57 (2000), 18-23.

Hindrichs, T., *Philipp de Monte (1521-1603). Komponist, Kapellmeister, Korrespondent* (Göttingen, 2002).

Hoeniger, F.D., 'How plants and animals were studied in the mid-sixteenth century', in idem and J.W. Shirley (eds.), *Science and the arts in the Renaissance* (Cranbury/London, 1985), 130-148.

Hoeniger, F.D., and J.W. Shirley (eds.), *Science and the arts in the Renaissance* (Cranbury/London, 1985).

Hofmann E., 'Verkieselte Hölzer der Vashegy- (Eisenberg-) Gruppe', *Vasvàrmegyei Mùzeumok Evkönyve*, 3 (1927-1929), 81-87.

Holtsmark, A., *Norsk biografisk leksikon*, vol. VI (1934), 444-445.

Honour, H., *L'Amérique vue par l'Europe* [Paris, 1976] [Exhibition catalogue, Grand Palais, Paris].

Hoogewerff, G.J., 'Henricus Corvinus', *Mededeelingen van het Nederlandsch Instituut te Rome*, 2 (1922), 113-118.

Horvàth E., 'Clusius Lithoxylonja ès lelöhelye a kèsöbbi szakirodalomban', *Vasi Szemle*, 27/4 (1973), 585-595.

Huizenga, E., *Tussen autoriteit en empirie. De Middelnederlandse chirurgieën in de veertiende en vijftiende eeuw en hun maatschappelijke context* (Hilversum, 2003).

Hunger, F.W.T., *Charles de l'Escluse (Carolus Clusius), Nederlandsch kruidkundige 1526-1609*, 2 vols. (The Hague, 1927-43).

Impey, O., and A. Macgregor (eds.), *The origins of museums. The cabinet or curiosities in sixteenth- and seventeenth-century Europe* (Oxford, 1985).

Iannaccone, Isaia, *Johann Schreck Terrentius. Le scienze rinascimentali e lo spirito dell' Accademia dei Lincei nella Cina dei Ming* (Naples, 1998) [Istituto Universitario Orientale Dipartimento di Studi Asiatici, Series minor, 54].

Istvánffi, Gy., *A Clusius-codex mykologiai méltatása adatokkal Clusius életrajzához* (Budapest, 1900).

—, *Études et commentaires sur le Code de l'Escluse augmentés de quelques notices biographiques* (Budapest, 1900).

Iványi, B., 'A körmendi Batthyány-levéltár reformációra vonatkozó oklevelei I: 1526-1625', in *Iványi Béla cikkei és anyaggyűjtése*, ed. L. Szilasi (Szeged, 1990) [Adattár a XVI-XVIII. századi szellemi mozgalmaink történetéhez, 29/1].

Ivins Jr., W.M., *Prints and visual communications* (London, 1953).

Jardine, N., J.A. Secord and E.C. Spary (eds.), *Cultures of natural history* (Cambridge, 1996).

Jeanplong, J., and I. Katona, 'Clusius in Westpannonien. Beziehungen zu Boldizsár Batthyány und István Beythe', in *Carolus Clusius' Fungorum in Pannoniis observatorum brevis historia et Codex Clusii. Mit Beiträgen von einer internationalen Autorengemeinschaft*, ed. S. A. Aumüller and J. Jeanplong, facsimile edn. (Budapest/Graz, 1983), 34-39.

Jones, W., *William Turner: Tudor naturalist, physician, divine* (London, 1988).

Jong, E. de, 'Nature and art. The Leiden hortus as 'Musaeum', in idem and L. Tjon Sie Fat (eds.), *The authentic garden. A symposium on gardens* (Leiden, 1991), 37-60.

Joret, Charles, 'Liste des noms de plantes envoyées par Peiresc à Clusius', *Revue des langues romanes*, 1893-94, 437-442.

Jørgensen, P.M., 'Byen er Bergen – faget er botanikk', *Bergens museums Årbok*, 2003, 37-39.

—, 'Tulipanen og Bergen i 400 år', *Bergens tidende*, 28 May, 1997.

Kamen, H., *Philip of Spain* (New Haven / London 1997).

Kemenes, P., 'Hogyan derìtette fel Jordàn Tamàs az 1577. èvi brünni jàrvàny okàt?', *Orvosi hetilap*, 131/44 (1990), 2438-2440.

Kickx, J., 'Esquisses sur les ouvrages de quelches anciens naturalistes Belges. II. François Van Sterbeeck', *Bulletin de l'Académie Royale des Sciences*, I, sér. 9, 2 (1842), 393-426.

Kinser, Samuel, *The works of Jacques-Auguste De Thou* (The Hague, 1966).

Kinzelbach, R.K., J. Hölzinger (eds), *Marcus zum Lamm (1544-1606). Die Vogelbücher aus dem Thesaurus Picturarum* (Stuttgart, 2000).

Klaniczay, T., 'Le mouvement académique à la renaissance et le cas de la Hongrie', *Hungarian studies*, 2/1 (1986), 13-34.

Klooster, W., *The Dutch in the Americas 1600-1800* (Providence, 1997).

Koltai, A., *Adam Batthyàny und seine Bibliothek* (Eisenstadt, 2002) [Burgenländische Forschungen, Sonderband 24].

Koreny, F., *Albrecht Dürer and the animal and plant studies of the Renaissance* (Boston, 1988) [Exhibition catalogue, Albertina, Vienna].

Kusukawa, S., *The transformation of natural philosophy. The case of Philip Melanchthon* (Cambridge, 1995).

—, 'Leonhart Fuchs on the importance of pictures', *Journal of the history of ideas*, 58/3 (1997), 403-427.

—, 'The uses of pictures in the formation of learned knowledge: The cases of Leonhard Fuchs and Andreas Vesalius', in eadem and I. Maclean (eds.), *Transmitting knowledge: Words, images, and instruments in early modern Europe* (Oxford, 2006), 73-96.

Lack, H.W., 'Lilac and horse-chestnut: Discovery and rediscovery', *Curtis's botanical magazine*, 6/17 (2000), 109-114.

Lange, C.A., *De norske klostres historie i Middelaldern* (Oslo, 1847).

Laureys, M. (ed.), *The world of Justus Lipsius: A contribution towards his intellectual biography* (Brussels, 1998).
Lestringant, F., 'Le déclin d'un savoir. La crise de la cosmographie à la fin de la Renaissance', *Annales E.S.C.*, 46/2 (1991), 239-260.
—, *Écrire le monde à la Renaissance* (Caen, 1993).
—, *L'Expérience huguenote au nouveau monde (XVIe siècle)* (Geneva, 1996).
Levine, D. (ed.), *Americas lost* (Paris, 1992) [Exhibition catalogue, Musée de l'Homme, Paris].
Longeon, C., *Conrad Gesner. Vingt lettres à Jean Bauhin fils (1563-1565)*, transl. Augustin Sabot, ed. and comm. Claude Longeon (Saint Etienne, 1976).
López de Gómara, F., *Historia general de las Indias*, ed. P. Guibelalde and E. M. Aguilera, 2 vols. (Barcelona, 1954).
López-Piñero, J.M., et al., *Bibliographia medica Hispanica, 1475-1950* (Valencia, 1987-89).
—, 'Introducción', in N. Monardes, *La historia medicinal de las cosas que se traen de nuestras Indias Occidentales (1565-1574)* (Madrid, 1989), 9-74.
— and M.L. López Terrada, *La traducción de Juan de Jarava de Leonhardt Fuchs y la terminología botánica castellana del siglo XVI* (Valencia, 1994).
— and —, *La influencia española en la introducción en Europa de las plantas Americanas (1493-1623)* (Valencia, 1998).
Lowood, H., 'The New World and the European catalog of nature', in K. Ordahl Kupperman (ed.), *America in European consciousness, 1493-1750* (Chapel Hill/London, 1995), 295-323.
Lugli, A., *Naturalia et mirabilia. Les cabinets de curiosités en Europe* (Paris, 1998; French version of the Italian edn. Milan, 1983, with an updated bibliography).
Lundquist, K., *Lilium martagon L. Krolliljans introduktion och tidiga historia i Sverige intill år 1795 – i en europeisk liljekontext* (Alnarp, 2005) [Acta Universitatis Agriculturae Sueciae, Doctoral thesis no. 2005:19].
Martin, Luis, *The intellectual conquest of Peru: The Jesuit College of San Pablo 1568-1767* (New York, 1968).
Martin Martin, Carmen, and José Luis Valverde, *La farmacia en la América colonial: el arte de preparar medicamentos* (Granada, 1995).
Mason, P., *Deconstructing America. Representations of the other* (London, 1990).
—, 'Escritura fragmentaria: aproximaciones al otro', in G.H. Gossen, J.J. Klor de Alva, M. Gutiérrez Estévez and M. León-Portilla (eds.), *De palabra y obra en el Nuevo Mundo*, vol. III: *La formación del otro* (Madrid, 1993), 395-430.
—, 'From presentation to representation: Americana in Europe', *Journal of the history of collections*, 6/1 (1994), 1-20.
—, *Infelicities. Representations of the exotic* (Baltimore/London, 1998).
—, *The lives of images* (London, 2001).
—, 'Faithful to the context? The presentation and representation of American objects in European collections', *Anuário antropológico*, 98 (2002), 51-95.
—, 'Il contributo dei Libri Picturati A. 32-38 alla comprensione dell'iconografia del Brasile olandese nei dipinti di Albert Eckhout e di Frans Post', in G. Olmi and G. Papagno (eds.), *La natura e il corpo. Studi alla memoria di Attilio Zanca* (Florence, 2006), 101-120.
—, 'A dragon tree in the Garden of Eden. A case study of the mobility of objects and their images in early modern Europe', *Journal of the history of collections* (in press).
Matthews, L.G., *The royal apothecaries* (London, 1967).
Mattirolo, O., *Le lettere di Ulisse Aldrovandi ai granduchi della Toscana Francesco I e Ferdinando I* (Turin, 1904).

Matricule de l'Université de Médecine de Montpellier (1503-1599), ed. Marcel Gouron (Geneva, 1957).

Mauss, M., *The gift. The form and reason for exchange in archaic societies* (London, 1990; originally published as 'Essai sur le don. Forme et raison de l'échange dans les sociétés archaïques', *L'Année sociologique*, 2nd ser., 1 (1923-24).

McCusker, J.J., and C. Gravestijn, *The beginnings of commercial and financial journalism. The commodity price currents, exchange rate currents and money currents in early modern Europe* (Amsterdam, 1991).

Meadow, M.A., 'Merchants and marvels: Hans Jacob Fugger and the origins of the Wunderkammer', in P.H. Smith and P. Findlen (eds.), *Merchants and marvels: Commerce, science and art in early modern Europe* (New York/London, 2002), 182-200.

Meier, Johannes (ed.), *Sendung – Eroberung – Begegnung. Franz Xaver, die Gesellschaft Jesu und die katholische Weltkirche im Zeitalter des Barock* (Wiesbaden, 2005) [Studien zur Außereuropäischen Christentumsgeschichte (Asien, Afrika, Lateinamerika) / Studies in the history of Christianity in the Non-Western world, 8].

Meyer, E.H.F., *Geschichte der Botanik*, 4 vols. (Königsberg, 1857).

Meyer, F.G., with E.E. Trueblood and J.L. Heller (eds.), *The great herbal of Leonhart Fuchs: De historia stirpium commentarii insignes, 1542*, 2 vols. (Stanford, 1999).

Mirek, Z., and A. Zemanek (eds.), *Studies in Renaissance botany* (Kraków, 1998) [Polish botanical studies, Guidebook series, 20].

Monok, I., P. Ötvös and E. Zvara, *Balthasar Batthyány und seine Bibliothek. Bibliotheken in Güssing in 16. und 17. Jahrhundert*, vol. II (Eisenstadt, 2004) [Burgenländische Forschungen, Sonderband, 26].

Montenegro, Pedro, *Materia médica misionera*, introd. Raúl Quintana (Buenos Aires, 1945).

Morford, M., *Stoics and neostoics. Rubens and the circle of Lipsius* (Princeton, 1991).

Morren, E., 'Charles de l'Escluse, sa vie et ses oeuvres', *Bulletin des Sociétés d'Horticulture de Belgique*, 184 (1875), 1-45.

—, *Charles de L'Ecluse, sa vie et ses oeuvres 1526-1609* (Liège, 1875).

Mossetti, U., 'Catalogo dell'Erbario di Ulisse Aldrovandi: I campioni ritrovati negli erbari di Giuseppe Monti e Ferdinando Bassi', *Webbia*, 44/1 (1990), 151-164.

Müller-Wille, S., 'Joining Lapland and the Topinambes in flourishing Holland: Center and periphery in Linnaean botany', *Science in context*, 16 (2003), 461-488.

Nave, F. de, and D. Imhof (eds.), *Botany in the Low Countries (end of the 15th century – ca. 1650)* (Antwerp, 1993) [Exhibition catalogue Museum Plantin-Moretus].

O'Brien, P., et al. (eds.), *Urban achievement in early modern Europe: Golden ages in Antwerp, Amsterdam and London* (Cambridge, 2001).

Ogilvie, B.W., 'The many books of nature. Renaissance naturalists and information overload', *Journal of the history of ideas*, 64 (2003), 29-40.

—, 'Image and text in natural history, 1500-1700', in Wolfgang Lefèvre, Jürgen Renn and Urs Schoepflin (eds.), *The power of images in early modern science* (Basle/Boston/Berlin, 2003), 141-166.

—, *The science of describing. Natural history in Renaissance Europe* (Chicago, in press).

Olmi, G., 'Ordine e fama: il museo naturalistico in Italia nei secoli XVI e XVII', *Annali dell'Istituto storico italo-germanico in Trento*, 8 (1982), 225-274.

—, *L'Inventario del mondo. Catalogazione della natura e luoghi del sapere nella prima età moderna* (Bologna, 1992).

—, 'Lettere di fra Gregorio da Reggio, cappuccino e botanico del tardo rinascimento', in M. Beretta, P. Galluzzi, C. Triarico (eds.), *Musa musaei, Studies on scientific instruments and collections in honour of Mara Miniati* (Florence, 2003), 117-139.

O'Malley, C.D., 'Andreas Vesalius' pilgrimage', *Isis*, no. 45 (July 1954), 138-144.

O'Malley, John W., *The first Jesuits* (Cambridge [Mass.], 1998⁶).

— et al. (eds.), *The Jesuits: Cultures, sciences, and the arts 1540-1773. Papers from the International Conference held late May 1997 in Boston* (Toronto/Buffalo/London, 1999; rpt. 2000).

Omont, Henri, 'Catalogue des manuscripts grecs de Guillaume Pellicier', *Bibliothèque de l'École des Chartes*, 46 (1885), 45-83, 594-624.

O'Neill, Charles Edwards, and Joaquín María Dominguez (dir.), *Diccionario histórico de la Compañia de Jesús: Biográfico-temático*, 4 vols. (Rome/Madrid, 2001).

Padgen, A., *European encounters with the New World: From Reniassance to Romanticism* (New Haven, 1993).

Papy, J., 'An antiquarian scholar between text and image? Justus Lipsius, humanist education and the visualisation of ancient Rome', *Sixteenth century journal*, 35 (2004), 97-131.

Pardo Tomás, J., 'Obras españolas sobre historia natural y materia médica americanas en la Italia del siglo XVI', *Asclepio*, 43 (1991), 51-94.

— and M.L. López Terrada, *Las primeras noticias sobre plantas americanas en las relaciones de viajes y crónicas de Indias (1493-1553)* (Valencia, 1993).

— and J.M. Lopez Pinero, *La influencia de Francisco Hernández (1515-1587) en la constitución de la botánica y de la materia médica modernas* (Valencia, 1996).

—, 'Le immagini delle piante americane nell'opera di Gonzalo Fernandez de Oviedo (1478-1557)', in G. Olmi, L. Tongiorgi and A. Zanca (eds.), *Natura-cultura. L'interpretazione del mondo fisico nei testi e nelle immagini* (Florence, 2000), 163-188.

—, '*Imago naturae*: historia natural, materia médica y nuevos mundos', *Historia*, 16/24 (2000), 25-40.

—, *El tesoro natural de América. Colonialismo y ciencia en el siglo XVI* (Madrid, 2002).

—, 'Tra "oppinioni" e "dispareri": la flora americana nell'erbario di Pier'Antonio Michiel (1510-1576)', in G. Olmi and G. Papagno (eds.), *La natura e il corpo. Studi alla memoria di Attilio Zanca* (Florence, 2006), 73-100.

Park, K., 'The criminal and the saintly body: autopsy and dissection in Renaissance Italy', *Renaissance quarterly*, 47 (1994), 1-33.

Parshall, P., '*Imago contrafacta*: Images and facts in the Northern Renaissance', *Art history*, 16 (1994), 554-579.

Pavord, A., *The tulip* (London, 1999).

Pelling, M., and C. Webster, 'Medical practitioners', in C. Wester (ed.), *Health, medicine and mortality in the sixteenth century* (Cambridge, 1979), 165-235.

Platter, Felix, *Beloved son Felix: The journal of Felix Platter, a medical student in Montpellier in the sixteenth century*, transl. Sean Jennett (London, 1961).

Pomian, Krzystof, *Collectors and curiosities: Paris and Venice 1500-1800* (London, 1990).

Popoff, Michel, *Index général des manuscrits décrits dans le Catalogue général des manuscrits des bibliothèques publiques de France* (Paris, 1993).

Prest, John, *The Garden of Eden: The botanic garden and the recreation of Paradise* (New Haven, 1981).

Ramón-Laca, L., 'The Spanish and American plants in Clusius' correspondence', *Polish botanical studies. Guidebook series*, 20 (1998), 135-160.

—, 'Las plantas americanas en la obra de Charles de l'Écluse', *Anales del Jardín Botánico de Madrid*, 57/1 (1999), 97-107.

—, 'Charles de l'Écluse y la flora ibérica', in idem and R. Morales Valverde (eds.), *Charles de l'Écluse de Arras, Descripción de algunas plantas raras encontradas en España y Portugal* (Castilla y León, 2005), 9-34.

Rance-Bourrey, A.-Joseph, *Jacques-Auguste de Thou, son Histoire universelle et ses démêlés avec Rome* (Paris, 1881).

Rath, G., 'Die Briefe Conrad Gessners aus der Trewschen Sammlung', *Gesnerus*, 7 (1950), 140-70; 8 (1951), 195-215.

Raugei, Anna-Maria (introd. and ed.), *Pinelli, Gian Vincenzo – Dupuy, Claude: Une correspondance entre deux humanistes* (Florence, 2001).

Reeds, Karen M., *Botany in medieval and Renaissance universities* (New York/London, 1991).

Reichardt, H.W., 'Carl Clusius' Naturgeschichte der Schwämme Pannoniens', *Festschrift zur Feier des fünfundzwanzigjährigen Bestehens der k.k. Zoologisch-Botanischen Gesellschaft in Wien*, (1876), 145-186.Rekers, B., 'Epistolario de Benito Arias Montano (1527-1598)', *Hispanofila*, 9 (1960), 25-37.

Renzi, S. de, 'Writing and talking of exotic animals', in M. Frasca-Spada and N. Jardine (eds.), *Books and the sciences in history* (Cambridge, 2000), 151-167.

Revillout, Ch., 'Les promoteurs de la Renaissance à Montpellier', *Mémoires de la société archéologique de Montpellier*, 2nd ser., 2 (1900), 14-383.

Rivolta, A., *Catalogo dei Codici Pinelliani dell'Ambrosiana latini* (Milan, 1933).

Roberts, R.S., 'The early history of the import of drugs into Britain', in F.N.L. Poynter (ed.), *The evolution of pharmacy in Britain* (London, 1965), 165-185.

Rodríguez-Marín, F., *La verdadera biografía del Doctor Nicolás Monardes* (Madrid, 1925).

Rossetti, L., *Matricula nationis Germanicae artistarum in gymnasio Patavino (1553-1721)* (Padua, 1986).

Roze, E., *Notes et souvenirs extraits de d'histoire des plantes rares de Charles de l'Escluse (Rariorum plantarum Historia 1601)* (Paris, 1898).

Sadler J., A növènytan törtènetei honunkban a 16-ik szàzadban', *Magyar Termèszettudomànyi Tàrsulat Evkönyvei*, 1 (1841-45), 78-118.

Salvesen, P.H., 'Levende kulturminner i Gamlehagen på Store Milde. Rosene', *Årringen*, 2002, 4-12.

Sánchez Rodríguez, C., *Perfil de un humanista: Benito Arias Montano (1527-1598)* (Huelva, 1996).

Schepelern, H.D., 'Natural philosophers and princely collectors: Worm, Paludanus and the Gottorp and Copenhagen collections', in O. Impey and A. MacGregor (eds.), *The origins of museums: The cabinets of curiosities in sixteenth and seventeenth-century Europe* (Oxford, 1985), 121-127.

Schierbeek, A., *Van Aristoteles tot Pasteur. Leven en werken der groote biologen* (Amsterdam, 1923).

Schlosser, J. von, *Die Kunst- und Wunderkammern der Spätrenaissance. Ein Beitrag zur Geschichte des Sammelwesens* (Braunschweig, 1978^2; 1st edn. 1923).

Schmidt, B., 'Exotic allies: The Dutch-Chilean encounter and the (failed) conquest of America', *Renaissance quarterly*, 52 (1999), 440-473.

Schnapper, A., *Le géant, la licorne et la tulipe. Collections et collectionneurs dans la France du XVIIe siècle* (Paris, 1988).

Schuler, Irmgard, *Das Florilegio medicinal von 1712 des Johann Steinhöfer, Jesuitenmissionar in Mexiko* (unpublished dissertation, Munich, 1973).

Sebastián de Elcano, J., et al., *La primera vuelta al mundo* (Madrid, 2003).

Secord, J.A., 'The crisis of nature', in N. Jardine, J.A. Secord and E.C. Spary (eds.), *Cultures of natural history* (Cambridge, 1996), 447-459.

Sievernich, Michael, and Günter Switek (eds.), *Ignatianisch. Eigenart und Methode der Gesellschaft Jesu* (Freiburg/Basel/Vienna, 1990^2).

Smith, P.H., and P. Findlen (eds.), *Merchants and marvels: Commerce, science and art in early modern Europe* (New York/London, 2002).

Smith, W.D., 'The function of commercial centres in the modernization of European capitalism: Amsterdam as an information exchange in the seventeenth century', *The journal of economic history*, 44/4 (1984), 985-1005.

Soldano, A., 'L'Erbario di Ulisse Aldrovandi. Volumi XV-XVI', *Atti Istituto Veneto di Scienze, Lettere ed Arti* (in press).

Sondervorst, F.A., *Histoire de la médecine belge* (Brussels, 1981).

Stien, R., 'Legitimis libri possessor – norske leger som boksamlere i eldre tid', *Axonet, Medlemsblad for norsk nevrologisk forening*, 1 (2002), 12-13.

Stirling, Johannes, *Lexicon nominum herbarum, arborum fruticumque linguae latinae* (Budapest, 1995).

—, *Magyar reneszánsz kertművészet a XVI-XVII. században* (Budapest, 1996).

Strong, R., *The Renaissance Garden in England* (London, 1979).

Swan, C., 'From blowfish to flower still life paintings: Classification and its images circa 1600', in P.H. Smith and P. Findlen (eds.), *Merchants and marvels: Commerce, science and art in early modern Europe* (New York/London, 2002), 109-136.

Szabó, A.T., I. Szabó and F. Wolkinger (eds.), *The beginnings of Pannonian ethnobotany: Stirpium nomenclator Pannonicus*, ed. S. Beythe (1583), C. Clusius (1584), D. Czvittinger (1711) (Graz/Güssing, 1992) [Collecta Clusiana, 2].

Szabó, M., and S. Tonk, *Erdélyiek egyetemjárása a korai újkorban 1521-1700* (Szeged, 1992) [Fontes rerum scholasticarum, 4].

Szlatky, M., 'A magyar nyelvű természettudományos és orvosi irodalom a XVI. században', *Orvostörténeti Közlemények*, 109-112 (1985), 91-97.

Takáts, S., *Rajzok a török világból*, 3 vols. (Budapest, 1917).

Talbott, J.H., *A biographical history of medicine: Excerpts and essays on the men and their work* (New York/London, 1970).

Terpstra, H., 'De Nederlandsch voorcompagnieën', in F.W. Stapel (ed.), *Geschiedenis van Nederlandsch Indië*, 5 vols. (Amsterdam, 1938-40), vol. II, 275-475.

Terwen-Dionisius, E.M., 'De eerste ontwerpen voor de Leidse hortus', in J.W. Marsilje et al. (eds.), *Uit Leidse bron geleverd. Studies over Leiden en de Leidenaren in het verleden, aangeboden aan drs. B.N. Leverland bij zijn afscheid als adjunct-archivaris van het Leidse Gemeentearchief* (Leiden, 1989), 392-400.

—, 'Date and design of the botanical garden of Padua', *Journal of garden history*, 14 (1994), 213-235.

Thomas, K., *Man and the natural world* (Harmondsworth, 1983).

Tjon Sie Fat, L.A., and E. de Jong (eds.), *The authentic garden. A symposium on gardens* (Leiden, 1991).

—, 'Clusius's garden: A reconstruction', in idem and E. de Jong (eds.), *The authentic garden. A symposium on gardens* (Leiden, 1991), 3-12.

Todorov, T., *La conquête de l'Amérique. La question de l'autre* (Paris, 1982).

Tongiorgi Tomasi, L., and F. Garbari, 'Carolus Clusius and the botanical garden of Pisa', in L. Tjon Sie Fat and E. de Jong (eds.), *The authentic garden. A symposium on gardens* (Leiden, 1991), 61-74.

Toni, G.B. de, 'Il carteggio degli italiani col botanico Carlo Clusio nella Biblioteca Leidense', *Memorie Regia Accademia, scienze, lettere e arti, Modena*, 10 (1911), 1-147.

—, *Spigolature Aldrovandiane. XIII. Un altro corrispondente di Ulisse Aldrovandi, il medico Giovanni Battista Balestri* (Leipzig, 1912).

Tosi, A., *Ulisse Aldrovandi e la Toscana. Carteggio e testimonianze documentarie* (Florence, 1989).

Traxler, G., 'Die burgenländischen Pflanzenstandorte bei Carolus Clusius', *Burgenländische Heimatblätter*, 35 (1973), 49-59.

Trevisan, R., 'Luigi Anguillara', in A. Minelli (ed.), *The botanical garden of Padua 1545-1995* (Venice, 1995), 57-59.

Ubrizsy-Savoia, A., 'Importance de l'oeuvre de Clusius dans l'histoire de la mycologie', *Bulletin de la Société Mycologique de France*, 91/4 (1975), 560-565.

—, 'Carolus Clusius kapcsolatai az olasz tudományos élettel', *Mikológiai Közlemények*, 1 (1975), 13-21.

—, 'Un aperçu de l'histoire de la mycologie jusqu'au Carolus Clusius', *Friesia*, 11/1 (1975), 31-40.

—, 'Les acquarelles mycologiques de Charles de l'Escluse', *Histoire et nature*, 7 (1975), 89-95.

—, 'Contribution à la connaissance des oeuvres de Clusius', *Revue d'histoire des sciences*, 28/4 (1975), 361-370.

—, 'Carolus Clusius és a termesztett növények', *Botanikai Közlemények*, 62/3 (1975), 223-226.

—, 'Il Codice di Clusius', *Rassegna di micologia ed ecologia Romana dell'Amer*, 3/7-8 (1976), 6-11.

—, 'Importance of Carolus Clusius' life-work in the history of mycology', *Acta agronomica Academiae Scientiarum Hungaricae*, 25 (1976), 400-417.

—, *Die Beziehungen des Lebenswerkes von Carolus Clusius zu Italien und Ungarn. Clusius' pilzkundliche Aquarelle* (Güssing/Vienna, 1977).

—, 'Clusius levelezotársai', *Vasi Szemle*, 32/78-1 (1978), 124-127.

—, 'I rapporti tra Carolus Clusius ed i naturalisti italiani del suo tempo', *Physis*, 20 (1978), 49-69.

—, 'Piante italiane in un'opera olandese del 1601', *Annali di botanica (Roma)*, 35-36 (1978 [= 1976-77]), 144-154.

—, 'Carolus Clusius munkásságát megelozo mikológiai ismeretek', *Mikológiai Közlemények*, 2 (1979), 89-98.

—, 'Il Codice micologico di Federico Cesi', *Rendiconti dell'Accademia Nazionale dei Lincei, Classe Sc. Fis. Mat. e Nat.*, ser. VIII, 68/2 (1980), 129-134.

—, 'Wissenschaftliche Beziehungen zu Italien. Der Maler der Pilzaquarelle im Clusius-Codex', in S.A. Aumüller and J. Jeanplong (eds.), *Carolus Clusius Fungorum in Pannoniis observatorum brevis historia et Codex Clusii* (Budapest/Graz, 1983), 54-56.

—, 'Magyar földrajzi nevek két XVI. sz.-i németalföldi könyvben', *Földrajzi Közlemények*, 31 (CVII) (1983), 130-132.

—, and J. Heniger, 'Carolus Clusius and American plants', *Taxon*, 32/3 (1983), 424-435.

—, 'I rapporti inglesi di Carolus Clusius; Piante inglesi nel 'Rariorum plantarum historia' di Carolus Clusius edito nel 1601', in Pècsi Akadèmiai Bizottsàg (ed.), *Studia phytologica nova in honorem jubilantis A.O. Horvàt* (Pècs, 1987).

—, 'Orvosbotanikai adatok Carolus Clusius néhány kevésbé ismert müvében', *Gyógyszerészet*, 32 (1988), 367-370.

—, 'The botanical garden of Padua in Guilandino's day', in A. Minelli (ed.), *The botanical garden of Padua 1545-1995* (Venice, 1995), 172-195.

—, *Rapporti italo-ungheresi nella nascita della botanica in Ungheria* (Pécs, 2002).

—, 'Il metodo sinottico, collante tra la *Syntaxis plantarum* di Aldrovandi e le *Tavole fitosofiche* di Cesi', in *Federico Cesi un principe naturalista, Atti del Convegno Internazionale di Studi, Acquasparta, September 29-30, 2003, Accademia Nazionale dei Lincei* (forthcoming).

Vandenbroeck, P., (ed.), *America bride of the sun* (Antwerp, 1992) [Exhibition catalogue, Royal Museum of Fine Arts, Antwerp].

Velsaco Lozano, A.M.L., 'El jardín de Itztapalapa', *Arqueología Mexicana* 57 (2000), 26-35.

Veress, E., *Olasz egyetemeken járt magy arországi tanulók anyakönyve és iratai (1222-1864)* (Budapest, 1941).

Verhaeren, H., *Catalogue de la bibliothéque du Pét'ang* (Beijing, 1949; rpt. Paris, 1969).
Voet, L., *The Golden Compasses: A history and evaluation of the printing and publishing activities of the Officiana Plantinana at Antwerp*, 2 vols. (Amsterdam/London, 1969-72).
—, *The Plantin press (1555-1589): A bibliography of the works printed and published by Christopher Plantin at Antwerp and Leiden*, 6 vols. (Amsterdam, 1980-83).
—, 'Christopher Plantin as a promoter of the science of botany', in F. de Nave and D. Imhof (eds.), *Botany in the Low Countries* (Antwerp, 1993), 39-45.
Waquet, F., *Latin, or the empire of a sign* (New York, 2001).
Wear, A., *Knowledge and practice in English medicine, 1550-1680* (Cambridge, 2000).
Wegener, Hans, 'Das grosse Bilderwerk des Carolus Clusius in der Preussischen Staatsbibliothek', *Forschungen und Fortschritte*, 12 (1936), 374-376.
White, B.M.I., *A Cast of Ravens. The strange case of Sir Thomas Overbury* (London, 1965).
Whitehead, P.J.P., 'Georg Markgraf and Brazilian zoology', in E. van den Boogaart (ed.), *Johan Maurits van Nassau-Siegen 1604-1679. A humanist prince in Europe and Brazil* (The Hague, 1979), 424-471.
Woodward, D., *Maps as prints in the Italian Renaissance: Makers, distributors and consumers* (London, 1996).
Zemanek, A., 'Renaissance botany and modern science', in Z. Mirek and A. Zemanek (eds.), *Studies in Renaissance botany* [Polish Botanical Studies, Guidebook Series 20] (Kraków, 1998), 9-47.
Zon, Stephen, *Petrus Lotichius Secundus: Neo-Latin poet* (Berne/New York, 1983).

List of illustrations

On the cover
Drawing of a daffodil, accompanying a letter from Carolus Clusius in Leiden to Mattea Caccini in Florence, dated 10 October 1608. Universiteitsbibliotheek Leiden, BPL 2414/14b.

1. Portrait of Carolus Clusius, at the age of 59. Canvas, painted in Vienna in 1585 by an unknown artist, possibly Jacob de Monte. Universiteitsbibliotheek Leiden.
2. Portrait of Carolus Clusius. Engraving by Martinus Rota, sixteenth-century. Universiteitsbibliotheek Leiden, PK I 152 Rot 1.
3. Watercolour of a clematis in *Libri Picturati*, vol. A 23, f. 18v. Courtesy of the Jagiellon Library, Kraków.
4. Portrait of princess Marie de Brimeu. Anonymous engraving from Jacques de Bie, *Livre contenant la généalogie et descente de ceux de la mayson de Croy* (Antwerp, c. 1615).
5. First page of the letter from Jacques Plateau in Tournay to Clusius in Leiden (8 February 1602). Universiteitsbibliotheek Leiden, VUL 101.
6. Guillaume Rondelet. Woodcut by an anonymous artist, from G. Rondelet, *Libri de piscibus [...]* (Lyon, 1554). Universiteitsbibliotheek Leiden, 667 A 16 (1).
7. Two woodcuts of fishes from G. Rondelet, *Libri de piscibus [...]* (Lyon, 1554), 175. Universiteitsbibliotheek Leiden, 667 A 16 (1).
8. Map of Hungary from Abraham Ortelius, *Theatrum orbis terrarum* (Antwerp, 1595), no. 96. Courtesy of the Library of the Hungarian Academy of Sciences, Budapest.
9. Portrait of Boldizsár Batthyány by unknown painter, seventeenth century. Courtesy of the Hungarian National Museum, Budapest.
10. A letter of Carolus Clusius written to Boldizsár Batthyány. Hungarian National Archives, Budapest.
11. Title-page of Rembert Dodoens, *Cruydt-boeck* (Antwerp, 1644). Photograph Bengt Melliander, UB-Media, Lund University Library.
12. Map of Scandinavia and Northern Europe from Abraham Ortelius, *Theatrum orbis terrarum* (Antwerp, 1570). Photograph Bengt Melliander, UB-Media, Lund University Library.
13. Ripe cloudberries (*Rubus chamaemorus*), growing on boggy ground in the mountains of the province of Jämtland, Sweden, close to the Norwegian border. Photograph Kjell Lundquist, August 2005.
14. The island of Ven with Uraniborg castle. Tycho Brahe's second map of the island. From Georg Braunius and Frans Hogenberg, *Civitates orbis terrarum*, part IV: *Liber quartus urbium praecipuarum totius mundi* (Cologne, 1586). Photograph Bengt Melliander, UB-Media, Lund University Library.
15. List of plants written by Clusius on the envelope to the first letter from Høyer of 9 April 1597. Universiteitsbibliotheek Leiden, VUL 101.

16. Scarlet Martagon (*Lilium chalcedonicum*), here called 'HEMEROCALLIS CHALCEDONICA umbellifera', on a contemporary painting by Daniel Rabel, the original of the engraved florilegium *Theatrum florae* (1622). Courtesy Bibliothèque Nationale de France, Dossier PF2315015, s. 75.
17. Extract from the third letter from Høyer to Clusius, 17 August 1597. Universiteitsbibliotheek Leiden, VUL 101.
18. Orange lily or Fire lily (*Lilium bulbiferum*). Photograph Kjell Lundquist, Enafors, Jämtland, Sweden, August 2005).
19. Cloudberry (*Rubus chamaemorus*) in flower and fruit. Woodcut from Johannes Palmberg, *Serta florea Svecana, eller Swenske örtekrantz* (Strängnäs, 1684). Photograph Bengt Melliander, UB-Media, Lund University Library.
20. Carolus Clusius. Painting, dated 1606 by Filippo Paladini after a woodcut portrait of Clusius at the age of 75 by Jacques de Gheyn II, published as a frontispiece to Clusius' *Rariorum plantarum historia* (Antwerp, 1601). Museo Botanico, Pisa.
21. Cercopithecus sagouin. From Clusius' *Exoticorum libri decem* (Leiden, 1605), 372. Universiteitsbibliotheek Leiden, 755 A 3 (2).
22. Manati. From *Exoticorum libri decem*, 133.
23. Sloth. From *Exoticorum libri decem*, 111.
24. Sloth. From *Exoticorum libri decem*, 373.
25. Rex bird of paradise. From *Exoticorum libri decem*, 362.
26. Bird of paradise. From *Exoticorum libri decem*, 360.
27. Echinomelocactus. From *Exoticorum libri decem*, 92.
28. Scaly horn. From *Exoticorum libri decem*, 109.
29. Bird (alcatraz?). From *Exoticorum libri decem*, 106.
30. Bixa orellana. From *Exoticorum libri decem*, 74.
31. Anser Magellanicus. From *Exoticorum libri decem*, 101.
32. Anser Magellanicus. From Barent Jansz Potgieter, *Wijdtloopigh verhael van tgene de vijf schepen (die int jaer 1598 tot Rotterdam toegherust werden, om door de Straet Magellana haren handel te drijven) wedervaren is [...]* (Amsterdam, 1600). Universiteitsbibliotheek, Amsterdam.
33. Anser Magellanicus. From Zacharias Heyns, *Dracht-thoneel* (Amsterdam, 1601), 145. Koninklijke Bibliotheek, The Hague.
34. Dutch penguin-hunting. From De Bry, *America*, vol. IX, part II (Frankfurt, 1601). Universiteitsbibliotheek, Amsterdam
35. Anser Magellanicus, detail of fig. 25.
36. Armadillo. From *Exoticorum libri decem*, 330.
37. Armadillo. From *Exoticorum libri decem*, 109.
38. Gian Lorenzo Bernini, Fontana dei Quattro Fiumi, 1651. Piazza Navona, Rome.
39. P. Bellonius, *Plurimarum singularium and memorabilium rerum [...] observationes*, tr. C. Clusius (Antwerp, 1589), 390. British Library, C.60.e.5. By permission of the British Library.
40. M. Lobelius, *Plantarum seu stirpium icones* (Antwerp, 1581), vol. II, 231. Universiteitsbibliotheek Leiden, 1371 E 29.
41. P. Bellonius' *Observationes* in C. Clusius, *Exoticorum libri decem* (Antwerp, 1605), 163. Bibliotheca Thysiana, Leiden, 2002.
42. P. Bellonius, *Plurimarum singularium and memorabilium rerum [...] observationes*, tr. C. Clusius (Antwerp, 1589), 220. British Library, C.60.e.5. By permission of the British Library.
43. Bellonius' *Observationes* in C. Clusius, *Exoticorum libri decem* (Antwerp, 1605), 94. Bibliotheca Thysiana, Leiden, 2002.

44. C. Clusius, *Curae posteriores* (Leiden, 1611), 57. Leiden University Library, 661 A 5.
45. N. Monardes, *Simplicium medicamentorum ex novo orbe delatorum, quorum in medicina usus est*, tr. C. Clusius (Antwerp, 1593), 364-365. British Library, C.60.e.5. By permission of the British Library.
46. N. Monardes, *Simplicium medicamentorum ex novo orbe delatorum*, in C. Clusius, *Exoticorum libri decem* (Antwerp, 1605), 326. Bibliotheca Thysiana, Leiden, 2002.
47. C. a Costa, *Aromatum et medicamentorum in Orientali India nascentium liber* in C. Clusius, *Exoticorum libri decem* (Antwerp, 1605), 267. Bibliotheca Thysiana, Leiden, 2002.
48. The funerary monument of Nicasius Ellebodius, Pozsony. From G. Galavics et al. (eds.), *Collectanea Tiburtiana. Tanulmányok Klaniczay Tibor tiszteletére* (Szeged, 1990), fig. 2. Photograph by Géza Galavics (detail).
49. Frontispiece of the manuscript *Watercolours of fungi, Caroli Clusii*, University of Oxford, Department of Plant Sciences, Sherard Collection Ms. 43.
50a. Table 6 in the *Clusius Codex*, Universiteitsbibliotheek Leiden, BPL 303.
50b. A page of the Oxford album, a copy after table 6 in the Leiden *Clusius Codex*. University of Oxford, Department of Plant Sciences, Sherard Collection, Ms. 43.
50c. Page n. 7 in the Brussels album. Royal Library Brussels, KBR Ms. 15475.
51a. Table 5 in the *Clusius Codex*, Universiteitsbibliotheek Leiden, BPL 303.
51b. A page of the Oxford album. University of Oxford, Library of the Department of Plant Sciences, Sherard Collection, Ms. 43.
51c. Page n. 6 in the Brussels album. Royal Library Brussels, KBR Ms. 15475.
52. Engraved title-page of C. Clusius, *Rariorum plantarum historia* (Antwerp, 1601). Universiteitsbibliotheek, Leiden, 661 A 3.
53. Title-page of Giovan Battista Ferrari, *Flora overo cultura di fiori* (Rome, 1638). Universitätsbibliothek Johann Christian Senckenberg, Frankfurt am Main. By friendly permission.
54. *Spartium junceum* L. Photograph Armin Jagel, Bonn.
55. *Hibiscus rosa-sinensis* L. Photograph Hilke Steinecke, Frankfurt am Main.
56. *Muscari botryoides* L. Photograph Hilke Steinecke, Frankfurt am Main.

About the contributors

Sabine Anagnostou is pharmacist and researcher at the Institut für Geschichte der Pharmazie at the Philipps-Universität in Marburg, Germany. Her fields of research are pharmacy in the missions (16th to 18th centuries), the history of medicinal plants, and ethnopharmacy. Recent publications are: *Jesuiten in Spanisch-Amerika als Uebermittler von heilkundlichem Wissen* (Stuttgart, 2000), 'Jesuits in Spanish America and their contribution to the exploration of the American materia medica', *Pharmacy in history*, 47 (2005), 3-17, and 'Pharmazie auf internationaler Ebene. Die Apotheke des Collegio Romano vom 16. bis 18. Jahrhundert', *Geschichte der Pharmazie*, 57 (2005), 57-63.

Irene Baldriga has a Ph.D in art history and collaborates with several Italian and international academic institutions. Her favourite field of research is Renaissance and Baroque art, the history of collecting and patronage and the relation art-science in early modern Europe. Among her recent publications is the book *L'occhio della Lince. I primi lincei tra arte, scienza e collezionismo, 1603-1630* (Rome, 2002).

Josep Lluis Barona is professor of history of science at the University of Valencia and visiting fellow at the Universities of Oxford and Bergen (Norway). He studied psychiatry and philosophy in Valencia and was Vice-Rector of the University from 1986 to 1992. His main fields of research are the history of medical thought in the early modern period and the history of contemporary public health, specially relating to children, international health organisations and health policies. Recent books: *Ciencia, salud pública y exilio. España 1875-1939* (Valencia, 2003), *Salud, tecnología y saber médico* (Madrid, 2004) and *Health and medicine in rural Europe* (Valencia, 2005), co-edited with Steven Cherry.

Dóra Bobory is soon to complete her dissertation at the Medieval Studies Department of the Central European University in Budapest. Her main interest is history of science in the early modern period, her doctoral work focuses on the scientific interests and correspondence of the Hungarian Count Boldizsár Batthyány (1537-1590).

Elizabeth Boutroue is researcher at the Institut de Recherche et d'Histoire des Textes in Paris. She is specialized in the transmission of – particularly – scientific texts from the classical era to the Renaissance. Having written her doctoral thesis on the transmission and reception of Pliny's *Historia naturalis* in the Renaissance, she is currently working on Renaissance herbals.

Florike Egmond is postdoc in the Clusius Project at the Scaliger Institute, University of Leiden (The Netherlands). She has published on the social and cultural history of crime, punishment, honour, the body and natural history during the early modern period. Author (with Peter Mason) of *The mammoth and the mouse. Microhistory and morphology* (Baltimore, 1997); co-editor with Peter Mason of *The whale book. Whales and other marine animals as described by Adriaen Coenen in 1585* (London, 2003), with Francisco Bethencourt of *Correspondence and cultural exchange in early modern Europe* (Cambridge, 2006), and with Rob Zwijnenberg of *Bodily extremities. Preoccupations with the human body in early modern European culture* (Aldershot, 2003).

Robert Visser is Teylers professor in the history of science at Leiden University and also teaches at Utrecht University. His research focuses on the relation between science and the public since 1800 and on the reception of Darwinism in the Netherlands.

Paul Hoftijzer is Director of the Scaliger Institute of Leiden University Library. In addition, he holds the P.A. Tiele chair in early-modern book history at the University of Leiden. He has published widely on the history of publishing, bookselling and reading in the Dutch Republic.

Sachiko Kusukawa is Tutor and Fellow in the History and Philosophy of Science at Trinity College, Cambridge. She is currently working on a monograph on the role of pictorial arguments in scientific knowledge in sixteenth-century Europe.

Gillian Lewis is a Senior Research Fellow of St. Anne's College in the University of Oxford. She is engaged upon a full-length study of Guillaume Rondelet in the context of medicine and natural history in Renaissance Montpellier.

Kjell Lundquist is landscape architect, Doctor of Agricultural Science, at the Department of Landscape Planning Alnarp, Swedish University of Agricultural Sciences. Lecturing: Courses in Garden and Park History, and in Restoration, Conservation and Maintenance of Gardens, Parks and Designed Landscapes. Fields of research: Garden, Park and Landscape History. The history of Garden Plants. Plant-introductions, within the Main Research Profile of the Faculty: Garden, Park and Landscape - History and Heritage. A complete list of his publications is available on: http://www.lpal.slu.se/ShowPage.cfm?OrgenhetSida_ID=6272

Peter Mason has written widely on the European cultural reception of the Americas. For several years he was a consultant in art and anthropology for the Taller Experimental Cuerpos Pintados, Santiago de Chile. He is Profesor Visitante at the Casa de América, Madrid, and is an international editorial adviser of the *Journal of the history of collections* (Oxford). His most recent publication (with Christian Báez Allende) is *En el jardín. Fotografías de fueguinos y mapuches en los zoológicos humanos europeos* (Santiago de Chile, 2006). E-mail: monti55@fastwebnet.it

José Pardo Tomás is researcher at the Department of History of Science in the Institute 'Mila i Fontanals' (CSIC, Barcelona, Spain). He has published on social and cultural history of medicine, natural history, and scientific publishing in the early modern period. He is the author of, among others, *Ciencia y censura* (Madrid, 1991); *El tesoro natural de América* (Madrid, 2002); and *El médico en la palestra* (Salamanca, 2004). Since 2002 he is vice-president of the Catalan Society of the History of Science and Technology.

Dr. Andrea Ubrizsy Savoia, of Hungarian origin, is employed at the chair of Ecology of the Plant Biology Department of the University of Rome 'La Sapienza', where she graduated in biology in 1970. Since 1973 she has dealing with the history of botany, especially with the history of mycology, of botanical gardens, of herbariums and of herbals. She discovered in 1979 at the Library of Institute de France in Paris and identified as the three volumes of the mycological codex of Federico Cesi, founder of the Accademia dei Lincei in 1603 in Rome. These volumes, believed lost since the end of the eighteenth century, contain the first scientific images of fungi, realized between the 1624 and 1630, observed with the microscope constructed by Galileo Galilei. She is also author of more than a hundred publications, most of them dealing with the activity and relationships by Carolus Clusius, Ulisse Aldrovandi and other Renaissance scholars.

Index

Abbondio, Antonio, 290
Acosta, Cristóbal (de) – see Costa, Christophorus A
Acosta, José de, 311-312
Adrian (gardener), 151
Aelian, 75, 83, 94
Aer, Ioannes Gorvertz van der, 241
Aicardo, Paolo, 270
Aicholtz Starzerin, Anna von, 29
Aicholz, 288
Alcalá, Duke of, 178
Aldino, Tobia, 290
Aldrovandi, Ulisse, 13-15, 33, 46, 249, 269, 274-275, 285-290
Alemán, Pedro, 104
Alessandrino, Giulio, 271, 273
Alfaro, Martin de, 177
Alfonso the Magnanimous, King, 104
Alpin, Prosper (Prospero Alpini), 62-63, 283
Althan, Lord, 139
Amelang, James, 174
Anagnostou, Sabine, 5, 9
André, Jacques, 60
Anguillara, Luigi, 286
António, Gaspar, 304
Apicius, 60
Arceo, Francisco de, 102
Aremberg, family, 263
Aremberg, Count of, 15, 20, 44
Arias Montano, Benito, 15, 100, 102, 105, 108-109, 181, 185
Aristotle, 66, 75, 83-84, 270
Arkel, Yzabeau van, 29
Arnold, Paul, 230
Arundel, Lord, 57, 58
Ashworth Jr., William B., 195, 204

Aubry, Jean, 125, 131
Aumüller, István (Stephan) A., 133, 276, 283
Avicenna, 67, 78-80, 242
Aviñón, Juan de, 178
Backerus, Stephan, 241
Bacon, Francis, 254
Baldriga, Irene, 5, 9, 18, 38, 112
Barbari, Jacopo de', 230
Barbaro, Ermolao, 66
Barbaro, G., 286
Barlay, Sz.Ö., 124, 126
Barona, J.L., 4, 198, 210
Barvitio, Ioannis, 284
Bátai, István, 123
Báthory, István, 127
Batthyány, family, 120, 123, 141, 269
Batthyány, Ádam, 125
Batthyány, Boldizsár (Balthasar), 4, 15, 119-128, 130-144, 268-271, 275, 277, 291
Batthyány, Ferenc, 120, 124
Batthyány, Gáspár, 120, 124
Batthyány, Kristóf, 130
Batthyány-Strattmann, Ladislaus, 132
Battistini, M., 285
Båtvik, J. Ingar I., 151
Bauhin, Jean, 74
Bauhin, Kaspar (Caspar), 26, 28, 60, 157, 203, 252, 254-256
Béjar, Duchess of, 178
Belleforest, François de, 55
Belon, Pierre (Petrus Bellonius), 13, 26, 50, 59, 61, 64, 66, 77, 84, 92, 215, 231-236, 299
Bembo, Orazio, 290
Benincasa, Giuseppe – see Giuseppe Casabona
Bernini, Gian Lorenzo, 217

Berzeviczy, Màrton, 270
Bethencourt, Francisco, 9
Betz, Jerome, 72
Bey, Ali, 140
Beyer, Absalon Pederssøn, 151
Beythe, András, 123, 142, 143
Beythe, István, 123, 142, 143, 268
Bèze, Théodore de (Theodorus Beza), 50
Biagioli, Mario, 34
Bie, Jacques de, 30
Biesius, Nicolas, 271
Bloeme, Hendrik, 37
Bloom, H., 175
Blotius (de Bloot, Blotz), Hugo, 127, 270, 272-274
Bobart I, Jacob, 277
Bobart II, Jacob, 277
Bobory, Dora, 4, 269
Bocaud, Jean, 81
Bock, Hieronymus, 1, 26
Boiastuau, Pierre, 255
Boisot, Jean, 29
Boisot, Louise de, 29
Bonhomme, Macé, 88
Bonnail d'Assas, Jacques Salomon de, 73
Bontius, Jacobus, 302
Boodt, Anselmus de, 16, 204
Borbàs, Vince, 269
Borcht, Peeter vander, 139, 223, 225-226, 231, 240
Bordes, Jacques des, 88
Boutroue, Marie-Elizabeth, 4, 87
Bouwman, André, 5
Boyle, Robert, 13
Brahe, Tycho, 149, 150, 152-153
Brancion, Jean de, 37, 101, 108, 189
Brandenburg, Elector of, 227
Brandmuller, J., 55
Braun, G., 153
Bright, Timothy, 242-243
Brimeu, Princess Marie de, 15, 20, 29-32, 42, 44-45
Briquet, Ch.M., 90
Brocas, Jean, 52
Browne, Janet, 9, 11
Broyardus, Franciscus, 290
Brueghel the Elder, Pieter, 123
Brunfels, Otto, 1, 24, 26, 60, 67, 229

Bruti, Ottavia Peverella de, 29
Bruto, Giovanni Michele (Joannes Michael Brutus), 127
Bry, De, family, 214
Bry, Johan Israel De, 213
Bry, Johan Theodore De, 213
Buchanan, George, 55-56
Burgkmair, Hans, 198
Burgos, Bernardino de (Duranus), 108
Busbecq(ue), Ogier Ghislain de (Augerius de Busbecke, Ogier van Busbeck), 16, 127, 199, 273
Busher, Joanna, 241
Buyck, Henricus, 241
Cabrol, Barthélemy, 83
Caccini, Matteo, iv, 15, 18, 72, 250-251, 253-254, 290
Caerl, Ioannes Ioannis F., 241
Calandrini, Giovanni, 290
Calzolari, Francesco, 244, 249, 290
Calzolari, Hieronimo, 33
Camerarius II, Joachim, 15, 29, 31, 44, 229, 239, 241, 244-245
Cantarini, Gregorio, 274
Cappa, Antonio, 290
Caretto, Philippus, 284
Cartagena, Bishop of, 180
Casabona, Giuseppe (Giuseppe Benincasa, Joseph Goedenhuize), 283, 286-287, 290
Castañeda, Juan (Joannes) de, 29, 100, 105-106, 108, 110-112, 210, 218, 241
Castellan, Honoré, 57, 81, 87, 96
Catelan, Laurent, 83
Catelli, Nora, 174-175
Cavendish, Thomas, 43
Cazoncin, 182
Celsus, 60
Cerenza (L'Acerenza), Duke of, 62
Cesi, Federico, 249, 251-265, 281
Charles V, Emperor, 206
Chiovenda, Emilio, 263
Choartus Buzenvallus, Paullus (Paul Choart de Buzenval), 240
Christianson, John Robert, 145, 147, 149
Cicero, Marcus Tullius, 127
Cieza de León, Pedro, 189-190, 203
Claesz, Cornelis, 202
Clerc, Franciscus le, 240, 241

Clusius, Carolus (Charles de l'Écluse), passim
Cole, Jacob (Jacob Colius Ortelianus), 43, 241
Coligny, Princess Louise de, 29
Colius Ortelianus, Jacobus – see Jacob Cole
Colonna, Fabio, 254, 256-258, 261, 264, 286, 290, 292
Columella, Lucius, 308
Comte de l'Escluse, Geneviève le, 29
Conti, Piero Ginori, 250, 285
Coornhardt, Volcardus (Volckert Coornhert), 240-241
Coornhert, Dietrich Clemensz, 204
Cordes, Simon de, 202, 211, 213
Cordus, Valerius, 66, 242
Corrozet, Giles, 59
Cortuso, Giacomo Antonio, 33, 274, 284, 286-287
Corvinus, Elias, 125-126, 130, 137-138, 271
Corvinus, Henricus, 254
Corzo, Juan Antonio, 180
Costa, Christophorus a (Christóbal Acosta), 100, 106, 108, 113, 224-226, 231, 243, 288, 295, 299, 302, 306-307, 309
Coudenberghe, Peeter (Pieter) van, 22-23, 242
Covarrubias, Sebastián de, 173-174
Crato (von Kraftheim), Johannes, 29, 89, 127, 271, 273
Csaba, József, 119
Csapody, I., 133
Daniel (gardener), 136
Darnley, Henry, 55
Darwin, Charles, 9, 10
David, Adrianus, 276
Dàvid, Ferenc, 283
Decembrio, Pietro Candido, 195
Dernschwam, Joannes, 124
Despota Sami, Jacobus, 53
Dioscorides, 60-61, 66-67, 71, 73, 75, 103-104, 183, 227, 229, 239, 245, 301
Diòszegi, Sàmuel, 268
Dodoens, Rembert (Rembertus Dodonaeus), 1, 13, 24, 26, 61, 64, 92, 96, 102, 127, 131, 146, 159, 222-224, 226-229, 242, 271, 273, 309
Domingus, Marlon, 5
Domitzer, Johann, 131
Dorn, Gherhardus, 131
Drake, Francis, 43, 99, 203, 234, 237

Drebbel, Cornelis, 263
Dudicius, Andrea(s) – see Andràs Dudith
Dudith, Andràs (Andreas Dudicius), 271, 273
Dupuy, brothers, 49
Durante, Castore, 290
Dürer, Albrecht, 230
Dyckman, Georgius, 286
Ecgk, Christian von, 33
Eckblad, Finn-Egil, 145-146, 162-164, 167, 169
Eckius, Johannes, 249, 251-255, 258, 263
Écluse, Charles de – see Carolus Clusius
Edward VI of England, King, 47, 58
Egenolff, Christian, 229-230
Egmont, Count Lamoraal of, 15, 57
Egmond, Florike, 4, 246, 249, 263
Elizabeth (I) of England, Queen, 47, 57
Ellebodius, Nicasius, 269-272, 274-275
Elzevier, 261
Erasmus, Desiderius, 36
Erasso, Francisco de, 51
Essen, Madeleine von, 145
Faber, G.B., 291
Faber, Johannes, 255, 257-258, 260, 264
Fabricius, Paulus, 273
Fabrizi d'Acquapendente, Girolamo, 284
Farfan, Agustín, 301
Fassardo, family, 57
Fazekas, A., 129
Fazekas, Mihàly, 268
Fejérpataky, Làszló, 132
Felizian von Herberstein, Count, 126
Ferdinand I of Hungary, King, 120, 128
Fernández de Oviedo y Valdés, Gonzalo, 84, 94, 176, 190-191, 208, 210, 214, 237
Ferrabosco, Pietro, 126
Ferrara, Duke of, 185, 289
Ferrari, Giovanni Battista, 308-309
Ficino, Marsilio, 270
Filiis, Anastasio de, 251
Findlen, Paula, 245, 256
Fontanon, François, 73
Forgách, Ferenc, 127
Forgách, Mihály, 273
Forgách, Simon, 127
Foss, Anders, 149-150, 152, 161, 167
Fosse, Thomas de la, 22, 37, 45
Fragoso, Juan, 191
Francesi, Domenico, 270

Francis I of France, King, 51, 80, 86
Francis II of France, King, 121
Franco, Cesare, 126
Frankovics (Frankovith), Gergely, 143
Frati, L., 287
Fraxinus – see Gáspár Szegedi Kőrös
Freundt, Nicolaus de, 148, 151-152
Frisius, Gemma, 102
Fuchs, Leonhard, 1, 24, 26, 68, 89, 221-224, 227, 238
Fugger, Alexander, 290
Fugger, Antoni, 99
Fugger, family, 124, 244
Fugger, Jacob, 20, 99-100, 226
Gaddus, Nicolaus, 288
Galen, 61, 66-67, 72, 78- 80, 83, 88, 93, 95, 301
Galilei, Galileo, 13, 253, 260, 264
Garbari, F., 285
Garcia da Orta (Garcia ab Horto), 40, 61, 99, 100, 106, 108-109, 113, 186, 191, 224-226, 231, 243, 288, 295, 298-299, 302, 307, 309
Garet, family, 16, 43, 44
Garet, Elizabeth, 241
Garet, Fernando, 241
Garet, Jacomina, 241
Garet Jr, James (Jacob), 16, 22-23, 25, 43, 190, 200, 203, 234, 241
Garet Sr, James, 16, 23, 39, 43
Garet, Mary, 241
Garet, P(i)eter, 16, 39, 40, 43-44, 210-211, 241
Gargatagli, Marietta, 174-175
Garlandius, Joannes, 131
Garth, Richard, 43-44, 190, 199-200, 241, 244
Gartner, Christian, 154
Gerard, John, 43
Gerretsen, Paul, 5
Gessner, Conrad, 13-14, 16, 26, 45, 47, 68, 94, 129, 227, 239
Gheyn II, Jacques de, 196, 264
Ghini, Luca, 83, 151
Gillon, Esaya le, 275-277, 284-285, 290
Giovio, Paolo, 55, 84
Gobelius, Severinus, 227-228
Goedenhuize, Joseph – see Giuseppe Casabona

Goltzius, Hendrick, 264
Goupyl, Jacques, 89, 93
Gregory XII, Pope, 180
Grolier, Jean, 272
Grotenhuys, Arnoldus, 241
Grotius, Hugo (Hugo de Groot), 16, 36
Gryll, Lorenz (Laurentius Gryllus), 75, 77, 290
Gualteri, 288
Guarinoni, Bartolomeo, 290
Guicciardini, Lodovico, 23
Guichard, Dr., 71
Guilandino, Melchiorre (Melchior Wieland), 270-271, 286
Guise, François de, 54
Haeften, Conillemette de, 29
Harris, Stephen A., 277
Hasselaer, Petrus, 241
Haton, Claude, 55
Heind(e)l, Andreas (András), 273
Hemal, Andrea – see Andreas Heind(e)l
Henri II of France, King, 211
Henry of Scotland, 55
Henry VIII of England, King, 47, 58
Herden, Balthazar ab, 290
Hernández, Francisco, 20, 111, 259-260, 306, 307
Héroard, Michel, 83
Hertel, Johannes, 275, 283-284
Heusenstain Starhemberg, Anna Maria von, 29, 31
Heyns, Zacharias, 213
Hiller, Erhardt, 125, 131
Hippocrates, 72, 78-80, 82, 95
Hoefnagel, Joris, 91
Hogenberg, F., 153
Hoghelande, Joannes (Jan) van, 29, 33, 39, 208
Hoghelande, Theobald van, 283
Hoierus, Henricus – see Henrik Høyer
Holstenius, Lucas, 261
Holtsmark, Anne, 168
Homelius, Dr., 126
Homelius, Joannes, 126, 130, 141
Homer, 127
Hooke, Robert, 13
Horace, 60
Hortensius, Lambertus, 239, 241, 245
Horváth, Tibor Antal, 127

Houtman, Cornelis de, 201
Høyer, Henrik (Henricus Hoierus), 4, 145-152, 154-155, 157-169, 241
Hudde, Hendricus, 241
Hulthem, Charles van, 277
Hunger, F.W.T., 16-17, 65, 85, 137, 147
Huygens, Christiaan, 16
Huygens, Constantijn, 16, 264
Hyperius, Andreas, 65, 68, 94
Idström, Christian, 145
Ignatius of Loyola, 293
Imhof, Dirk, 221
Imperato, Ferrante, 33, 62, 249, 259, 270, 272, 286, 290-291
Isengrin, Michael, 221-223
Isidore of Sevilla, 60
Istvànffi, Gyula, 276, 277
Istvànffy, Miklòs (Nicholas), 127, 132-133, 137-138, 268, 271-273
Iványi, Béla, 120, 124, 130
Ivins, W.M., 221
Jackson, Elizabeth, 241
Jansen, Zacharias, 263
Jansz, Rutger, 204
Jeanplong, J., 132-133
John the Baptist, 123
Jong, E. de, 200
Jordán de Asso, Ignacio, 107-108, 112
Jordanus, Thomas (Tamàs Jordàn), 271, 274-275
Jørgensen, Per Magnus, 145, 152, 161
Joubert, Laurent, 57, 72, 84, 87, 89, 95
Juhász, Péter Melius, 142
K'ang-hsi, Emperor, 303
Kallimachos, 154
Kamel, Georg Joseph, 305-306
Kampen, Gerard Janssen van, 139, 223
Katona, I., 121, 132-133
Kepler, Johannes, 254
Kerti, István, 129
Kickx, J., 277
Kircher, Athanasius, 18, 217
Kis, István Szegedi, 123
Klaniczay, T., 124
Koltai, A., 124
Kőrös (Fraxinus), Gáspár Szegedi, 129, 142
Kramer, Johann, 207

Kusukawa, Sachiko, 5
Laca, Luis, 103
Laet, Johannes de, 261
Laguna, Andrés de, 301
Lalaing, Anna de, Domina de Marquette, 29
Lamm, Marcus zum, 213
Lancisi, L.M., 291
Lange, Johan, 145, 151
Langhe, Johannes, 239
Languetus, Hubertus, 34, 134
Laurin, brothers, 33
Laurin, Guy, 25
Lechner, Andreas, 303
Léry, Jean de, 191-192, 215
Lesley, John, 56
Lewenklau, Joannes, 33
Lewis, Gillian, 4, 24
Ligozzi, Jacopo, 91
Lindhout, Henricus a, 255
Linnaeus, Carolus, 9, 27, 185
Linschoten, Jan Huygen van, 40, 201
Linz, Roderic, 198
Linz, Sebald, 198
Lipsius, Justus, 15-16, 23, 29, 32-33, 36, 245, 273, 284
Listhius, Joannes (János Liszthi), 123, 128, 273
Lobel, Paul de, 46
Lobelius, Matthias (Matthieu de L'Obel, de Lobel), 23, 43, 46, 87, 190, 227-230, 233, 241-242, 252, 254, 275
Loe, Hendrik van der, 224
Loe, Jan van der, 222-224, 227, 229
Loew, K.F., 269
López de Gómara, Francisco, 190, 192, 225, 237
Lotichius II, Petrus, 50, 51, 95
Louis II of Hungary, King, 120
Ludwig I, Duke of Wurttemberg, 15
Ludwig VI, Elector of the Palatinate, 15
Lugo, Juan de, 303
Lullus, Raimundus, 131
Lumley, Lord, 57
Lumnitzer, István, 269
Lundquist, Kjell, 4
Lusenet, Yola de, 5
Lusitanus, Amatus, 286
Lyte, Henry, 224

Macer, 61
Maclean, I., 246
Maes, Jan (Jean) de, 29, 202
Magellan, Ferdinand, 205
Mahu, Jacques, 202, 211, 213
Malocchi, Francesco, 287, 290
Manlius, Joannes, 123, 132, 143
Manutius, Aldus, 66
Manuzio, Paolo, 270
Marchetis, Jacobus de, 53-54, 57
Margaret of Austria, 206
Marinus, 61
Marnix de Saint Aldegonde, Philip de, 37, 47
Marsigli, L.F., 290
Martín, Hipólito, 100, 108
Martín, Pedro, 100, 108-109
Mary of Hungary, Queen, 124, 128
Mary Tudor, Queen of England, 47, 58
Mary Stuart, Queen of Scots, 47, 55-57, 121
Mascardi, Giacomo, 264
Mason, Peter, 5-6, 9, 175, 239-240, 249
Matthias of Austria, Archduke, 284
Mattioli, Pietro (Pier) Andrea (Petrus Andreas Matthiolus), 60, 66-67, 142, 239, 245, 273, 274, 286, 301
Maximilian II, Emperor, 240
Medici, Catherine de', 211
Melanchthon, Philip, 65, 68, 94
Melissus, Paulus Schedius, 290
Menéndez de Avilés, 189
Mera, Giovanni de, 286
Mercator, Gerard, 102
Mercuriale, Girolamo, 270
Merian, Maria Sybilla, 32
Mersenne, Marin, 36
Michiel, Pier' Antonio, 198, 290
Moffet, Thomas, 43
Monardes, Nicolás, 5, 99-100, 105-106, 108, 113, 131, 173-174, 176-192, 201, 214-215, 224-225, 231, 234, 237-238, 243, 288, 295, 298-299, 306-307
Monok, István, 124-125
Montaigne, Michel de, 91, 200
Monte, Jacob de, 2
Monte, Antonio del, 295
Monte, Filippo di, 29, 270, 290
Monte, Giacomo di, 290
Montenegro, Pedro, 301-302

Morandi, J.B., 291
Mordentius Salernitanus, Fabricius, 240
Moretus, Jan, 101, 229, 233
Morgan, Hugh (Hugo), 16, 22, 43, 190, 241
Morren, E., 276
Morton, A., 28
Moulins, Jean Des, 88
Mouton, Jean, 16, 22, 45-46,
Müller (Molitor), Theophilo, 263
Muller, Cornelius, 223
Münster, Sebastian, 51, 57
Nádasdy, family, 125, 127-130, 142
Nádasdy, Ferenc, 126
Nádasdy, Tamás, 129
Nave, Francine de, 221
Neck, Jacob van, 43, 201, 245
Newton, Isaac, 13
Nicander, 61
Nicolai, Arnold, 223, 225
Nidaros, Bishop of, 166
Nieremberg, 261
Noort, Olivier van, 202
Norfolk, Duke of, 57-58
Nunez, Hector, 225
Núñez de Herrera, Juan, 178
Nygaard, T., 150
Ogilvie, B., 240
Olàh, Nicolaus, 273
Oldendorp, Joannes, 65
Olmi, Giuseppe, 9, 46
Olsen, Venke Åsheim, 145
Ommen, Kasper van, 5, 221
Oppian, 83
Orange, William of, 147
Orlers, Jan Jansz, 255
Örneholm, Urban, 145
Orsini, Fulvio, 61
Orta, Garcia da (ab) – see Garcia da Orta
Ortelius, Abraham, 74, 102, 121, 148
Os, Theodoricus ab, 241, 245
Osma, Pedro de, 180-182, 188
Ötvös, Péter, 123-125
Overbury, Thomas, 46
Oviedo, Gonzalo Fernández de – see Gonzalo Fernández de Oviedo
Paaw, Pieter (Petrus Pavius, Pawius), 149, 152-153, 166, 207, 210, 241, 255
Paaw, Reynerus, 241

Pacquet de l'Escluse, Catherine, 29
Pál (servant), 128
Paladini, Filippo, 196
Palatinate, Prince-Electors of the, 213
Paleotti, Gabriele, 214
Palingenius, Marcellus, 136
Palmberg, Johannes, 167
Paludanus, Bernardus, 39-40, 105, 110, 201-202, 210, 241, 255
Pancio (Pantius), Alfonso, 29, 185, 286
Paracelsus, Theophrastus, 126, 131, 227
Paradiso di Sette Monti, Arnoldo, 290
Pardo Tomás, José, 5, 9
Parduyn(us), Simon, 201, 241, 244
Pathay, István, 123
Patricius, 54
Pavius, Petrus – see Paaw, Pieter
Pavord, A., 147
Pedersdotter, Anne, 149
Pederssøn, Geble (Gjeble Pedersen), 149, 151
Peiresc, Nicolas-Claude Fabri de, 49-50, 58, 61-63, 217-218
Pelagonius, 60
Pellicier, Guillaume, 51, 52, 56, 75, 80-81, 84, 86-88, 91, 95-96
Pena, Pierre, 87, 227, 241
Penny, Thomas, 22, 43, 167
Peréz de Morales, Catalina, 177
Peréz de Morales, García, 177
Petivier, James, 305
Peucer, Casper, 67
Peutinger, Conrad, 207
Peverello, Baldasarre, 290
Peverello de Bruti, Octavia, 290
Pfefferkorn, Ignaz, 300
Philip II of Spain, King, 20, 51, 102, 112, 179, 185-186
Pietersen, Sylvertus, 241
Pigafetta, Antonio, 205-206
Pinelli, Giovanni Vincenzo, 29, 49, 61-63, 144, 259, 269-270, 272, 284, 286, 289
Piombo, Sebastian del, 195
Piso, Willem, 302, 306
Pistalotius, Nicolaus, 126
Plantin, Christopher (Christoffel Plantijn), 16, 101-102, 108, 112, 138, 181, 185, 198, 223-231, 242

Plateau, Jacques (Jacob), 29, 40-42, 44, 208, 214-215, 217-218, 240
Platter, Felix, 15, 71, 74, 83, 91
Plaza, Juan, 100-101, 103-106, 108-109
Pliny, 52, 56, 60-61, 66, 75, 83-84, 86-88, 103, 131, 173, 208, 301, 308
Pomagaics, Mihály, 130
Pona, Giovanni, 241, 244, 260, 289, 299
Popoff, Michel, 49
Popper, Ioannes, 241
Porret, Christian (Christianus Porretus), 16, 22, 217, 241, 244
Posthius, Johan, 65
Potgieter, Barent Jansz, 202, 212-213
Pozzo, Cassiano dal, 217
Prussia, Duke of, 228
Purfoot, Thomas, 227
Purkircher, György (Georg), 130, 270-274, 292
Quattrami, Evangelista, 289
Quercetanus, Josephus, 134, 136
Rabel, Daniel, 156
Rabelais, François, 56, 61, 79-80, 86
Radéczy, István (Steven), 123, 127, 270-273
Raffio, N., 286
Raleigh, Walter, 200
Raphael, 197
Raphelengius, Franciscus I, 16, 229, 231, 233
Raphelengius, Franciscus II, 50, 101
Raphelengius, Justus, 219
Rauwolff, Leonhart, 91
Ray, John, 305-306
Recchi, Nardo Antonio, 111-112, 259-261, 265
Rediger, Thomas, 177, 186
Reggio, Gregorio da, 218, 250, 253, 289
Reneaulme, Paul, 49, 58, 63
Révay, Péter, 273
Reverdy, Georges, 85
Rhazes, 79
Riccoboni, Antonio, 270
Rich, John, 43
Richier de Belleval, Pierre, 63
Ridolfi, Roberto di, 57
Rimay, János, 273
Robin, Jehan, 40, 42, 44
Roderiguez, Franciscus, 239
Roelsius, Tobias, 241, 244, 299
Rogers, Daniel, 50

Rojo, José, 303
Rondelet, Guillaume, 13, 16, 50-52, 54, 56-57, 61, 68-78, 80-96, 227
Rosenkrantz, Erik, 168
Rota, Martinus, 10
Roussel, Claude de, 283
Rubens, P.P., 263
Rudolf II, Emperor, 204, 207-208, 284
Ruel (Ruelle), Jean, 66, 104, 309
Ruppertsdatter, Marine, 149
Rutgerus, Joannes, 244
Ruysch, Rachel, 32
Sadler, Jòzsef, 276
Saint Omer, Charles de, Lord of Moerkerke and Dranoutre, 15, 20, 23, 25, 44, 91, 97
Saint-Hilari, 54
Salviani, Ippolito, 84, 286
Salzburg, Archbishop of, 206
Sambucus, Johannes (János Zsámboky), 124, 127, 270-273
Sanravius, Joannes, 54
Saporta, Antoine, 81
Sárkány, O., 126
Savile, Henry I, 241
Savile, Henry II, 241
Savile, Jane, 241
Savile, John, 241
Saxo Grammaticus, 149
Scaliger, Josephus Justus, 16, 55-56, 59, 62-63, 253, 283, 286
Scaliger, Jules-César, 79
Schaller, Farkas, 139
Scharm, Johannes, 241
Scharm, Stephen Jan, 201
Scheurleer, H., 55
Schott, Hans, 229
Schreck, Johann (Joannes Terrentius), 260, 298, 306-307
Schultz, Georg, 303
Schwendi, Làzàr, 274
Schyron, Jean, 79-80, 82, 96
Scutellari, Giacomo, 290
Sebeok, Sàndor, 269
Seijen, Arnoldus, 276-277
Sepp, Anton, 300
Severinus, Petrus, 131
Seville, Archbishop of, 178-179

Sinapius, David, 201
Sivry, Philippe de, 203
Smetius, Martinus, 74
Soldano, Adriano, 289
Solis, Virgil, 139
Soop, Thyri Anfinnsdatter, 150
Staden, Hans, 215
Steenhuis, Maarten, 5
Steinhöfer, Johann, 302
Stelluti, Francesco, 251, 258
Sterbeeck, Franciscus van, 276-278, 281
Stirling, Johannes, 60, 133, 136, 141
Stork, Agnes, 147
Swan, Claudia, 256
Sweert(s) (Sweertius), Emanuel, 201, 204-205, 230, 241
Sydney, Philip, 15, 23
Sylvius, Jacobus, 82
Syvertz, Walichius, 241
Szabó, A., 142
Szőnyi, Gy.E., 124
Takáts, S., 129, 136
Tertullian, 127
Theophrastus of Eresos, 28, 59-61, 66, 301, 308
Thevet, André, 191, 203, 215
Thott, Peder, 149
Thou, Jacques Auguste de, 49-59, 63, 86-87
Thury, E., 124
Thurzós, family, 125
Tilmannus, Hendricus, 241
Tongiorgi Tomasi, Lucia, 9, 285
Toni, G.B. De, 249, 285
Tournon, François de, 82, 84
Tovar, Simón de, 16, 100-102, 105, 108-110, 190, 198, 241
Tragus, Hieronymus, 67
Transylvanus, Maximilian, 206
Trigault, Nicolas, 298, 307
Turenne, Vicomte de, 82
Turner, William, 46
Turre, G., 290
Tuscany, Grand Duke of, 91, 288
Ubrizsy Savoia, Andrea, 5, 132, 138, 251, 285
Udine, Giovanni da, 198
Uffele, Johannes van, 218
Ujlaki, 273

Ungnadin, Eva, Baroness von Sonnegk, 29
Urban VIII, Pope, 308
Urne, 168
Valleriola, Francesco, 74
Vellekoop, Jan, 5
Venerius, Joachim, 29
Ventura, Laurentius, 131
Verancsics, Antal, 270, 273
Verweij, Michiel, 277
Vesalius, Andreas, 52-53, 238
Villanova, Arnoldus De, 255
Viviani, Giovanni, 290
Voet, Leon, 221, 228
Vorstius, Aelius Everhardus, 72, 190, 214, 241
Vulcanius, Bonaventura, 16, 267
Waehle, Espen, 208
Waesberghe, Jan van, 202
Walton, Izaak, 283
Warwijck, Wybrant van, 43, 201
Weely, Jehan (Johannes) de, 201, 207
Weert, Sebald de, 211
Wegener, Hans, 90
Weiditz, Hans, 67
Weidner de Bilterburg, Ferdinand, 290
Welser, Marcus, 260
Widmar, 125
Wieland, Melchior – see Guilandino, Melchiorre
Wilhelm IV of Hessen-Kassel, 20
Winther, Jean, 82
Withoos, Maria, 32
Wotton, Henry, 283, 290
Würzburg, Bishop of, 130
Wyetfleet, Aleidis, 29
Zamorano, Rodrigo, 100, 105-106, 108, 110-112
Zárate, Agustin de, 191
Zeitler, Joseph, 303
Zemanek, A., 28
Zennig, Francisco, 189
Zouche, Lord Edward, 15, 20
Zrínyi, family, 127
Zrínyi, Dorica, 131
Zrínyi, Miklós, 131
Zsámboky, János – see Sambucus, Johannes